T0324166

KALANCHOE (CRASSULACEAE) IN SOUTHERN AFRICA

KALANCHOE (CRASSULACEAE) IN SOUTHERN AFRICA

Classification, Biology, and Cultivation

GIDEON F. SMITH
Department of Botany, Nelson Mandela University, Port Elizabeth, South Africa

ESTRELA FIGUEIREDO
Department of Botany, Nelson Mandela University, Port Elizabeth, South Africa

ABRAHAM E. VAN WYK
H.G.W.J. Schweickerdt Herbarium, Department of Plant and Soil Sciences, University of Pretoria, Pretoria, South Africa

ACADEMIC PRESS
An imprint of Elsevier

Academic Press is an imprint of Elsevier
125 London Wall, London EC2Y 5AS, United Kingdom
525 B Street, Suite 1650, San Diego, CA 92101, United States
50 Hampshire Street, 5th Floor, Cambridge, MA 02139, United States
The Boulevard, Langford Lane, Kidlington, Oxford OX5 1GB, United Kingdom

Notices
Knowledge and best practice in this field are constantly changing. As new research and experience broaden our understanding, changes in research methods, professional practices, or medical treatment may become necessary.

Practitioners and researchers must always rely on their own experience and knowledge in evaluating and using any information, methods, compounds, or experiments described herein. In using such information or methods they should be mindful of their own safety and the safety of others, including parties for whom they have a professional responsibility.

To the fullest extent of the law, neither the Publisher nor the authors, contributors, or editors, assume any liability for any injury and/or damage to persons or property as a matter of products liability, negligence or otherwise, or from any use or operation of any methods, products, instructions, or ideas contained in the material herein.

Library of Congress Cataloging-in-Publication Data
A catalog record for this book is available from the Library of Congress

British Library Cataloguing-in-Publication Data
A catalogue record for this book is available from the British Library

ISBN: 978-0-12-814007-9

For information on all Academic Press publications
visit our website at https://www.elsevier.com/books-and-journals

Front cover: The leaves of the iconic *Kalanchoe luciae* are soup plate-sized, and often glaucous and strikingly red-infused. Photograph: Gideon F. Smith.
Back cover: Close-up of the flowers of *Kalanchoe leblanciae*. Photograph: Gideon F. Smith.
Title page: The leaves of *Kalanchoe sexangularis* var. *sexangularis* turn bright red when it grows in fully exposed positions. Photograph: Gideon F. Smith.

Publisher: Charlotte Cockle
Acquisition Editor: Nancy Maragioglio
Editorial Project Manager: Michael Lutz
Production Project Manager: Maria Bernard
Cover Designer: Mark Rogers

Typeset by SPi Global, India

Working together
to grow libraries in
developing countries

www.elsevier.com • www.bookaid.org

Dedication

This book is dedicated to

Rosette Batarda Fernandes (Portugal) [1 October 1916 (Redondo, Portugal)–28 May 2005 (Coimbra, Portugal)] at the age of 51. In 1982 and 1983, Rosette Fernandes authored the treatments of the Crassulaceae, including of *Kalanchoe*, for the *Conspectus florae angolensis* and *Flora zambesiaca*, respectively. Image kept in the private collection of Dr António Coutinho, University of Coimbra, Portugal. Reproduced with his permission.

Hellmut Richard Toelken (Namibia, South Africa, and Australia) [01 September 1939 (Windhoek, Namibia)] at the age of 69. In 1985, Dr Toelken authored the treatment of the Crassulaceae, including of *Kalanchoe*, for the *Flora of Southern Africa*. Photograph: H.R. Toelken.

Contents

Foreword

An inflorescence of the African and Socotran *Kalanchoe rotundifolia* carries a very rare double flower. Just such a deviation from the regular-flowered form of the Madagascan *Kalanchoe blossfeldiana* became the basis of the Double Flaming Katy Group or Rosebud(-flower) kalanchoes, which today represents a lucrative, global horticultural industry. Photograph: Gideon F. Smith.

Kalanchoe is a much neglected genus of leaf succulents in the stonecrop family (Crassulaceae). It was first named in 1763 by the French botanist Michel Adanson with the only monograph dating from 1907 to 1908 being woefully out of date and completely unillustrated. Over the past 50 years, generally well-illustrated works on *Kalanchoe* have appeared: in 1964, a survey was undertaken of the plants growing in the fabulous private botanical garden of *Les Cèdres* on the French Riviera, and a review of the Malagasy species took place in 1995, beautifully illustrated with watercolour paintings. Finally, in 2003, a summary treatment of the genus was published as part of the six-volume work, the *Illustrated Handbook of Succulent Plants*. For this, the French botanist Bernard Descoings recognised 144 species, together with a handful of named hybrids. In summary, *Kalanchoe* in the literature has a patchy history with most of the significant work being generally inadequately illustrated.

The genus does not deserve its 'Cinderella' status since it includes many horticulturally attractive plants. They are often exceptionally easy to propagate because they proliferate readily from seed, cuttings, bulbils produced on the leaf margins, and in many cases even severed leaves. One species, *Kalanchoe blossfeldiana*, is commercially important as the basis of plant breeding programmes that have resulted in, among others, the 'Flaming Katy' series of cultivars propagated on a vast scale in the horticultural trade. Members of the genus have also been used extensively as experimental subjects in physiological studies on the control of flowering by different light treatments and on the mechanism of crassulacean acid metabolism (CAM), one of three metabolic pathways found in the photosynthetic tissues of vascular plants.

Kalanchoes are naturally widespread from Africa and Madagascar eastward through Arabia to Southeast Asia. The diversity within the genus is impressive, ranging from diminutive dwarf plants a mere 20 cm tall to trees up to 8 m tall. There is much here to interest the grower, the professional horticulturalist, and plant scientist.

The scene is set for a modern treatment of *Kalanchoe*. Gideon Smith, Estrela Figueiredo, and Braam van Wyk, the authors of this book, are well placed to do so since together they have an impressive track record, having published several books and a formidable number of scientific papers on succulents. In preparation for this book, a number of their recent publications have focussed on southern African species. This book provides a sound foundation for the appreciation and understanding of *Kalanchoe*, and I highly recommend it.

Colin C. Walker, PhD
President of the British Cactus & Succulent Society,
Honorary Research Associate,
The Open University, United Kingdom

1

Preface

FIG. 1.1 *Kalanchoe crundallii* is one of the rarest of the southern African kalanchoes. It is restricted to a few locations on the Soutpansberg in South Africa's Limpopo province, where it invariably grows in the understory of dense bushveld in deep or dappled shade. Photograph: Gideon F. Smith.

This book provides the first comprehensive, illustrated account for members of the succulent plant genus *Kalanchoe* Adans. in southern Africa. While several other genera of Crassulaceae in southern Africa have had books dedicated to them, the genus *Kalanchoe*

has suffered some neglect. There are several possible reasons for this:

- With just less than 12% of the known global species diversity of *Kalanchoe* occurring in southern Africa,

1

the genus is not as well represented in the subcontinent as other locally much more speciose crassuloid genera, such as the ubiquitous *Crassula*.

- When in bloom in autumn and winter, their clusters of small, tubular, cigar-shaped flowers are arguably not as eye-catching as the thickly flowered, brightly coloured candles and candelabras of other succulents such as aloes that more brightly lighten up the drab winter landscape.
- In some instances having opted for biennial or multi-annual habits that often culminate in the plants that have flowered dying off, some kalanchoes tend to be comparatively short-lived, also in cultivation.
- Kalanchoes do not yet have a collective, locally well-known and widely used, descriptive common name that allows anyone to refer to them with ease, enabling informal communication about the plants by non-specialists.
- Perhaps most importantly, the local species are absent from the high-density succulent plant areas in southern Africa, for example, the Klein Karoo, parts of the Groot Karoo, Knersvlakte, and Richtersveld.

Yet, in their typical bushveld (savanna) and grassland habitats, the smooth-edged or scallop-margined succulent leaves arranged in low-growing rosettes make kalanchoes striking and noticeable, and when they flower, their thin, often much-branched, or club-shaped to cylindrical, densely flowered inflorescences provide some respite from the otherwise muted-coloured autumn to late-winter landscapes. These waterwise plants are exceedingly tough and resilient; they thrive where they often grow under sparse or dense tree canopies (Fig. 1.1), interspersed among dense swathes of grasses, or survive in thin soils that settled in rock crevices on well-drained slopes.

Southern Africa, the main study area covered in this book, is here defined as comprising the countries Namibia, Botswana, Swaziland, Lesotho, and South Africa (see Fig. 1.2). Together, these countries comprise the so-called *Flora of Southern Africa* (FSA) region, which is abutted by the *Conspectus florae angolensis* region (Angola only) in the northwest, and the *Flora zambesiaca* region (Mozambique, Zimbabwe, Botswana, Zambia, and Malawi) in the centre and northeast. Note that Botswana is included in both the *Flora of Southern Africa* and *Flora zambesiaca* regions. Throughout the text, the part of Africa immediately north of the FSA region and south of the equator is referred to as south-tropical Africa.

This book includes overviews of the family Crassulaceae and the genus *Kalanchoe* in global and subcontinental contexts, followed by information on the taxonomic history of the genus in southern Africa, important collectors and researchers, kalanchoes in locally published botanical art, and the recorded common names of kalanchoes.

The characters and ecology of the species are discussed, including the distribution ranges where they occur, as well as whether they are widespread or range-restricted, for example, their habitat preferences, and from where the species were formally recorded for the first time, also mentioning type localities, which are in some cases well beyond the northern border of southern Africa.

The morphological characters and phenology, physiology and anatomy, biocultural significance and toxicity of southern African kalanchoes are also discussed. Uses of the southern African and some introduced kalanchoes in garden settings (Fig. 1.3), as well as the tendency of nonnative species to become invasive, are also noted, discussed, and illustrated.

For each of the indigenous, naturalised, garden-escape, and cultivated *Kalanchoe* species (Fig. 1.4), comprehensive nomenclatural, taxonomic, descriptive, and other information is provided, including maps of their geographical distribution ranges in southern Africa and slightly beyond. Detailed information is supplied on the type specimens of the species names—that is, those specimens that fix the application of the scientific names of the kalanchoes. All the naturalised and introduced, horticulturally popular *Kalanchoe* species in southern African are indigenous to Madagascar, the Red Island or Ile de Rouge, a major centre of present-day diversity for the genus.

As part of an ongoing reassessment of the taxonomic status of southern African kalanchoes, we comprehensively investigated the infrageneric classification of the southern African kalanchoes and also provide, where it proved necessary, updates of the taxonomy of the locally naturalised, invasive, and cultivated kalanchoes. For the southern African species included in this work, we applied a cut-off date of 31 December 2016. All the southern African species that had been known on that date are treated in this book. Most names in *Kalanchoe* are burdened by many synonyms; these are discussed in detail to unambiguously establish the application of the names that should be applied to kalanchoes in southern Africa.

Figueiredo & Smith (2015) argued strongly for the typification of plant names, given that the application of names is determined by types. During the course of studies on *Kalanchoe*, it was necessary to typify several names that either lacked types or where previous typifications were not effective. This applies to the names of both indigenous southern African *Kalanchoe* taxa and to the Madagascan ones that have become locally naturalised. Correctly identifying and therefore naming species that are naturalised or even invasive are as important as documenting the indigenous diversity (Pyšek et al., 2013). The typifications made during the course of this study and several scientific papers that deal with the taxonomy of the *Kalanchoe* species included in this book were

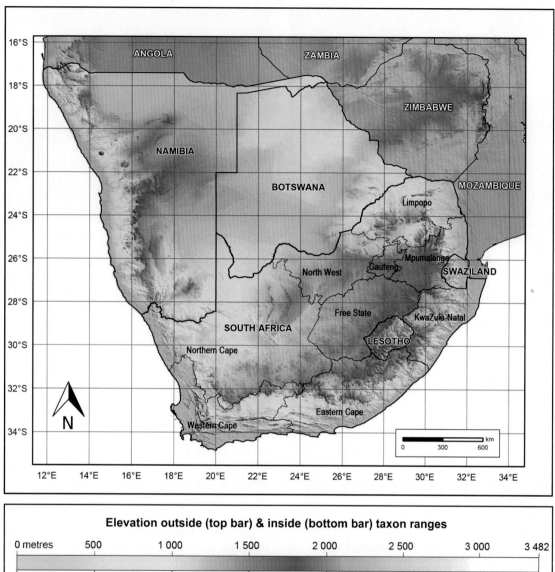

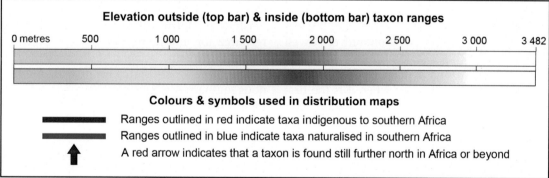

FIG. 1.2 Map of the African continent south of 16°S, serving as a key to the maps used in this book. We cover those species of *Kalanchoe* indigenous to or cultivated/naturalised in southern Africa, a geographical region here defined as comprising Namibia, Botswana, Swaziland, Lesotho, and South Africa. The relief of the study area is shown in colour. Provincial boundaries and names are supplied for South Africa only. Bordering countries (Angola, Zambia, Zimbabwe, and Mozambique) are depicted with a grey relief and are broadly treated as part of south-tropical Africa. In the distribution maps supplied in this book, the ranges of indigenous (outlined in red) and naturalised (outlined in blue) species are shown with brighter relief colours, hence the additional bar of elevation colours in the map legend. In these maps, the ranges of only those indigenous species known to extend outside southern Africa are shown in the grey area, either as a red outline or with a red arrow indicating that a taxon is found further north in Africa or beyond. Note that the range maps depict probable and not definite distribution. We also do not map any of the species that are indigenous to south-tropical Africa but absent from the main study area, southern Africa.

FIG. 1.3   Some of the Madagascan kalanchoes that are commonly grown in gardens in southern Africa have escaped and become naturalised in the subcontinent; seven such species are included in this book. Here, the regular and white-leaved forms of *Kalanchoe fedtschenkoi* are cultivated together in a border in a garden in Pretoria, Gauteng, South Africa. Photograph: Gideon F. Smith.

FIG. 1.4   Potted specimens of *Kalanchoe blossfeldiana*, such as this one that belongs to the double-flowered Rosebud or Roseflower Group, are widely sold in nurseries and florist shops in southern Africa. In this book, we treat and discuss six noninvasive, exotic kalanchoes that are widely cultivated in the subcontinent. Photograph: Gideon F. Smith.

deliberately first published in the scientific literature; all the relevant papers are cited in this book. No new species are described here, and no new typifications are done.

Author citations of the names of the kalanchoes treated here are given in the 'Contents' page, as well as where the species are taxonomically treated in Chapter 12. In the rest of the text, author citations of the names are largely not given, except if there is a specific reason to do so. However, in the case of *Kalanchoe* species mentioned, but not treated in the taxonomic part of this book, either as accepted species or as synonyms, author citations are provided at their first reference in a chapter. For the format of author citations, we follow Brummitt & Powell (1992).

In the case of species and infraspecific names, the genus name *Kalanchoe* is spelled out when it is the first word in a sentence or at first mention in a paragraph, after which it is abbreviated to "*K.*" However, to prevent confusion between, for example, *Kalanchoe* and *Kitchingia* Baker, which is at the genus rank a synonym of *Kalanchoe*, these genus names are at times spelled out in full throughout parts of the text. *Kitchingia* is the basionym of *Kalanchoe* subgen. *Kitchingia* (Baker) Gideon F.Sm. & Figueiredo.

Etymological information on people commemorated in the names of succulent plants, including kalanchoes, as well as the derivation of other specific epithets, is given in, among others, Korevaar et al. (1983) and Eggli & Newton (2004). However, in some instances, information that we sourced differed from that given in these references; we provide information on the origin of specific epithets where the species are treated in Chapter 12.

With very few exceptions, place names in South and southern Africa referenced in this book were standardised according to Raper et al. (2014), which elaborated on the earlier works of Raper (1987, 1989, 2004). Leistner & Morris (1976) was an additional invaluable resource for information on ¼-degree grids and place names in the subcontinent.

The conservation status we provide for all the indigenous southern African kalanchoes is based on natural history observations made during fieldwork over many years and following discussions with colleagues. Our suggested listings closely coincide with those given by the *Red List of South African Plants* (redlist.sanbi.org). We did not express an opinion on the conservation status of the species of *Kalanchoe* indigenous to Madagascar that are included in this book because of their local horticultural appeal only or invasiveness in southern Africa.

CHAPTER

2

Crassulaceae of the World

FIG. 2.1   A naturalised specimen of *Aeonium arboreum* (L.) Webb & Berthel. growing at Cabo Sardão on the Alentejo coast, southwestern Portugal. Photograph taken on 29 December 2017 by Gideon F. Smith.

INTRODUCTION

Along with the Agavaceae (century plants), Aizoaceae (mesembs), Apocynaceae (carrion flowers), Asphodelaceae (aloes), and Cactaceae (cacti), the Crassulaceae (plakkies or stonecrops) are one of the best known succulent plant families globally. In terms of the number of species, it is the third largest family of succulent plants, after the mesembs and cacti. All the members of the Crassulaceae, even the few aquatic species, display leaf succulence, to a greater or lesser degree, and some are additionally stem succulents. Characters that define the Crassulaceae include:

- a comparatively primitive flower structure;
- regular, usually pentamerous flowers;
- isomerous flowers (having floral whorls with the same number of parts);
- free carpels; and
- a single or double whorl of stamens.

Apart from being predominantly leaf succulents, the Crassulaceae are further characterised by usually having small, star-shaped flowers borne in dense clusters. Most Crassulaceae species are shrubby, subshrubby, or low-growing, long-lived perennials or short-lived multiannuals. A few, such as *Sedum mucizonia* (Ortega) Raym.-Hamet from southwestern continental Europe (the Iberian Peninsula: Portugal and Spain), and northern Africa (Algeria and Morocco), are annuals, however, and are quick-growing, completing their vegetative and reproductive life cycles within a single growing season (Smith et al., 2016b). These annuals, also called therophytes, therefore survive unfavourable growing conditions in the form of seed that is often produced in great abundance. The majority of the world's succulents are perennials, growing and flowering rather predictably year after year. The annual life form is fairly uncommon among succulents, given their ability to survive unfavourable growing conditions using, inter alia, moisture accumulated in their leaves, stems, or roots.

At the other end of the growth form spectrum, a very few representatives of the Crassulaceae become small trees; these are included in the predominantly southern African genera *Tylecodon* Toelken (one species) and *Crassula* L. (two species and one subspecies). In *Crassula*, only *C. arborescens* (Mill.) Willd. subsp. *arborescens*, *C. arborescens* subsp. *undulatifolia* Toelken, and *C. ovata* (Mill.) Druce have been accorded National Tree List numbers. The fourth southern African arborescent crassuloid taxon is *Tylecodon paniculatus* (L.f.) Toelken. The Mexican *Sedum frutescens* Rose with its thick stems and peeling bark has a similar vegetative appearance to and shows remarkable convergence with *Crassula arborescens*, *C. ovata*, and *Tylecodon paniculatus* (Smith et al., 2017c).

The circumscription of the Crassulaceae is reasonably stable, but as with most large families, in some instances, classification at genus and infrageneric ranks remains fluid. Over the past more or less 20–30 years, the Crassulaceae and several of its genera have been the subject of a number of phylogenetic studies and other genus- and family-level investigations (see, e.g. Eggli, 1987, 1988; Van Ham, 1994; Mes, 1995; Stevens, 1995; 't Hart & Eggli, 1995; and Gontcharova & Gontcharov, 2007, and references therein). The family studies showed that the Crassulaceae are a monophyletic unit that represents a 'natural' group. The Crassulaceae are mostly interpreted as closely related to the Saxifragaceae, a predominantly northern temperate group that also extends to subartic zones (Mabberley, 2017: 244; Harding, 1992; Köhlein, 1984; McGregor, 2008). The Penthoraceae either have been included in the Crassulaceae or in the Saxifragaceae (most often) or treated as a distinct family (Mabberley, 2017: 694).

GENUS AND SPECIES DIVERSITY IN, AND DISTRIBUTION OF, THE CRASSULACEAE

At present, about 1400 species are included in the family Crassulaceae, the most species-rich family in the order Saxifragales (Gontcharova & Gontcharov, 2007). This large diversity of species displays a bewildering range of growth forms, leaf shapes, and flower morphologies and colours. The species are arranged into between 35 and 45 genera that range in the number of included species from monotypic ones (one species only) such as *Meterostachys* to large ones with several hundred species, such as *Sedum* and *Crassula*. The number of genera recognised varies depending on the taxonomic concept followed, with several genera being segregates of, especially, *Sedum*. The genera often recognised in the Crassulaceae are summarised in Table 2.1.

Phytogeographically, the Crassulaceae are widespread and occur on all the continents, mostly in warm and dry regions. They are often found in climatically harsh natural habitats that span a vast range, from seasonal to year-round aquatic pools to hot, fully exposed rocks, and the understory of deeply shaded forests and savannas. Some species are frost-hardy, while others grow best in the tropics and mild subtropics. With a global geographical distribution range that straddles the Old and New Worlds, the family Crassulaceae has a significant presence in continental North America, especially in Mexico, in Eurasia, and in sub-Saharan Africa, and in patches in South America but is underrepresented

TABLE 2.1 An alphabetical catalogue of subfamilies, tribes, and genera of the Crassulaceae of the world, based on and mostly following Eggli (2003) and Thiede & Eggli (2007). Species numbers (third column) are approximations. The subdivision of Crassulaceae (subfam. Sempervicoideae) tribe Sedeae into clades is not reflected in the Table. See also Rowley (1978: 118–119) for an older subfamily classification of the Crassulaceae

No.	Genus	Number of species	Distribution
I	Crassulaceae subfam. Crassuloideae		
I.1	*Crassula* L. (incl. *Tillaea* L.)	195	Arabia, southern Africa and rest of Africa, a few species near cosmopolitan
I.2	*Hypagophytum* A.Berger	1	Ethiopia
II	Crassulaceae subfam. Kalanchooideae (as Kalanchoideae)		
II.3	*Adromischus* Lem.	30	Southern Africa
II.4	*Cotyledon* L.	11	Southern Africa, one outlier to the north
II.5	*Kalanchoe* Adans.	150	Arabia, Central and East Africa, Madagascar, Southeast Asia, southern Africa
II.6	*Tylecodon* Toelken	50	Southern Africa
III	Crassulaceae subfam. Sempervivoideae		
III.A	Tribe Aeonieae		
III.A.7	*Aeonium* Webb & Berth.	36	Macaronesia, North Africa, Yemen
III.A.8	*Aichryson* Webb & Berth.	14	Macaronesia
III.A.9	*Monanthes* Haw.	9	Macaronesia
III.B	Tribe Sedeae		
III.B.10	*Afrovivella* A.Berger	1	Ethiopia
III.B.11	*Cremnophila* Rose	2	Mexico
III.B.12	*Dudleya* Britton & Rose	47	Mexico, USA
III.B.13	*Echeveria* DC.	139	Central and South America, Mexico, USA
III.B.14	*Graptopetalum* Rose	18	Mexico, USA
III.B.15	*Lenophyllum* Rose	7	Mexico, USA
III.B.16	*Pachyphytum* Link, Klotzsch & Otto	15	Mexico
III.B.17	*Pistorinia* DC.	3	Iberian Peninsula, North Africa
III.B.18	*Prometheum* (A.Berger) H.Ohba	8	Armenia, Caucasus, Greece, Iran, Turkey
III.B.19	*Rosularia* (DC.) Stapf	17	Inner Asia, Mediterranean, Near East
III.B.20	*Sedella* Britton & Rose (incl. *Parvisedum* R.T.Clausen)	3	USA
III.B.21	*Sedum* L.[a]	420	North America, Europe, North Africa, Near East, Asia, South America, Central and Eastern Africa
III.B.22	*Thompsonella* Britton & Rose	6	Mexico
III.B.23	*Villadia* Rose	21	Central America, Mexico, USA
III.C	Tribe Semperviveae		
III.C.24	*Petrosedum* Grulich	7	Balkans, Europe, Mediterranean
III.C.25	*Sempervivum* L. (incl. *Jovibarba* Opiz)[b]	63	Balkans, Carpathians, Caucasus, Europe, Iran, North Africa, Peninsula, Russia, Turkey
III.D	Tribe Telephieae		
III.D.26	*Hylotelephium* H.Ohba	27	Caucasus, East Asia, Europe, North America, Siberia
III.D.27	*Kungia* K.T.Fu	2	China

Continued

TABLE 2.1   An alphabetical catalogue of subfamilies, tribes, and genera of the Crassulaceae of the world, based on and mostly following Eggli (2003) and Thiede & Eggli (2007). Species numbers (third column) are approximations. The subdivision of Crassulaceae (subfam. Sempervicoideae) tribe Sedeae into clades is not reflected in the Table. See also Rowley (1978: 118–119) for an older subfamily classification of the Crassulaceae—cont'd

No.	Genus	Number of species	Distribution
III.D.28	*Meterostachys* Nakai	1	China, Japan, Korea
III.D.29	*Orostachys* Fischer	11	China, Japan, Kazakhstan, Korea, Mongolia, Russia
III.D.30	*Perrierosedum* (A.Berger) H.Ohba	1	Madagascar
III.D.31	*Sinocrassula* A.Berger	7	Bhutan, China, India, Nepal, Pakistan, Tibet
III.E	Tribe Umbiliceae		
III.E.32	*Phedimus* Raf.	18	Asia, eastern Europe
III.E.33	*Pseudosedum* (Boiss.) A.Berger	12	Central Asia
III.E.34	*Rhodiola* L. (incl. *Tolmachevia* Á.Löve & D.Löve)	58	East Asia, Europe, North America, Siberia
III.E.35	*Umbilicus* DC. (incl. *Chiastophyllum* Stapf)	13	Arabia, Caucasus, Europe, Macaronesia, near East, North Africa

[a] *Sedum* here including *Altamiranoa* Rose ex Britton & Rose, *Amerosedum* Á.Löve & D.Löve, *Ohbaea* V.V.Byalt & I.V.Sokolova, *Poenosedum* Holub, and *Sedastrum* Rose.
[b] We here regard *Jovibarba* Opiz as a genus distinct from *Sempervivum* L.

in Australia (Kapitany, 2007), South America, and Polynesia [a subregion of Oceania consisting of about 1000 islands] (Van Ham, 1994: 23).

In Europe, crassuloid species grow in low-lying, near-maritime niches on the Iberian Peninsula in the west (Smith & Figueiredo, 2013), to much higher elevations in the Caucasus, and eastern continental European countries such as Slovakia and the Czech Republic (Hadrava & Miklánek, 2007a, b), and further east, for example, in Siberia, where winter snow is a common occurrence ('t Hart & Eggli, 1995; Smith et al., 2015a). With the exception of the Sahara, the Crassulaceae occur virtually throughout the African continent, both south and north of this sandy desert, with a marked concentration of genera and species in southern Africa.

From a utilitarian economic point of view, the Crassulaceae are mostly important in the horticultural industry through the production, sometimes on an enormous scale, of material for outdoor cultivation in mild and even some continental-type climates (see, e.g. Horvath, 2014 on *Sedum*, and Walker, 2017 on *Phedimus*) and in harsher climates for the indoor-plant trade, through species such as *Kalanchoe blossfeldiana*, a typical windowsill plant. However, the Crassulaceae do not include any crop species.

PICTORIAL GALLERY OF SOME GENERA OF THE CRASSULACEAE NOT INDIGENOUS TO SOUTHERN AFRICA, AND OF THE CLOSELY RELATED SAXIFRAGACEAE

FIG. 2.2 *Aeonium balsamiferum* Webb & Berthel., a Canary Islands endemic growing in the Shoenberg Temperate House in the Missouri Botanical Garden, St. Louis, USA. Photograph: Gideon F. Smith.

FIG. 2.5 *Dudleya saxosa* (M.E.Jones) Britton & Rose subsp. *collomiae* (Rose) Moran prefers rocky outcrops in central Arizona, USA. Photograph: Gideon F. Smith.

FIG. 2.3 *Aeonium urbicum* (C.Sm. ex Hornem.) Webb & Berthel. occurs in a wide area of Tenerife, Canary Islands. Photograph: Estrela Figueiredo.

FIG. 2.6 Close-up of the flowers of *Dudleya saxosa* subsp. *collomiae*. Photograph: Gideon F. Smith.

FIG. 2.4 *Dudleya farinosa* Britton & Rose growing on coastal cliffs near San Francisco, California, USA. Photograph: Gideon F. Smith.

FIG. 2.7 The leaves of *Echeveria agavoides* Lem. are often strongly red-margined. Photograph: Gideon F. Smith.

FIG. 2.8   *Echeveria purpusorum* A.Berger has an aloe-like, rosulate growth form. Photograph: Gideon F. Smith.

FIG. 2.11   *Graptopetalum paraguayense* (N.E.Br.) E.Walther carries small, compact rosettes at the ends of short, thin branches. Photograph: Gideon F. Smith.

FIG. 2.9   Close-up of the bright orange and yellow flowers of *Echeveria purpusorum*. Photograph: Gideon F. Smith.

FIG. 2.12   The flowers of *Graptopetalum paraguayense* are white with small red spots. Photograph: Gideon F. Smith.

FIG. 2.10   *Graptopetalum bartramii* Britton & Rose subsp. *arizonica* (Rose) S.L.Welsh growing against a shady cliff in southern Arizona near Nogales, south of Tucson, USA. Photograph: Gideon F. Smith.

FIG. 2.13   *Hylotelephium cauticola* (Praeger) H.Ohba occurs naturally in Japan. Photograph: Gideon F. Smith.

FIG. 2.14 *Hylotelephium spectabile* (Boreau) H.Ohba grown as a bedding plant in the garden of the Golden Phoenix Hotel, Beijing, China. Photograph taken on 25 July 2017 by Gideon F. Smith.

FIG. 2.17 *Orostachys boehmeri* (Makino) H.Hara is a small-growing, rosulate species from Japan. Photograph: Gideon F. Smith.

FIG. 2.15 Close-up of the pinkish white flowers of *Hylotelephium spectabile*. Photograph Gideon F. Smith.

FIG. 2.16 *Jovibarba heuffelii* (Schott) Á.Löve & D.Löve, a small, rosette-forming species from Eastern Europe, is sometimes treated as *Sempervivum heufellii* Schott. Photograph: Gideon F. Smith.

FIG. 2.18 Close-up of the spikelike inflorescence of *Orostachys boehmeri*. The flowers are white with prominently purplish red anthers. Photograph: Gideon F. Smith.

FIG. 2.19   *Petrosedum forsterianum* (Sm.) Grulich, here growing near Alvados in Portugal, has bright yellow flowers that are densely clustered together in forked but nevertheless often head-shaped inflorescences. Photograph: Estrela Figueiredo.

FIG. 2.22   *Phedimus spurius* (M.Bieb.) 't Hart is native to the Caucasus, Iran, and Turkey. Photograph: Gideon F. Smith.

FIG. 2.20   The blue-leaved *Petrosedum sediforme* (Jacq.) Grulich growing near Alcanena in central Portugal. Photograph: Gideon F. Smith.

FIG. 2.23   *Rhodiola integrifolia* Raf., also sometimes regarded as *Sedum integrifolium* (Raf.) A.Nelson, occurs sympatrically with *R. rhodantha* (see Fig. 2.24) in a wetland in the Rocky Mountains above the treeline near Summit Lake along the Mount Evans road, about 100 km west of Denver, Colorado, USA. Photograph: Gideon F. Smith.

FIG. 2.21   *Phedimus kamtschaticus* (Fischer & C.A.Meyer) 't Hart has a very wide geographical distribution range in Russia, China, Japan, and Korea. Photograph: Gideon F. Smith.

FIG. 2.24 *Rhodiola rhodantha* (A.Gray) H.Jacobsen, variously treated as *Sedum rhodanthum* A.Gray, or as *Clementsia rhodantha* (A.Gray) Rose ex Britton & Rose, growing in an ankle-deep, perennial wetland at an elevation of about 4000 m on Mount Evans, the highest summit of the Chicago Peaks in the Front Range of the Rocky Mountains, Colorado, USA. Photograph: Gideon F. Smith.

FIG. 2.25 The small *Sedella pumila* Britton & Rose, sometimes treated as *Parvisedum pumilum* (Benth.) R.T.Clausen, growing in a very moist habitat in the Hetch Hetchy Valley in California's Yosemite National Park, USA. Photograph taken on 22 April 2011 by Estrela Figueiredo.

FIG. 2.26 The small, low-growing *Sedum albertii* Regel, sometimes treated as *Pseudosedum affine* (Schrenk) A.Berger, occurs in Siberia and Asia. Photograph: Gideon F. Smith.

FIG. 2.27 *Sedum dasyphyllum* L. is a small-leaved species from Europe. Photograph: Gideon F. Smith.

FIG. 2.28 The bright yellow-flowered *Sedum lanceolatum* Torr., sometimes treated as *Amerosedum lanceolatum* (Torr.) Á.Löve & D.Löve, growing in the Rocky Mountains, Colorado, USA. Photograph: Gideon F. Smith.

FIG. 2.29   One of the an annual species included in the genus, *Sedum mucizonia* (Ortega) Raym.-Hamet here grows against a vertical rock face along the Fórnea Trail near Alvados, central Portugal. Photograph: Estrela Figueiredo.

FIG. 2.30   *Sedum pachyphyllum* Rose, a Mexican species, has intensely red-tipped leaves. Photograph: Gideon F. Smith.

FIG. 2.31   The shrubby *Sedum treleasei* Rose is a native of Mexico. Photograph: Gideon F. Smith.

FIG. 2.32   *Sempervivum arachnoideum* L. has cobwebby rosettes. Photograph: Gideon F. Smith.

FIG. 2.33 *Sempervivum tectorum* L., the common houseleek of Europe. Photograph: Gideon F. Smith.

FIG. 2.34 Close-up of an inflorescence of *Sempervivum tectorum*. Photograph: Estrela Figueiredo.

FIG. 2.35 *Umbilicus rupestris* (Salisb.) Dandy growing on a rock wall near Serra de Santo António in the Parque Natural das Serras de Aire e Candeeiros in central Portugal. Photograph taken on 18 May 2015 by Gideon F. Smith.

FIG. 2.36 The family Saxifragaceae is often interpreted as a close relative of the Crassulaceae. The growth form of *Saxifraga hostii* Tausch subsp. *rhaetica* (Kerner) Br.-Blanq. is reminiscent of that of a number of crassuloid species. Photograph: Gideon F. Smith.

FIG. 2.37 Close-up of the small, white flowers of *Saxifraga hostii* subsp. *rhaetica*. Photograph: Gideon F. Smith.

FIG. 2.38 The low-growing *Saxifraga paniculata* Mill. carries its flowers in densely clustered inflorescences. Photograph: Gideon F. Smith.

FIG. 2.39 The leaves of *Saxifraga rotundifolia* L. are reminiscent of those of *Umbilicus rupestris* (Salisb.) Dandy, the navelwort, which is a member of the Crassulaceae. Photograph: Gideon F. Smith.

FIG. 2.40 *Saxifraga stolonifera* Curtis is a popular house plant and arguably the best known of the *Saxifraga* species. The open rosettes produce numerous, long runners at the ends of which plantlets are formed. The small, red-spotted, white flowers are zygomorphic, but nonetheless reminiscent of those of some representatives of the Crassulaceae. Photograph: Gideon F. Smith.

Crassulaceae in Southern Africa

FIG. 3.1   *Crassula perfoliata* L. var. *minor* (Haw.) G.D.Rowley. Eastern Cape, South Africa. Photograph: Gideon F. Smith.

INTRODUCTION

In southern Africa, the Crassulaceae are represented by five of the *c.* 34 genera (Thiede & Eggli, 2007) recognised in the family. Therefore, about 15% of the crassuloid *genera* occur in the region (see Meyer et al., 1997: 69 and Smith et al., 1997: 56–63). These are: (1.) *Adromischus* Lem.; (2.) *Cotyledon* L.; (3.) *Crassula* L.; (4.) *Kalanchoe* Adans.; and (5.) *Tylecodon* Toelken (see Table 3.1). Two of these genera (*Adromischus* and *Tylecodon*) are endemic to the region, while the vast majority of the species of *Cotyledon* and *Crassula* occur in southern Africa. In contrast, just over 10% of the known species of *Kalanchoe* occur in southern Africa.

In southern Africa, these five genera are represented by about 250 species, which account for 18% of the *c.* 1400 known crassuloid *species* of the world.

TABLE 3.1 Subfamilies and genera of Crassulaceae that occur naturally in southern Africa (Namibia, Botswana, Swaziland, Lesotho, and South Africa)

No.	Genus	Number of species	Percentage of global total (%)	Distribution
1.	Crassulaceae subfam. Crassuloideae			
1.1	*Crassula* L.	*c.* 150	75	Winter- and summer-rainfall regions
2.	Crassulaceae subfam. Kalanchooideae			
2.1	*Adromischus* Lem.	*c.* 30	100	Winter-rainfall region
2.2	*Cotyledon* L.	*c.* 11	100	Winter- and summer-rainfall regions
2.3	*Kalanchoe* Adans.	*c.* 17	11	Summer-rainfall region
2.4	*Tylecodon* Toelken	*c.* 50	100	Winter-rainfall region

Four of the southern African crassuloid genera (*Adromischus*, *Cotyledon*, *Kalanchoe*, and *Tylecodon*) are included in Crassulaceae subfam. Kalanchooideae A.Berger (Berger, 1930, as Kalanchoideae), while *Crassula* is classified in Crassulaceae subfam. Crassuloideae.

Adromischus and *Tylecodon* are distributed mostly in the western, winter-rainfall parts of southern Africa, while *Kalanchoe* is a summer-rainfall genus. Representatives of *Crassula* and *Cotyledon* occur in both rainfall regions (Jürgens, 1995).

BRIEF SUMMARY OF THE GENERA

For the first 30 to 50 post-Linnean years (1753 to *c.* 1800), many species, from South Africa and beyond, with a general *plakkie* appearance (variously shaped, fat, but often flattened leaves) were at first described as belonging in the genus *Cotyledon*. It took a few decades before diversity at the genus rank in the Crassulaceae was better appreciated, but by the beginning of the 19th century, there was increasing recognition of the superficiality of the one- or two-genus [*Cotyledon* and *Crassula*] Linnean classification system that predominated during the mid- to late-18th century. *Kalanchoe* was the first genus name to be published after the appearance of Linnaeus' *Species plantarum* in 1753, followed by *Adromischus* in 1852, and, much more recently, *Tylecodon* in 1978.

1. CRASSULACEAE SUBFAM. CRASSULOIDEAE

1.1 *Crassula* L., *Sp. pl.* 1: 282 (1753)

The genus *Crassula* occurs naturally in Europe, the Americas, Australia, New Zealand, and some southern islands but with the largest species-level diversity found in southern Africa (Jürgens, 1995). In this regard, the genus is the southern hemisphere counterpart of the very diverse *Sedum* L. ('t Hart, 1978). Of the five crassuloid genera that occur naturally in southern Africa, *Crassula* is by far the most diverse (Figs. 3.1–3.4). Growth forms vary immensely, with some species attaining treelike dimensions, while other species are miniature, soil-hugging creepers, or even perennial hydrophytes. *Crassula* displays a bewildering array of floral morphologies and flower colours. Most species have small, rather dull-coloured flowers, which are often arranged in dense, head-shaped inflorescences, but a few others have large, brightly coloured, loosely packed, almost aloe-like flowers, *Crassula coccinea* being an example.

The taxonomy of *Crassula*, at least in southern Africa, is not yet stable and still undergoes refinement: new species are continuously discovered and described, while other, older names are sometimes resurrected for species long synonymised. Within *Crassula*, several, often specialised, sections are recognised. Not all of these are natural groups though.

Illustrated reference: Rowley (2003).

FIG. 3.2 *Crassula arborescens* (Mill.) Willd. subsp. *undulatifolia* Toelken. Eastern Cape, South Africa. Photograph: Estrela Figueiredo.

FIG. 3.3 *Crassula coccinea* L. Southwestern Cape, South Africa. Photograph: Gideon F. Smith.

FIG. 3.4 *Crassula corallina* Thunb. subsp. *corallina*. Western Cape, South Africa. Photograph: Gideon F. Smith.

2. CRASSULACEAE SUBFAM. KALANCHOOIDEAE

2.1 *Adromischus* Lem., *Jard. Fleur. 2, Misc.* 59 (1852)

Adromischus, with about 30 recognised species, is endemic to southern Africa, being mostly restricted to South Africa. Species occur predominantly in the Klein Karoo and parts of the southwestern Groot Karoo, with significant outliers along the west coast, stretching northwards to the Richtersveld, while to the east, several species are found in the Eastern Cape. Very few species extend into the climatically severe interior beyond the Great Escarpment.

Species mostly grow as small shrublets that carry fat, mottled, bloated-looking leaves on thin wiry stems (Figs. 3.5–3.7). Flowers are tubular and often not very conspicuously coloured.

Illustrated reference: Pilbeam et al. (1998).

FIG. 3.6 *Adromischus triflorus* (L.f.) A.Berger. Klein Karoo, South Africa. Photograph: Gideon F. Smith.

FIG. 3.5 *Adromischus maculatus* (Salm-Dyck) Lem. Southwestern Cape, South Africa. Photograph: Gideon F. Smith.

FIG. 3.7 *Adromischus umbraticola* C.A.Sm. subsp. *umbraticola*. North-West province, South Africa. Photograph: Gideon F. Smith.

2.2 *Cotyledon* L., *Sp. pl.* 1: 429 (1753)

The highest species diversity in *Cotyledon* is found in South Africa, with the vast majority of the species being endemic to the country. One species, *C. barbeyi* Schweinf. ex Baker, extends from northeastern South Africa northwards through eastern Africa and the Horn of Africa to the Arabian Peninsula, while in western south-tropical Africa, the widespread *C. orbiculata* L. enters southern Angola. *Cotyledon* species are generally shrubby and much-branched, with often quite large leaves arranged at intervals along sturdy or weak stems and branches. Flowers are usually quite showy, very large, and aloe-like with variously reflexed corolla lobes (Figs. 3.8–3.11).

Although a few new *Cotyledon* species have been described from South Africa in recent years, the genus is well studied. However, the taxonomy of the immensely variable *Cotyledon orbiculata* L., in which five varieties are recognised, requires resolution.

Illustrated reference: Van Jaarsveld & Koutnik (2004).

FIG. 3.8 *Cotyledon barbeyi* Schweinf. ex Baker var. *barbeyi*. Mpumalanga, South Africa. Photograph: Gideon F. Smith.

FIG. 3.10 *Cotyledon orbiculata* L. var. *dactylopsis* Toelken. Free State, South Africa. Photograph: Gideon F. Smith.

FIG. 3.9 *Cotyledon campanulata* Marloth. Eastern Cape, South Africa. Photograph: Gideon F. Smith.

FIG. 3.11 *Cotyledon orbiculata* L. var. *oblonga* (Haw.) DC. Eastern Cape, South Africa. Photograph: Gideon F. Smith.

2.3 *Kalanchoe* Adans., *Fam. pl.* 2: 248 (1763)

With only 17 species recorded, *Kalanchoe* is the smallest of the crassuloid genera found in southern Africa. Other significant present-day centres of species diversity include western south-tropical Africa, eastern Africa, and Madagascar, where the genus has diversified most extensively. The genus has also been recorded from the east, including India and China. The infrageneric classification of *Kalanchoe* is best reflected at the subgenus rank, with three subgenera recognised: *K.* subg. *Kalanchoe*, *K.* subg. *Bryophyllum*, and *K.* subg. *Kitchingia*. The latter two are naturally endemic to Madagascar but with several species and interspecific hybrids of *K.* subg. *Bryophyllum* having become naturalised in mild-climate parts of the world.

At species rank, the taxonomy of *Kalanchoe* is reasonably stable, but over the past few years, at least in southern Africa, several new species have been described, some infraspecific taxa raised to species rank, and range extensions recorded.

The present work is the first-ever comprehensive treatment of the kalanchoes of southern Africa.

2.4 *Tylecodon* Toelken in *Bothalia* 12: 378 (1978)

The genus *Tylecodon* was split from *Cotyledon*—the name is an anagram of 'Cotyledon'—and established for the soft-, deciduous-leaved species that were in the past included in the genus *Cotyledon* (Figs. 3.12 and 3.13). *Tylecodon* species are endemic to southern Africa and occur mostly in the western, winter-rainfall region of South Africa, with outliers in parts of the arid, western interior.

New species of *Tylecodon* are sporadically described, but the genus taxonomy is quite stable.

Illustrated reference: Van Jaarsveld & Koutnik (2004).

FIG. 3.12 *Tylecodon paniculatus* (L.f.) Toelken. Worcester-Robertson Karoo, South Africa. Photograph: Gideon F. Smith.

FIG. 3.13 *Tylecodon reticulatus* (L.f.) Toelken var. *reticulatus*. Groot Karoo, South Africa. Photograph: Gideon F. Smith.

4

The Genus *Kalanchoe*

FIG. 4.1   The terrestrial or epiphytic *Kalanchoe porphyrocalyx* (Baker) Baill., a Madagascan endemic, has large, purplish red tubular flowers. Photograph: Gideon F. Smith.

DERIVATION OF THE NAME *KALANCHOE*

Various possibilities have been offered as to the derivation of the name "Kalanchoe". Eggli & Newton (2004: 125) suggest two such possibilities:

- It is a phonetic transcription of the Chinese "Kalan Chauhuy", which means "that which falls and grows", although, as they remark, no bulbiliferous taxa are native to China; or
- It is derived from ancient Indian "kalanka-" (= spot, rust) and "chaya" (= gloss), perhaps referencing the glossy and perhaps sometimes reddish leaves of *Kalanchoe laciniata* that occurs in India.

Harvey (1862: 378) was of the view that the name "…is from the Chinese term for one of the species", a suggestion with which Johnson & Smith (1982: 58) agreed. More recently, Clarke & Charters (2016: 179) stated that "Kalanchoe" is the Chinese name, "jiā lán cài" (meaning "temple vegetable"), for *Kalanchoe ceratophyllum*, correctly *K. ceratophylla* Haw., a name treated by Descoings (2003: 144) as being of uncertain application. The protologue associated with the name *K. ceratophylla* was published in Haworth (1821: 23), and not in Haworth (1819), as stated by Descoings (2003: 144). The origin of the material on which Haworth (1821: 23) based the name was recorded as "*Habitat in Sina*". In the *Flora of China*, the name *K. ceratophylla* is today accepted as applicable to a species that occurs in "Fujian, Guangdong, Guangxi, Taiwan, Yunnan [India, SE Asia]" (Fu et al., 2001). The following taxa are treated as occurring in the *Flora of China* region: *K. ceratophylla*, *K. garambiensis* Kudô, *K. integra* (Medik.) Kuntze, *K. spathulata* var. *annamica* (Gagnep.) H.Ohba, and *K. tashiroi* Yamam. The common name given for representatives of the genus is "伽蓝菜属 jia lan cai shu".

SUPRA- AND INFRAGENERIC CLASSIFICATION AND NOMENCLATURE OF THE GENUS *KALANCHOE*

Along with the mostly southern African genera *Adromischus* Lem., *Cotyledon* L., and *Tylecodon* Toelken, as well as the Madagascan *Bryophyllum* Salisb., *Kalanchoe* was included in the tribe *Kalanchoeae* 't Hart in the family Crassulaceae ('t Hart, 1995: 167). Note though that the split of *Bryophyllum* from *Kalanchoe* sensu stricto is

nowadays largely regarded as artificial, with *Bryophyllum* being preferentially included in *Kalanchoe* (see below.) Perhaps rather paradoxically, representatives of the exclusively Madagascan *Bryophyllum* are generally better known globally than most of the African and Asian kalanchoes, given that a number of them tend to naturalise, some becoming weedy and invasive, in countries with a mild (tropical, subtropical, and Mediterranean) to temperate or even near-continental-type climate (Smith et al., 2015b). In contrast, species of *Kalanchoe* sensu stricto occur from Asia, through Madagascar and other Indian Ocean islands, as well as eastern and central Africa to southern Africa ('t Hart & Eggli, 1995; Thiede & Eggli, 2007). Recognition of *Kalanchoe* sensu lato (i.e. including *Kalanchoe* sensu stricto, *Bryophyllum*, and *Kitchingia*) (Fig. 4.1) was further supported by molecular data that indicated that a broadly conceived *Kalanchoe* is a monophyletic group ('t Hart, 1995; Van Ham & 't Hart, 1998; Mort et al., 2001).

In a somewhat revised classification of the Crassulaceae, Thiede & Eggli (2007: 110) proposed the inclusion of *Kalanchoe* (now with the Madagascan *Bryophyllum* and *Kitchingia* included in *Kalanchoe*) as part of the derived *Kalanchoe* clade, in the subfamily Kalanchooideae A.Berger (Berger, 1930; as Kalanchoideae), rather than in the tribe *Kalanchoeae* (as proposed by 't Hart, 1995: 167). The other genera included in the Kalanchooideae are *Adromischus*, *Cotyledon*, and *Tylecodon*. This suggested classification is nowadays widely followed.

Smith & Figueiredo (2018a) provided a comprehensive analysis of the history of the infrageneric classification of *Kalanchoe*. Their findings are summarised here.

The genus *Kalanchoe* includes plants that were at one time or another considered to belong to three separate genera: *Kalanchoe*, *Bryophyllum*, and *Kitchingia*. *Kalanchoe* (Figs. 4.2–4.4) was the first to be described (Adanson, 1763), followed by *Bryophyllum* (Salisbury, 1805), and *Kitchingia* (Baker, 1881). Shortly thereafter Baillon (1885: 468) considered *Kitchingia* to be an unranked subdivision (§) of *Kalanchoe*. In his revision of the group, Hamet (1907) was the first to treat *Kalanchoe*, *Bryophyllum*, and *Kitchingia* as a single genus; this view was widely accepted until Berger (1930), in the influential *Die Natürlichen Pflanzenfamilien*, provided a synopsis of the group and recognised the three taxa as distinct genera. The recognition of *Bryophyllum* as distinct from *Kalanchoe* was later followed by Resende (1956), and supported by Lauzac-Marchal (1974)

FIG. 4.2 *Kalanchoe marmorata* Baker occurs naturally in eastern and central Africa. The species has light green to glaucous green leaves that are ornamented with irregularly shaped, purple blotches. Photograph: Gideon F. Smith.

FIG. 4.3 *Kalanchoe millottii* Raym.-Hamet & H.Perrier is a small to medium-sized Madagascan species. Its velvety leaves, although much smaller, are reminiscent of those of the much more commonly cultivated, arborescent *K. beharensis*. Photograph: Gideon F. Smith.

FIG. 4.4 Often, all that remains visible after monocarpic *Kalanchoe* species have flowered are the dry, brittle leaves and stems. This specimen of *K. salazarii* Raym.-Hamet grew wedged between two rocks near Lubango, in southern Angola. Photograph: Gideon F. Smith.

and Byalt (2008), as well as by those authors who treated the Crassulaceae for regional African Floras, for example Fernandes (1982a, 1983) for the *Conspectus florae angolensis* and *Flora zambesiaca* projects, respectively; Tölken (1985) for the *Flora of Southern Africa*; and Wickens (1987) for the *Flora of Tropical East Africa*.

Hamet (1907, 1908a) proposed 13 unranked and unnamed groups (Table 4.1) for 54 species of *Kalanchoe*, to which he later added an additional group when he described *K. luciae* (Hamet, 1908b). Koorders (1918–1920) was the first to propose a subdivision of the genus in two ranked groups (subgenera): "*Kalanchoe* subg. *Eukalanchoe*" and *K.* subg. *Bryophyllum* (Salisb.) Koorders. However, the combination "*K.* subg. *Eukalanchoe*" was not validly published. The correct name for the typical subgenus of *Kalanchoe* is *K.* subg. *Kalanchoe*, as epithets for subdivisions of a genus cannot consist of the name of the genus with the prefix "*Eu-*" (McNeill et al., 2012: 49, Article 21.3.), and a subdivision of a genus that includes the type of the genus must have the same name as the genus (McNeill et al., 2012: 51, Article 22.1.).

With recent improvements in the understanding of the genus *Kalanchoe* as a whole, it is clear that the boundaries of *Kalanchoe* sensu stricto and *Bryophyllum* are artificial (Eggli et al., 1995; Descoings, 2003; Smith & Figueiredo, 2018a, b), and Thiede & Eggli (2007) rightly treated *Bryophyllum* as included in *Kalanchoe*. Chernetskyy (2012: 15) similarly argued that the existence of "intermediate" species makes it impossible to distinguish separate genera, a notion earlier supported by Mort et al. (2001) who argued for the inclusion of *Bryophyllum* in *Kalanchoe*, based on strong support from *matK* sequence analyses. Geographically, taxa previously regarded as belonging in *Bryophyllum* occur in Madagascar only, representatives of *Kalanchoe* sensu stricto occur throughout Africa, Madagascar, Arabia, and Asia, with particularly high species diversity in Madagascar (Allorge-Boiteau, 1996), and south-central, eastern, and southern Africa. Gehrig et al. (2001) also regarded *Kalanchoe* as best divided into three groups that would correspond to *Kalanchoe*, *Bryophyllum*, and *Kitchingia*.

In his treatment of the Crassulaceae, Berger (1930: 404) created 10 named, but unranked, groups (§) in the genus *Kalanchoe* sensu stricto that corresponded approximately to Hamet's groups (Table 4.2).

Boiteau (1947) proposed three sections in *Kalanchoe* for, respectively, the genera *Kalanchoe* sensu stricto, *Kitchingia*, and *Bryophyllum*, as *Kalanchoe* sect. *Eukalanchoe*, *K.* sect. *Kitchingia* (Baker) Boiteau, and *Kalanchoe* sect. *Bryophyllum* (Salisb.) Boiteau. "*Kalanchoe* sect. *Eukalanchoe*" was not validly published, the correct name being *K.* sect. *Kalanchoe*. Boiteau (1947) further proposed 14 subsections (§) (see Table 4.3), seven of which were included in *K.* sect. *Bryophyllum*, five in "*K.* sect. *Eukalanchoe*", and two in *K.* sect. *Kitchingia*.

Boiteau (1947) did not describe these subsections and did not provide Latin diagnoses or descriptions for the new groups; therefore, the names were not validly published (McNeill et al., 2012: 85, Article 39). However, two of the subsections, *K.* subsect. *Tricanthae* and *K.* subsect. *Integrifoliae*, being new combinations of Berger's unranked taxa, were validly published. These are cited as *K.* sect. *Kalanchoe* subsect. *Tricanthae* (A.Berger) Boiteau [= *K.* sect. *Kalanchoe* (unranked) *Tricanthae* A.Berger] and *K.* sect. *Kalanchoe* subsect. *Integrifoliae* (A.Berger) Boiteau [= *K.* sect. *Kalanchoe* (unranked) *Integrifoliae* A.Berger]. In subsequent papers, Boiteau & Mannoni (1948, 1949) described the subsections, up to subsection number 8, in French. However, they did not furnish these with Latin diagnoses or descriptions, and the new subsections were therefore still not validly published. This classification was later adopted by Boiteau & Allorge-Boiteau (1995: 16 [page unnumbered], 33–38), with the same three sections [*Kalanchoe* sect. *Kalanchoe*, *K.* sect. *Kitchingia* (Baker) Boiteau, and *Kalanchoe* sect. *Bryophyllum* (Salisb.) Boiteau] and 15 unranked and informal but named groups ("Groupe I–XV").

Descoings (2003) accepted two sections in *Kalanchoe* [*K.* sect. *Kalanchoe* and *K.* sect. *Bryophyllum*, which included *Kitchingia*]. Later, Descoings (2006) proposed an altogether new infrageneric classification, recognising three subgenera: *K.* subg. *Kalanchoe* (with 58 species), *K.* subg.

TABLE 4.1 Infrageneric classification of *Kalanchoe* according to Hamet (1907, 1908a, b)

Group no.	Distribution	Number of species
1	Madagascar	9
2	Madagascar	1
3	Madagascar	1
4	Madagascar	1
5	Madagascar	1
6	Madagascar	1
7	Madagascar	1
8	Madagascar	3
9	Madagascar, Comores, and one widespread species [*K. pinnata* (Lam.) Pers.]	9
10	Madagascar	7
11	Socotra	1
12	Socotra, Angola	1
13	Africa, Yemen, Arabia, India, and one widespread (*K. laciniata*, the type species of the genus, in sensu lato)	18
14	Africa	1

TABLE 4.2 Infrageneric classification of *Kalanchoe* by Berger (1930). Names accepted for southern African kalanchoes are given in bold, italic font in the fifth column; synonyms are given in nonbold, italic font in the same column

Genus	Infrageneric classification ("Gruppen" sensu Berger, 1930: 404)	Infrageneric classification ("Groupe" sensu Hamet, 1907: 877–879)	Distribution	Southern African species included
Kalanchoe	§ 1. *Stellatopilosae*	Group 8	Madagascar	–
	§ 2. *Crenatae*[a]	Group 13	West, East, and Tropical Africa [Sierra Leone, Angola, Congo, Somalia, Eritrea, Abyssinia, Tanzania, Zanzibar, Zambesi, Nyassa]; South Africa; Madagascar; Comoros; Socotra; Arabia; India; Yemen; China; Formosa; Malaca; Java; Brazil	***K. brachyloba*** (= *K. baumii* Engl. & Gilg) ***K. crenata*** (= *K. brittenii* Raym.-Hamet) ***K. hirta*** ***K. laciniata*** ***K. lanceolata*** (= *K. platysepala* Welw. ex Britten) ***K. longiflora*** ***K. rotundifolia*** (= *K. seilleana* Raym.-Hamet) ***K. sexangularis*** (= *K. vatrinii* Raym.-Hamet)
	§ 3. *Scapigerae*	Group 12	Angola; Socotra	–
	§ 4. *Tricanthae*	Groups 4 and 6	Madagascar	–
	§ 5. *Pubescentes*	Group 7	Madagascar	–
	§ 6. *Linearifoliae*	Group 10	Madagascar	–
	§ 7. *Socotranae*	Group 11	Socotra	–
	§ 8. *Transvaalenses*[b]	Group 14	South Africa	***K. luciae*** ***K. thyrsiflora***
	§ 9. *Integrifoliae*	Group 2	Madagascar	–
	§ 10. *Pumilae*	Group 3	Madagascar	–
Kitchingia		Group 1	Madagascar	–
Bryophyllum		Groups 5 and 9	Madagascar	–

[a] *Kalanchoe waterbergensis* would likely have been included in § 2. *Crenatae*, should it have been known at the time that Berger (1930) proposed his classification.
[b] *Kalanchoe montana* and *K. winteri* would have been included in § 8. *Transvaalenses*, should they have been known at the time that Berger (1930) proposed his classification.

TABLE 4.3 Infrageneric classification of *Kalanchoe* proposed by Boiteau (1947). The subsection names that were not validly published by Boiteau are given in nonitalic font in the third column. The names of the two subsections proposed by Berger (1930: 404) (§ 12 and § 13) that Boiteau (1947) validated are given in italic font

Genus	Section	Subsection
Kalanchoe	*Kitchingia*	§ 1. Sylvaticae § 2. Campanulatae
	Bryophyllum	§ 3. Centrales § 4. Epidendrae § 5. Scandentes § 6. Bulbilliferae § 7. Suffrutescentes § 8. Streptanthae § 9. Proliferae
	'Eukalanchoe'	§ 10. Lanigera § 11. Alpestres § 12. *Tricanthae* (A.Berger) Boiteau § 13. *Integrifoliae* (A.Berger) Boiteau § 14. Occidentales

TABLE 4.4   Placement of the southern African species of *Kalanchoe* in the classification of Descoings (2006)

No.	Southern African species	K. subg. *Kalanchoe*	K. subg. *Calophygia*	K. subg. *Bryophyllum*
1.	*Kalanchoe alticola* Compton		x	
2.	*Kalanchoe brachyloba* Welw. ex Britten	x		
3.	*Kalanchoe hirta* Harv.	Regarded as a synonym of *K. crenata*		
4.	*Kalanchoe crenata* (Andrews) Haw.		x	
5.	*Kalanchoe crundallii* I.Verd.		x	
6.	*Kalanchoe laciniata* (L.) DC.	x		
7.	*Kalanchoe lanceolata* (Forssk.) Pers.		x	
8.	*Kalanchoe leblanciae* Raym.-Hamet	x		
9.	*Kalanchoe longiflora* Schltr. ex Wood	x		
10	*Kalanchoe luciae* Raym.-Hamet		x	
11.	*Kalanchoe montana* Compton	Treated as a subspecies of *K. luciae*		
12.	*Kalanchoe neglecta* Toelken	x		
13.	*Kalanchoe paniculata* Harv.	x		
14.	*Kalanchoe rotundifolia* (Haw.) Haw.	x		
15.	*Kalanchoe sexangularis* N.E.Br.	x		
16.	*Kalanchoe thyrsiflora* Harv.		x	
17.	*Kalanchoe winteri* Gideon F.Sm., N.R.Crouch & Mich.Walters	Only published in 2016		
18.	*Kalanchoe waterbergensis* Van Jaarsv.	Only published in 2017		

Bryophyllum (with 26 species), and *K.* subg. *Calophygia* Desc. (with 66 species). The latter comprises the species that have characters (for example flower position, calyx and corolla structure, stamen insertion, shape and size of squamae, pistil structure) intermediate between the other two subgenera and that were previously placed in either *K.* subg. *Kalanchoe* or *K.* subg. *Bryophyllum*. Previously, the genera (or subgenera/sections) *Bryophyllum* and *Kitchingia* included only species from Madagascar. In this proposed classification of Descoings (2006), the delimitation of the infrageneric groups was amended, and the species occurring in Africa were classified either in *K.* subg. *Kalanchoe* or in *K.* subg. *Calophygia* (see Table 4.4), while the species from Madagascar were split among all three subgenera.

In terms of infrageneric classification, our concept of *Kalanchoe* accepts three subgenera that coincide with the three genera traditionally recognised, *K.* subg. *Kalanchoe*, *K.* subg. *Bryophyllum* (Salisb.) Koorders, and *K.* subg. *Kitchingia* (Baker) Gideon F.Sm. & Figueiredo. Placing some of the southern African species in *Kalanchoe* subg. *Calophygia*, as proposed by Descoings (2006), results in having some taxa that are likely related in different subgenera. Examples are: *K. rotundifolia* and *K. alticola*, and *K. laciniata* and *K. lanceolata*; the species in these pairs should be included in the same subgenus (see Table 4.4).

Kalanchoe in Southern Africa

FIG. 5.1   The flowers of *Kalanchoe longiflora* are some of the largest in the genus in southern Africa. The species was described from South Africa's KwaZulu-Natal province, to which it is endemic, by Medley Wood. Photograph: Gideon F. Smith.

BACKGROUND

In the *Flora of Southern Africa* (*FSA*) region [Namibia, Botswana, Swaziland, Lesotho, South Africa] 17[1] species of *Kalanchoe* are recognised, therefore, just over 10% of the global species diversity in the genus. In the subcontinent, this essentially Old World genus is largely absent from the western and south-central, succulent-rich, winter-rainfall and arid karroid parts; the 17 species occur predominantly in the more mesic, summer-rainfall, bush-veld (savanna) region, with significant outliers in typical grasslands (Figs. 5.1–5.7). *Kalanchoe* is therefore reasonably well represented in the subcontinent, where genera such as *Adromischus* Lem., *Cotyledon* L., *Crassula* L., and *Tylecodon* Toelken have noteworthy present-day centres of diversity (Tölken, 1977a, b, 1985), especially in the more western and southern regions. All 17 southern African *Kalanchoe* species belong to the typical subgenus, *K.* subg. *Kalanchoe*. The other two subgenera, *K.* subg. *Bryophyllum* (Salisb.) Koorders and *K.* subg. *Kitchingia* (Baker) Gideon F.Sm. & Figueiredo, are Madagascan groups (Boiteau & Allorge-Boiteau, 1995; Smith & Figueiredo, 2018a).

Prior to the recent description of *Kalanchoe winteri* in 2016, the last upheld *Kalanchoe* species described from the *Flora of Southern Africa* (*FSA*) region was *K. neglecta* from Maputaland, which was described about 40 years

FIG. 5.2   In the dry season (especially late autumn and winter), the leaves of *Kalanchoe sexangularis* var. *sexangularis*, Mpumalanga province, South Africa, turn bright red, especially if plants are exposed to direct sunlight. The species was described in England in the early 1900s by Brown based on material collected near Barberton by Thorncroft. Photograph: Gideon F. Smith.

[1] Note that an 18th species of southern African *Kalanchoe*, *K. waterbergensis* Van Jaarsv., was described after this book manuscript was finalised.

FIG. 5.3   *Kalanchoe rotundifolia* coming into flower in late autumn near Pretoria, Gauteng province, South Africa. This species was first collected by James Bowie in the Eastern Cape province of South Africa. Photograph: Gideon F. Smith.

FIG. 5.6   A strikingly beautiful form of *Kalanchoe luciae* with bluish green leaves that are purplish red-infused along the margins. The species was described by Raymond-Hamet. Photograph: Gideon F. Smith.

FIG. 5.4   *Kalanchoe paniculata* growing well camouflaged among tall grasses at the edge of a bush clump in savanna vegetation, North-West province, South Africa. This species was described by Harvey when he jointly worked on the *Flora capensis* in the mid-1800s. Photograph: Gideon F. Smith.

FIG. 5.7   *Kalanchoe lanceolata*, which is common in eastern and northern southern Africa, such as here in the Marakele National Park, Thabazimbi, North-West province, South Africa, was first described from the Arabian Peninsula. Photograph: Gideon F. Smith.

ago. This is indicative of how taxonomic activity in the genus has ebbed and flowed over the past more or less 150 years, with each period of study invariably leading to the description of new taxa. The monographic work of Raymond-Hamet (early-1900s to mid-1900s), and Flora treatments of Harvey (mid-1800s), Compton (1950–1970), and Toelken (1960s–1985) in particular have invariably led to an improvement of our understanding of the diversity and phytogeography of the kalanchoes of southern Africa (Table 5.1).

The indigenous southern Africa kalanchoes can be divided into two broad groups based on inflorescence and vegetative morphology. One group, consisting of *Kalanchoe luciae*, *K. montana*, *K. thyrsiflora*, and *K. winteri*,

FIG. 5.5   *Kalanchoe alticola* growing in northwestern Swaziland. This species was first recorded and described by Compton when he worked on the Flora of Swaziland. Photograph: Gideon F. Smith.

TABLE 5.1 Chronology of the description of *Kalanchoe* species that occur naturally in southern Africa

No.	Taxon	Date of description	Author(s) of the name
1.	*K. laciniata*	1753 (in *Cotyledon*)	Carl Linnaeus
		1802 (transferred to *Kalanchoe*)	Augustin Pyramus de Candolle
2.	*K. lanceolata* (see Fig. 5.7)	1775 (in *Cotyledon*)	Pehr Forsskål
		1805 (transferred to *Kalanchoe*)	Christiaan Hendrik Persoon
3.	*K. crenata*	1798 (in *Vereia*)	Henry Charles Andrews
		1812 (transferred to *Kalanchoe*)	Adrian Hardy Haworth
4.	*K. rotundifolia* (see Fig. 5.3)	1824 (in *Crassula*)	Adrian Hardy Haworth
		1825 (transferred to *Kalanchoe*)	Adrian Hardy Haworth
5.	*K. hirta*	1862	William Henry Harvey
6.	*K. paniculata* (see Fig. 5.4)	1862	William Henry Harvey
7.	*K. thyrsiflora*	1862	William Henry Harvey
8.	*K. brachyloba*	1871	Friedrich Welwitsch ex James Britten
9.	*K. longiflora* (see Fig. 5.1)	1903	Friedrich Richard Rudolf Schlechter ex John Medley Wood
10.	*K. luciae* (see Fig. 5.6)	1908	Raymond-Hamet
11.	*K. leblanciae*	1912	Raymond-Hamet
12.	*K. sexangularis* var. *sexangularis* (see Fig. 5.2)	1913	Nicholas Edward Brown
13.	*K. crundallii*	1946	Inez Clare Verdoorn
14.	*K. montana*	1967	Robert Harold Compton
15.	*K. alticola* (see Fig. 5.5)	1975	Robert Harold Compton
16.	*K. neglecta*	1978	Hellmut Richard Toelken
17.	*K. winteri*	2016	Gideon F. Smith, Neil R. Crouch & Michele Walters

bears very distinctive cylindrical inflorescences in which the flowers are densely arranged in a thyrse borne by a stout peduncle and the leaves are large, paddle-shaped, and somewhat to markedly rounded. All the other southern African *Kalanchoe* taxa have open, rather diffuse,

flat-topped inflorescences and smaller, often marginally crenate leaves. *Kalanchoe crundallii* is an exception in that it has characters somewhat reminiscent of both groups.

Table 5.1 provides a chronology of the description of *Kalanchoe* species that occur naturally in southern Africa, while information about the type localities and type specimens of *Kalanchoe* species indigenous to southern Africa is summarised in Table 5.2.

TABLE 5.2 Type localities and type specimens of *Kalanchoe* species that occur naturally in southern Africa

No.	Taxon	Type specimen(s)	Type locality
1.	*K. alticola*	*Compton 32107* (NBG NBG0076374-0 holo-; K000232855, PRE0390372-0, PRE0523573-0 iso-)	Swaziland, Mukusini Hills, *c.* 1300 m above sea level, 2631 (Mbabane) (-AB); 29 May 1964
2.	*K. brachyloba*	*Welwitsch 2486* (LISU LISU218888 lecto-!). *Welwitsch 2486* designated by Fernandes, 1980: 329 (as "holotype"), specimen LISU218888 designated by Smith & Figueiredo (2017a: 8) in a second-step lectotypification	Lower Guinea [West Africa], [Angola], Benguela, Huilla [probably Huíla province]
3.	*K. crenata*	*s.c. s.n.* (BM BM000649706, holo-)	"Hort. Vere 1798 (e Sierra Leone)"
4.	*K. crundallii*	*A.H. Crundall s.n. sub PRE 27157*, barcode PRE0457400-0, (PRE holo-!)	[South Africa], Transvaal [Limpopo province], Zoutspansberg [Soutspansberg] district, Mount Lejuma, no date
5.	*K. hirta*	*K.L.P Zeyher s.n.* (S S-G-3472, holo-)	[South Africa, Eastern Cape province], Olifantshoek, Uitenhage, [C.P.; Cape province]
6.	*K. laciniata*	*Hortus Cliffortianus 175* (BM BM000628567, lecto-). Designated by Fernandes (1980: 376)	Cultivated plant from Africa, "Cotyledon afra, folio crasso, lato laciniato, flosculo aureo laciniata"
7.	*K. lanceolata*	*Herb. Forsskål 689* (C C10002080, holo-)	Arabia, [Yemen], Kurma
8.	*K. leblanciae*	*H. Junod 443* (G holo-G00418140!, in 2 sheets; G G00418141!; P not seen, image not available online, BR BR0000008886750!; Z Z-000101954!).	[Mozambique], Delagoa Bay, 1893

Continued

TABLE 5.2 Type localities and type specimens of *Kalanchoe* species that occur naturally in southern Africa—cont'd

No.	Taxon	Type specimen(s)	Type locality
9.	*K. longiflora*	*J. Medley Wood 4439* (NH NH0005206-0, holo-; K K000232852 iso-)	[South Africa.], Natal [Kwazulu-Natal province].—Near the brook Dumbeni, Weenen District [—2830 (Dundee): between Greytown and Weenen, (–CD)], 15 April 1891
10.	*K. luciae*	*[H.] Junod s.n.* (G G00418142 holo-!; P)	South Africa, Limpopo province, Soutpansberg district, Spelonken
11.	*K. montana*	*Compton 29471* (NBG lecto-! [Tölken, 1978: 89]; NBG0099565-0 designated as lectotype in a second-step lectotypification by Smith et al., 2016a)	Swaziland, near Devil's Bridge, Pigg's Peak district, Emlembe Mountains, altitude 5500' [*c.* 1676 m] above sea level
12.	*K. neglecta*	*Vahrmeijer & Tölken 835* (PRE PRE0523752-0, holo-)	[South Africa], Natal [KwaZulu-Natal province], (the) Ubombo (region), Sordwana [Sodwana] Bay, 5 May 1965
13.	*K. paniculata*	*K.L.P. Zeyher 671* (S S-G-3475 lecto-; BM without barcode, K K000232854, GRA GRA0001092-0, SAM SAM0036012-1 & SAM0036012-2 isolecto-). Designated by Tölken (1985: 66)	[South Africa], Orange Free State [Free State], Vetrivier
14.	*K. rotundifolia*	Plate of *Kalanchoe rotundifolia* dated "Octob.[-er] 1st. 1823" prepared by Thomas Duncanson of sterile material (a stem in leaf). Below, the painting is written as follows: "Raised from seeds + roots forwarded by Mr [James] Bowie in 1822". (K neo-). Designated by Tölken (1985: 62), as lectotype; corrected to neotype by Figueiredo & Smith (2017b: 111)	Bowie likely collected material of *K. rotundifolia* during a journey he undertook to the eastern region of the Cape Colony (the vicinity of Uitenhage, Port Elizabeth, and Grahamstown) from early 1820 to 29 January 1821 (Figueiredo & Smith, 2017b). By 1822 the material was already in cultivation at Kew and had finished flowering by the time it was described by Haworth (1824)
15.	*K. sexangularis* var. *sexangularis*	*Thorncroft s.n.* (K K000232850, holo-)	[South Africa], Barberton [annotated in pencil] [Mpumalanga province, 2531 (–CC)], without a date
16.	*K. thyrsiflora*	*Ecklon & Zeyher 1953* (S S-G-10758 lecto-; SAM SAM0036022-0 pro parte isolecto-). Designated by Tölken (1985: 69)	Cape, Katrivier near Philipstown
17.	*K. winteri*	*P.J.D. Winter 4430*, (PRE holo-; BNRH, PRU iso-)	South Africa, Wolkberg, Limpopo province, Thabakgolo Escarpment, Sedibeng sa Lebese Mountain, west of Strasburg, 10 September 2000

HISTORICAL OVERVIEW

James Bowie (*c.* 1789–2 July 1869 [Claremont, Cape Town, South Africa]), who professionally collected plants for the Royal Botanic Gardens, Kew, in the early 1800s (Smith & Van Wyk, 1989), is acknowledged as having introduced *Kalanchoe rotundifolia* (Fig. 5.3), the first indigenous southern African kalanchoe with a type locality in South Africa, into cultivation in the United Kingdom (Table 5.1). This material originated from the Eastern Cape province of South Africa (Figueiredo & Smith, 2017b), likely from near the coastal city of Port Elizabeth. Thomas Duncanson, a botanical artist employed by Kew, illustrated two of Bowie's *K. rotundifolia* specimens. The first plate was of sterile material "Raised from seeds + roots forwarded by Mr [James] Bowie in 1822" and is the neotype of the name *K. rotundifolia* (Figueiredo & Smith, 2017b). The second plate was made of material in flower and produced on 23 March 1829, some seven years later.

Shortly after Bowie's material of *Kalanchoe rotundifolia* was described from South Africa (Table 5.2), initially in 1824 in error as a species of *Crassula* by Adrian Hardy Haworth (19 April 1768 [Hull, Yorkshire, England]–24 August 1833 [Chelsea, England]), but soon transferred to *Kalanchoe* by Haworth (1825) himself, Haworth (1829) produced a "remodelled" global treatment of the genus *Kalanchoe*. This work included only 10 species and was published four years before his death from cholera. This *Kalanchoe* treatment of Haworth (1829) was the first attempt at a global-scale monograph of the genus. *Kalanchoe rotundifolia* was the only southern African species collected in South Africa that was included in the treatment; it was recorded as having originated from "*Habitat* ad Caput Bonae Spei, ubi legit amicus Dom. Bowie". The other three *Kalanchoe* species that occur in southern Africa that Haworth treated had originated as follows:

1. *Kalanchoe laciniata*: "*Habitat* in India. Ægypto? ";
2. *Kalanchoe crenata*: "*Habitat* in Africâ, prope Sierram Leonam."; and
3. *Kalanchoe lanceolata*: "Habitat in Arabiâ" (Fig. 5.7).

From the 1790s to the first two decades of the 1800s, Haworth contributed extensively to the taxonomy of a number of southern African succulent plant families, including the Aizoaceae, Asphodelaceae, and Crassulaceae (Stearn, 1965, 1971).

After the publication of Haworth's *Kalanchoe* treatment in 1829, herbarium specimens of the other southern African species that are widespread in the subcontinent were for the first time prepared, mostly in the 1830s and 1840s, by early botanical collectors, including Carl (Karl) Ludwig Philipp Zeyher (1799–1858) and Christian Frederick (Friedrich) Ecklon (1795–1868) jointly with Zeyher. Ecklon and Zeyher together and individually amassed considerable sets of herbarium accessions while collecting material in South Africa during the first half of the 19th century (see Gunn & Codd, 1981 for information on their collecting activities).

In volume 2 of the *Flora capensis* publication series (Harvey, 1862), only six species of *Kalanchoe* were recognised for South Africa, mostly because the exploration of the northern and eastern parts of the country was still in its infancy. These regions of South Africa are host to the greatest diversity of kalanchoes in the country. Four of these species are still upheld, with a further one, *K. oblongifolia* Harv., which was described "from an imperfect specimen" (Harvey, 1862: 379), and *K. hirta*, later being included in the synonymies of *K. paniculata* and *K. crenata*, respectively. Note, however, that in our work, we regard *K. hirta* as distinct from *K. crenata* (Smith et al., 2018a).

Harvey (1862: 380) inevitably made extensive use of the herbarium collections of Ecklon and Zeyher, and based *Kalanchoe paniculata* (Fig. 5.4), *K. hirta*, and *K. thyrsiflora* on specimens that Ecklon and Zeyher, collectively or separately, collected. For example, *Zeyher 671* is the type of the name *K. paniculata*, which was collected in South Africa's Free State province [previously the Orange Free State], at Vetrivier. The Vetrivier, which literally translates into English as 'Fat River', arises between Marquard and Clocolan in the eastern Free State, near the border with Lesotho, and is a westward-flowing tributary of the substantial Vaalrivier in central South Africa. This area experiences very cold winters, and only two kalanchoes are known to occur in the area. Dyer (1947) noted that this particular Zeyher specimen (*671*) on which the name *K. paniculata* is based was collected en route during an expedition that Zeyher and another early collector, Joseph Burke (1812–1853), undertook to the South African interior, from 1839 to 1840. Burke and Zeyher only reached the Vetrivier on

23 February 1841. It was likely around this date that the type specimen was collected. This particular joint expedition that Burke and Zeyher undertook was the first one during which parts of the northern provinces of South Africa (formerly the Transvaal province, now split into the North-West, Gauteng, Limpopo, and Mpumalanga provinces) were investigated for botanical and other natural history specimens (Gunn & Codd, 1981: 111).

Predictably, some anomalies can arise when dealing with specimens collected by early travellers in southern Africa. One example concerns the type locality of *Kalanchoe thyrsiflora*, another species that Harvey (1862: 380) described. The type specimen on which this name is based was recorded as having originated from "[South Africa], Cape, Katrivier near Philipstown", with reference to *Ecklon & Zeyher 1953* (S S-G-10758 lecto-; SAM SAM0036022-0 pro parte isolecto-), the lectotype designated by Tölken (1985: 69). The specimen states: "Katrivier haud procul a Philipstown" (English: Katrivier not far from Philipstown) and "prope fontes fluminis Katrivier and Ceded territory" (English: near the source of the Katrivier and Ceded territory). However, the present-day "Katrivier" is not that near "Philipstown" at all. "Katrivier" is in the Eastern Cape (Leistner & Morris, 1976: 204), and "Philipstown" in the Northern Cape, and the Katrivier, a tributary of the Great Fish River, is only about 150 km long. It could be that the district of Philipstown had a very large size in the 1800s, but it remains a stretch to imagine these two locations as being "near" each other. Alternatively, other water courses in the vicinity of Philipstown may also have been named "Katrivier" as the same name was often used as a place name in different parts of South Africa or for distinct, widely separated, natural features in the landscape (see for example Leistner & Morris, 1976). However, we were unable to locate a Katrivier at or near Philipstown. Given that the species is well known from the northern parts of the Eastern Cape, we regard the type locality as most likely having been near the source of the present-day Katrivier [*c.* 32°34'17"S 26°45'34"E].

About 10 years after the publication of Harvey's *Flora capensis* treatment in 1862 of the Crassulaceae, *Kalanchoe brachyloba* was described (Britten, 1871) from material that the indefatigable Friedrich Welwitsch collected in Angola, where the species is not nearly as plentiful as in northeastern South Africa, for example. At this time, that part of the country had not yet been explored botanically. *Kalanchoe brachyloba* has a very broad geographical distribution range south of the equator that stretches across the African continent in a broad, west-east direction, in southern, south-tropical, and parts of tropical Africa.

After the publication of Harvey (1862) and Britten (1871), parts of two early African Floras that included coverage of the genus *Kalanchoe*, a further 30 years passed before the collecting activities in the subtropical eastern parts of the subcontinent, lead to the discovery of further species of this genus with type localities in South Africa: John Medley Wood (1 December 1827 [Mansfield, Nottinghamshire, England]–26 August 1915 [Durban, KwaZulu-Natal, South Africa]) collected and described *K. longiflora* (Fig. 5.1) from the KwaZulu-Natal midlands, while Henri-Alexandre Junod (17 May 1863 [St Martin, Val de Ruz, Switzerland]–22 April 1934 [Genève, Switzerland]), a Swiss Romande missionary, collected material of *K. luciae* (Fig. 5.6) in northeastern South Africa and *K. leblanciae* in Mozambique. Junod's material found its way to the Herbarium in Genève, Switzerland, with Raymond-Hamet, who was based in Paris, eventually describing both species. Junod was commemorated in *K. junodii* Schinz, which is generally considered conspecific with *K. lanceolata* (Forssk.) Pers.

George Thorncroft (1857 [Kent, England]–19 July 1934 [Barberton, South Africa]), an English trader and plant collector who immigrated to South Africa in 1882 (Gunn & Codd, 1981), eventually moved to Barberton in present-day Mpumalanga, after initially spending five years in KwaZulu-Natal. During his collecting activities near Barberton, Thorncroft came across material of a kalanchoe that he sent to the Cambridge Botanic Garden, and this was eventually described as a new species, *Kalanchoe sexangularis* (Fig. 5.2), by N.E. Brown (11 July 1849 [Redhill, Surrey, England]–25 November 1934 [Kew Gardens, London]) who was based at the Royal Botanic Gardens, Kew. In the protologue of the name *K. sexangularis*, Brown stated that he was describing the new species from "…a living plant sent by Mr Thorncroft to Cambridge Botanic Garden and communicated to Kew by R.I. Lynch" (Brown, 1913).

Shortly after WWII Inez Verdoorn (15 June 1896 [Pretoria, South Africa]–2 April 1989 [Pretoria, South Africa]) described material collected by Arthur Crundall in the late 1930s as *Kalanchoe crundallii*. This species has a very narrow distribution range along the Soutpansberg in the Limpopo province of South Africa. Verdoorn wrote one of the first Afrikaans textbooks on botany in South Africa (Verdoorn, 1942). In this work, Robert Allen Dyer (21 September 1900 [Pietermaritzburg, KwaZulu-Natal, South Africa]–26 October 1987 [Johannesburg, Gauteng, South Africa]), who was the director of the Botanical Research Institute from 1944 to 1963, a forerunner of the South African National Biodiversity Institute, wrote a chapter on the cultivation of succulents, including reference to *Kalanchoe* species and other representatives of the Crassulaceae (Dyer, 1942). This early Afrikaans work of Verdoorn (1942) followed on from early South African botany textbooks published in English, such as those of Edmonds & Marloth (1909) and Storey & Wright (1916; see p. 171 for reference to *Kalanchoe*).

R. Harold Compton (1886–1979), who retired as the director of the then National Botanical Gardens of South Africa in 1953, shortly thereafter started work on a Flora of Swaziland (Compton, 1966), the country where he temporarily took up residence. During his collecting activities in Swaziland, he came across material of several *Kalanchoe* entities that he later described as new species, for example, *K. montana* (Smith et al., 2016a) and *K. alticola* (Fig. 5.5). His *K. decumbens*, however, was synonymised with *K. rotundifolia* by Tölken (1985: 62).

Hellmut R. Toelken (1939–) revised the Crassulaceae for the *Flora of Southern Africa* project and substantially improved our understanding of the family, including of *Kalanchoe*. He described *Kalanchoe neglecta* in 1978.

Most recently, Gideon F. Smith, Neil R. Crouch, and Michele Walters (Crouch et al., 2016b) described *Kalanchoe winteri* from material that Pieter Winter collected in Wolkberg, Limpopo province, in 2000.

Smith, Crouch, and Estrela Figueiredo for the first time illustrated most of southern Africa's kalanchoes in their field guide to the succulents of southern Africa (Smith et al., 2017b).

ORIGIN OF THE TYPE MATERIAL OF SOUTHERN AFRICAN *KALANCHOE* SPECIES

The majority of the *Kalanchoe* species indigenous to southern Africa were described for the first time from material collected in the subcontinent, or slightly to the north in south-tropical Africa (Table 5.2). Three very widespread species, *K. crenata*, *K. laciniata*, and *K. lanceolata*, have type localities well beyond the borders of the subcontinent, as material was clearly first collected from places where they occur naturally in the northern hemisphere. In the case of *K. laciniata*, the type of the name was prepared from a cultivated plant that originated from [likely northern] Africa, while the type of *K. lanceolata* is from Yemen and that of *K. crenata* from Sierra Leone. In contrast, living material from which the illustration that serves as the type of the name *K. rotundifolia* was prepared was collected in South Africa's Eastern Cape province. *Kalanchoe rotundifolia* is the fourth species of southern African *Kalanchoe* that has an exceedingly broad distribution range that straddles the equator. The type of the name *K. hirta*, a predominantly southern African species, was similarly collected in the Eastern Cape, and it is likely that the origin of the type material of the name *K. thyrsiflora* was also in the Eastern Cape.

Kalanchoe brachyloba was for the first time collected from the Huíla province in southern Angola. However, this species does not occur north of the equator and is essentially a south-tropical and southern African species with a broad west-east geographical distribution range.

The majority of the southern African kalanchoes were for the first time described, with types from, the eastern, subtropical parts of southern Africa, especially South Africa's KwaZulu-Natal [*Kalanchoe longiflora, K. neglecta*] and Mpumalanga [*K. sexangularis*] provinces, Swaziland [*K. alticola, K. montana*], and Mozambique [*K. leblanciae*].

One species, *Kalanchoe paniculata*, has its type locality in South Africa's climatically severe, centrally located Free State province, while further north in South Africa, in the Limpopo province, again in a more subtropical part of the subcontinent, three species [*K. crundallii, K. luciae, K. winteri*] were collected for the first time.

RESEARCHERS ON SOUTHERN AND SOUTH-TROPICAL AFRICAN *KALANCHOE*

A NOTE ON THE ORTHOGRAPHY OF THE NAMES OF THREE RESEARCHERS ON SOUTHERN AFRICAN CRASSULACEAE

- Selmar Schonland. Schonland's surname was originally spelled with an 'umlaut' on the 'o', as in "Schönland", but Schonland himself later dropped the diacritical sign, preferring the spelling "Schonland" (Gunn & Codd, 1981: 318). Rather confusingly, the standardised form of his surname in the authorship of novel plants he described or in plants he reclassified is "Schönland", with the 'umlaut' therefore retained (Brummitt & Powell, 1992: 573).
- Raymond-Hamet. As far as we could ascertain, it was shortly after *Kalanchoe leblanciae* was described (Hamet, 1912a) that Raymond-Hamet changed his name to "Raymond-Hamet", therefore, by hyphenating his given name and surname. Shaw (2008: 29) notes that this may have been done (perhaps officially) in 1935 only, as "Hamet" was still listed as such in the annual list of members published in the *Bulletin de la Société Botanique de France*, in volume 82 (1935), and after that as "Raymond-Hamet". In our book, his publications are cited under either "Hamet, R.", or "Raymond-Hamet", depending on which version of his name is used in the publication referenced. Raymond-Hamet's abbreviation as an author in plant nomenclature is "Raym.-Hamet" (Figs. 5.8 and 5.9).
- Hellmut Toelken. Although Hellmut Toelken published numerous works using the spelling "Tölken" of his surname, he prefers the "ö" being transcribed as "oe" (H.R. Toelken personal communication dated 05 December 2016 to one of us [G.F. Smith]). We cite his publications under the spelling of his surname as it appeared in his various publications. Toelken's abbreviation as an author in plant nomenclature is "Toelken". Hellmut's officially correct first name, as on

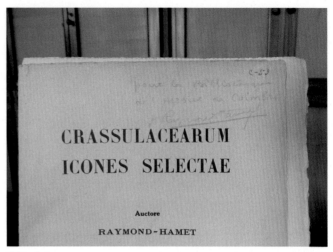

FIG. 5.8 This unbound copy of "Fasciculus Tertius", the third fascicle, of *Crassulacearum icones selectae* (*CIS*) that Raymond-Hamet self-produced in Paris, France, in 1958, is kept in the library of the Herbarium (COI) of the University of Coimbra, Coimbra, Portugal. Raymond-Hamet signed and sent a full set of all five parts ("Fasciculus Primus–Fasciculus Quintus") of his *CIS* to the Herbarium, where Rosette Fernandes worked on the Crassulaceae, including on *Kalanchoe*. Photograph: Gideon F. Smith.

FIG. 5.9 For *Kalanchoe germanae* Raym.-Hamet ex Raadts, Raymond-Hamet included, inter alia, a photograph ("Tab. 23") of a herbarium specimen in Fasciculus 2, dated 1956, of his *Crassulaceaearum icones selectae* (*CIS*). The name *K. germanae* was proposed by Raymond-Hamet, but as with many other names included in his *CIS*, not validly published by him. Raadts (1984: 377) eventually validly published the name for this Tanzanian species. Photograph: Gideon F. Smith.

his birth certificate, is Helmut (spelled with one "l"). This only transpired when he left South Africa for Australia in 1979. However, for personal use, he continues to use Hellmut (spelled with two "l"s) as he had always done (H.R. Toelken personal communication dated 06 March 2018 to one of us [G.F. Smith]).

In these three cases, when referring to the individuals, and not their publications, we use the names "Raymond-Hamet", "Schonland", and "Toelken", to honour their wishes.

William Henry Harvey (5 February 1811 [Summerville, Ireland]–15 May 1866 [Torquay, England]).

Between 1835 and 1842, while in his mid-20s and early-30s, William Harvey (Fig. 5.10), who had an interest in natural history from a young age, spent about four years in South Africa. This exposure to the floristic riches of the country prepared him to eventually take on the onerous task of contributing to a Flora of one of the colonies of Great Britain, the *Flora capensis* [*being a systematic description of the plants of the Cape Colony, Caffraria, & Port Natal*]. By 1836, he was appointed Colonial Treasurer for the Cape, a post theretofore briefly filled by his brother, Joseph, who died in mid-1836. Such were Harvey's commitments to his botanical activities (for which he was not employed at the Cape) that, by 1838, he had authored *The genera of South African plants*, the first substantial botanical book to have been published in South Africa (Victor et al., 2016).

FIG. 5.10 William Henry Harvey (1811–66). Photo: Ipswich Museum, via Wikimedia Commons.

By 1856, Harvey accepted the position of Chair of Botany at the University of Dublin. Earlier, in the mid-1840s, he had been awarded an honorary MD degree by the Senate of the Trinity College, Dublin. Together with Otto W. Sonder (13 June 1812 [Oldesloe, Schleswig-Holstein, Germany]–21 November 1881 [Hamburg, Germany]) of Hamburg, Harvey agreed to undertake the writing of the monumental *Flora capensis*, for which he contributed most of the first three volumes between 1860 and 1865. He therefore wrote the first floristic work that covered the genus *Kalanchoe* in South Africa, with three of the species he described (*K. hirta*, *K. paniculata*, and *K. thyrsiflora*) still being upheld. In 1866, one year after the third volume of *Flora capensis* appeared, Harvey died at the age of 55. His floristic works had a lasting impact on botany in South Africa, and many taxonomists who followed him were enthused by the prospects of finally having a descriptive and identification manual on the richest temperate flora in the world (Gunn & Codd, 1981: 179).

Selmar Schonland (15 August 1860 [Frankenhausen, Germany]–22 April 1940 [Grahamstown, South Africa]).

Soon after having obtained his PhD from Kiel University in Germany, Selmar Schonland (Fig. 5.11) became interested in the Crassulaceae and contributed an account of the family, including of *Kalanchoe*, to A. Engler & K. Prantl's *Die natürlichen Pflanzenfamilien* (Schönland, 1891).

Schonland emigrated to South Africa in 1889, accepting a position as head of the Albany Museum in Grahamstown in the Eastern Cape. Schonland maintained his interest in the Crassulaceae while he was the director of the Museum and published scores of new species in the genus *Crassula* and one species of *Kalanchoe*, *K. pyramidalis* Schönland, which is today included in the synonymy of *K. brachyloba* (Schönland, 1903a, b, 1907, 1929).

Raymond-Hamet (25 March 1890 [Dijon, Côte D'Or, France]–2 October 1972 [Paris, France]).

What little is known about Raymond-Hamet concerns his precociousness and eccentricity. He was born in Dijon, France, son of Étienne François Donatien Hamet and Ernestine Amique, and named Raymond Hamet. He was a pharmacologist, botanist, chemist, and doctor *honoris causa* in medicine (Marion, 1973; Schmitt, 1974), as well as a barrister at the court of appeal.

Remarkably, he started publishing scientific papers when he was a teenager; his first paper, describing a new species, was published in 1905 when he was 15 years old. Two years later, well before he was 20, he published the first part of a revision of *Kalanchoe*, the second part following one year later. In these two papers, he cited an astounding number of herbarium specimens and literature references and was clearly familiar with taxonomic methodology. After graduating from university, Raymond-Hamet wanted to research the subjects he was interested in without constraints, so he installed a laboratory in his house at Rue de la Glacière in Paris, time of his death in 1972. Drawings of plants, with proposed names but no descriptions, to which he apparently still intended to add descriptive text, were collected into five fascicles published by himself in Paris as *Crassulacearum icones selectae* (Figs. 5.8 and 5.9). Accordingly, Fernandes (1978–1980), Tölken (1978), and others, independently, re-evaluated some of these (for *Kalanchoe*), validating some names and discarding others.

Robert Harold Compton (06 August 1886 [Tewkesbury, England]–11 July 1979 [Cape Town, South Africa]).

Harold Compton was the second director of the National Botanic Gardens of South Africa, one of the forerunners of the present-day South African National Biodiversity Institute (SANBI) (Fig. 5.12). Compton was appointed to this administrative post shortly after WWI and additionally made the first Harold Pearson professor of botany at the University of Cape Town (Gunn & Codd, 1981), after the untimely death in 1916 of Prof. H. Harold W. Pearson, the first director of the Botanic Gardens. Compton held both positions from March 1919 until his retirement in 1953, a remarkable 34-year period. Compton was, inter alia, responsible for the establishment of

FIG. 5.11   Selmar Schonland (1860–1940). Photo: Selmar Schonland Herbarium, Albany Museum.

France, and pursued his research at his own expense. There, he developed pharmacological studies, conducting experiments on dogs (Marion, 1973). He is perhaps mostly known for his contributions to chemistry, especially our knowledge of the properties of alkaloids (Shaw, 2008, 2018).

With an interest in Crassulaceae since he was a teenager, Raymond-Hamet became a renowned specialist on the family, especially of the genera *Kalanchoe* and *Sedum* L. Raymond-Hamet is often acknowledged as rather idiosyncratically coining the epithet *'mitejea'* for a species of *Kalanchoe*. The epithet was an anagram for 'je t'aime', and the species is attributed to Leblanc & Raym.-Hamet. Alice Leblanc, about whom nothing is known, was also commemorated by him in *Kalanchoe aliciae* Raym.-Hamet, *K. leblancae* Raym.-Hamet, and *Sedum celiae* Raym.-Hamet, the latter epithet also being an anagram.

Raymond-Hamet died at home at the age of 82. Probably due to his many interests, he had not validly published a number of the names for his concepts by the

FIG. 5.12   Robert Harold Compton (1886–1979). Photo: South African National Biodiversity Institute.

the *Journal of South African Botany* in 1935. He edited this journal until his retirement and favoured it as the vehicle for disseminating his own research findings. The Compton Herbarium (acronym NBG), located today in the Kirstenbosch Research Centre of SANBI in Cape Town, is named in his honour.

After retiring from the directorship of the Botanic Gardens, Compton, who apparently had more than a passing interest in *Kalanchoe*, moved to Swaziland and worked on a Flora for this eastern southern African country for the next 20-odd years. During his research activities in Swaziland, he collected and described a number of new *Kalanchoe* species, two of which are upheld. By the time his Flora of Swaziland appeared in 1976, he had already returned to Cape Town [in 1971] where he lived until his death in 1979 (Gunn & Codd, 1981: 121).

Rosette Mercedes Saraiva Batarda Fernandes (1 October 1916 [Redondo, Portugal]–28 May 2005 [Coimbra, Portugal]).

Rosette Fernandes was a taxonomist at the Herbarium (COI) of the University of Coimbra, Portugal (Fig. 5.13). She graduated from the University of Lisbon in 1941

FIG. 5.13 Rosette Batarda Fernandes (1916–2005). Image kept in the private collection of Dr. António Coutinho, University of Coimbra, Portugal. Reproduced with his permission.

and later moved to Coimbra, starting her career as a naturalist in the Herbarium in 1947 and staying there until she retired as a Principal Researcher.

Fernandes was arguably the most prolific taxonomist in Portugal in the 20th century with hundreds of taxon names published, mostly in the families Anacardiaceae, Crassulaceae, Lamiaceae, Melastomataceae, and Verbenaceae, including *c.* 50 new taxa, three being new genera. She published over 250 papers (Coutinho, 2007) and produced treatments for several Floras, including *Flora europaea*, *Flora iberica*, *Flora de Moçambique*, *Conspectus florae angolensis*, and *Flora zambesiaca*.

The treatments of the Crassulaceae, including those of *Kalanchoe*, for the regions covered by the *Conspectus florae angolensis* (Fernandes, 1982a) and *Flora zambesiaca* projects (Fernandes, 1983), were published in quick succession. This afforded Fernandes the opportunity to compare herbarium collections across a broad, contiguous swathe of Africa south of the equator, from Angola in the west to Mozambique in the east. She additionally expressed views on the taxonomy of some southern African crassuloid species; at that time, Hellmut Toelken worked on the Crassulaceae treatment for the *Flora of Southern Africa* project. Inevitably, there are differences between their treatments of the family for these abutting and overlapping (Botswana) geographical areas. This is exampled by Fernandes (1983: 72) recognising *Cotyledon oblonga* Haw. at species rank, while Tölken (1985: 14) treated it at the varietal rank, as *Cotyledon orbiculata* L. var. *oblonga* (Haw.) DC.

Rosette Fernandes is known to have communicated and exchanged reprints of published papers with Raymond-Hamet in Paris, France, as there is at least a partial record of the reprints she mailed to him. However, her correspondence with Raymond-Hamet and other researchers for that matter is not deposited in the Herbarium of the University of Coimbra where she worked nor as far as we know elsewhere in the University archives; the correspondence is either held privately or was destroyed.

Hellmut Richard Toelken (01 September 1939 [Windhoek, Namibia]–).

Hellmut Toelken completed his first degree at the Stellenbosch University and thereafter successively registered for and obtained BSc (Hons), MSc, and PhD degrees from the University of Cape Town (Fig. 5.14). As the topic of his PhD dissertation, Toelken studied the genus *Crassula*, the results of which were published in two parts in the journal *Contributions from the Bolus Herbarium*; together, the two parts took up nearly 600 printed pages (Tölken, 1977a, b). Toelken was therefore well placed to complete a treatment of the entire family Crassulaceae in southern Africa for the *Flora of Southern Africa* project; the outcomes of his *FSA* work on the family appeared in print eight years after his PhD was published (Tölken, 1985).

FIG. 5.14 Hellmut Richard Toelken (1939–). Photograph: H.R. Toelken.

FIG. 5.15 Neil Robert Crouch (1967–). Photograph: Tanza Crouch.

novelty, and the reassessment of the status and rank of various genus members. The resulting reports have regularly appeared in the scientific press since the late 1990s.

From 1963 until 1967, Toelken was employed by the Botanical Research Institute (BRI), one of the forerunners of the South African National Biodiversity Institute (SANBI), after which he lectured at the University of Cape Town until 1973. From September 1973 until 1979, he was again attached to the BRI, including completing a stint as South Africa's Botanical Liaison Officer at the Royal Botanic Gardens, Kew, from June 1974 until July 1976. In 1979, Toelken emigrated to Adelaide, Australia (Gunn & Codd, 1981: 351).

In Australia, Toelken pursued his research on the local Crassulaceae, *Kunzea* Rchb. (Myrtaceae), and *Hibbertia* Andrews (Dilleniaceae). He also contributed to and was coeditor of *Flowering Plants in Australia* and *Flora of South Australia*.

Neil Robert Crouch (29 January 1967 [Bulawayo, Zimbabwe]–).

Neil Crouch, a PhD graduate of the University of Natal (now the University of KwaZulu-Natal) and subsequently employed by the National Botanical Institute, one of the forerunners of SANBI, has worked on several aspects of *Kalanchoe* taxonomy, biology, geography, and ethnobotany (Fig. 5.15). His work has resulted in new southern African country records, the description of a

KALANCHOES IN SOUTHERN AFRICAN BOTANICAL ART

Since it was first published in 1920, the *Flowering Plants of Africa* (*FPA*) has been the primary journal that deals with and promotes botanical art in southern Africa (Arnold, 2001). To date (2017), 2340 plates dealing with a vast range of, almost exclusively, flowering plants have been published in the 65 volumes of the series. With the emphasis strongly on paintings of visually attractive material, botanical art of a wide diversity of succulents has been included. Up to now, 15 plates of kalanchoes have graced the pages of *FPA*. Eight of these are of taxa that occur in southern African (therefore, *c.* 50% of the indigenous species), two are of African species that do not occur naturally in southern African, two are Socotran, and three depict Madagascan species.

As from volume 25 [1945 (published in 1946)] the *Flowering Plants of South Africa* (*FPSA*) was renamed the *Flowering Plants of Africa* (*FPA*), for a while with "*The*" as part of the title, so broadening the scope of the material that qualifies for inclusion (Killick & Du Plessis, n.d.). Madagascar was included in the broad concept of what constituted "Africa", especially as several staff members, notably David S. Hardy, of the Institute that published the journal, and nonstaff collaborators, such as Gilbert

W. Reynolds, had a research and horticultural interest in the Red Island. Since 1946, numerous Madagascan plant taxa have been illustrated in *FPA*, especially a range of succulents, including many aloes. To date, only three Madagascan *Kalanchoe* species have been featured in the journal, two of which are treated in this book as they either are invasive (*K. tubiflora*) or show such tendencies (*K. gastonis-bonnieri*) (Table 5.3).

Among the African species illustrated, only *Kalanchoe marmorata* has achieved a low level of popularity in domestic and amenity horticulture in southern Africa. The other African species, the two from Socotra and *K. beauverdii* from Madagascar, are largely unknown in cultivation in the subcontinent.

A wide range of botanical artists has contributed plates to *FPA* (Condy & Rourke, 2001), including of kalanchoes,

with five (33⅓%) of the kalanchoe plates having been done by Gill Condy.

Apart from paintings of about 50% of the kalanchoes having appeared in *FPA* over the past *c.* 90 years, these striking plants have been included in several other works that use botanical art to illustrate the flora of specific parts of southern Africa, or provinces of South Africa. Of these, the paintings executed by Anita Fabian (Condy & Rourke, 2001: 193) that were included in Germishuizen & Fabian (1982, 1997) are perhaps the best known. Five species that all occur in the northern provinces of South Africa were featured in this work. Fabian also contributed the art work to Hardy & Fabian (1992), where different paintings of some of the same species included in Germishuizen & Fabian (1982, 1997) were featured.

TABLE 5.3 Chronology of southern African, other African, and Madagascan *Kalanchoe* taxa illustrated in the *Flowering Plants of South Africa* and the *Flowering Plants of Africa*. The name under which the plants were illustrated is given in parentheses in the second column

No.	Taxon	Volume no.	Plate no.	Artist	Author	Date
Indigenous southern African taxa						
1.	*K. thyrsiflora*	9	341	Cythna Letty	E.P. Phillips	July 1929
2.	*K. crundallii*	25	967	E. (Edith) K. Burges	I.C. Verdoorn	January 1945[a]
3.	*K. paniculata*	26	1007	R. Brown[b]	R.A. Dyer	January 1947
4.	*K. laciniata* (=*K. brachycalyx*)	27	1052	Artist not stated	E.A Bruce	January 1949
5.	*K. lanceolata*	47[c]	1848	Rita Weber	H.R. Tölken	September 1982
6.	*K. sexangularis* var. *sexangularis*	47	1878	Rosemary Holcroft	H.R. Tölken	June 1983
7.	*K. crenata* subsp. *crenata*	61	2249	Gill Condy	N.R. Crouch & G.F. Smith	June 2009
8.	*K. leblanciae*	65	2328	Gill Condy	G.F. Smith, E. Figueiredo & N.R. Crouch	June 2017
Other African species						
9.	*K. marmorata*	27	1049	Edith K. Burges	E.A Bruce	September 1948
10.	*K. densiflora*	28	1089	Edith K. Burges	E.A Bruce	April 1950
Socotran species						
11.	*K. farinacea*	34	1329	P.R.O. Bally	P.R.O. Bally	March 1960
12.	*K. robusta*	45	1783	Rhona Collett	R.A. Dyer	June 1979
Madagascan species						
13.	*K. tubiflora* (=*Bryophyllum delagoense*)	49	1938	Gill Condy	H.R. Tölken & O.A. Leistner	September 1986
14.	*K. beauverdii* (=*Bryophyllum beauverdii*)	52	2050	Gill Condy	P.B. Phillipson	May 1992
15.	*K. gastonis-bonnieri*	52	2051	Gill Condy	W. Rauh	May 1992

[a] Published in July 1946.
[b] Rhona Collett (née Brown) (Condy & Rourke, 2001: 190).
[c] Plate later reproduced in Smith et al. (1997: x).

COMMON NAMES OF *KALANCHOE* IN SOUTHERN AFRICA

In this treatment, 126 records of common names of *Kalanchoe* species were compiled for the *Flora of Southern Africa* region (see Table 5.4). These records refer to 123 distinct names in the following languages: English (56), Afrikaans (39), isiXhosa (6), isiZulu (10), Setswana (6), siSwati (2), Sesotho (1), Tshiveṇḍa (2), and one in an unknown language. The most common name that applies to several species and to some degree also used for *Kalanchoe* as a whole, as well as representatives of other genera of the Crassulaceae, is the Afrikaans name 'plakkie'.

TABLE 5.4   Alphabetical list of the common names of *Kalanchoe* species compiled for the *Flora of Southern Africa* region. For the sake of accuracy, the author citations of the *Kalanchoe* species names are given in the third column

Common name	Language	Species
air plant	English	*Kalanchoe pinnata* (Lam.) Pers.
alligator plant	English	*Kalanchoe daigremontiana* Raym.-Hamet & H.Perrier
bergplakkie	Afrikaans	*Kalanchoe montana* Compton
blomkalanchoe	Afrikaans	*Kalanchoe blossfeldiana* Poelln.
blooming boxes	English	*Kalanchoe prolifera* (Bowie ex Hook.) Raym.-Hamet
boksplakkie	Afrikaans	*Kalanchoe prolifera* (Bowie ex Hook.) Raym.-Hamet
bolatsi	Setswana	*Kalanchoe paniculata* Harv. *Kalanchoe rotundifolia* (Haw.) Haw.
bosveldplakkie	Afrikaans	*Kalanchoe sexangularis* N.E.Br.
bush kalanchoe	English	*Kalanchoe hirta* Harv.
bushveld kalanchoe	English	*Kalanchoe sexangularis* N.E.Br.
Canterbury bells	English	*Kalanchoe pinnata* (Lam.) Pers.
cathedral bells	English	*Kalanchoe pinnata* (Lam.) Pers.
chandelier plant	English	*Kalanchoe tubiflora* (Harv.) Raym.-Hamet
chinese lantern (kalanchoe)	English	*Kalanchoe prolifera* (Bowie ex Hook.) Raym.-Hamet
Christmas tree kalanchoe	English	*Kalanchoe laciniata* (L.) DC.
Christmas tree plant	English	*Kalanchoe laciniata* (L.) DC.
coin kalanchoe	English	*Kalanchoe alticola* Compton
common kalanchoe	English	*Kalanchoe rotundifolia* (Haw.) Haw.

TABLE 5.4   Alphabetical list of the common names of *Kalanchoe* species compiled for the *Flora of Southern Africa* region. For the sake of accuracy, the author citations of the *Kalanchoe* species names are given in the third column—cont'd

Common name	Language	Species
crowned kalanchoe	English	*Kalanchoe crenata* (Andrews) Haw.
Crundall's kalanchoe	English	*Kalanchoe crundallii* I.Verd.
Crundall-se-plakkie	Afrikaans	*Kalanchoe crundallii* I.Verd.
curtain plant	English	*Kalanchoe pinnata* (Lam.) Pers.
devil's backbone	English	*Kalanchoe daigremontiana* Raym.-Hamet & H.Perrier
donkey's ear	English	*Kalanchoe beharensis* Drake
donkie-oor	Afrikaans	*Kalanchoe beharensis* Drake
elephant's ear kalanchoe	English	*Kalanchoe beharensis* Drake
flaming katy	English	*Kalanchoe blossfeldiana* Poelln.
flipping flapjacks	English	*Kalanchoe luciae* Raym.-Hamet
flower dust kalanchoe	English	*Kalanchoe pumila* Baker
geelplakkie	Afrikaans	*Kalanchoe thyrsiflora* Harv.
green mother of millions	English	*Kalanchoe prolifera* (Bowie ex Hook.) Raym.-Hamet
hasie[s]oor	Afrikaans	*Kalanchoe paniculata* Harv.
idambisa	isiZulu	*Kalanchoe rotundifolia* (Haw.) Haw.
idlebe lenkau	isiZulu	*Kalanchoe montana* Compton
indabulaluvalo	siSwati	*Kalanchoe paniculata* Harv.
indunjane	isiZulu	*Kalanchoe tubiflora* (Harv.) Raym.-Hamet
ipewula	isiXhosa	*Kalanchoe rotundifolia* (Haw.) Haw.
kalanchoe	English	*Kalanchoe lanceolata* (Forssk.) Pers.
kandelaarplant	Afrikaans	*Kalanchoe tubiflora* (Harv.) Raym.-Hamet
krimpsiektebos	Afrikaans	*Kalanchoe brachyloba* Welw. ex Britten
krimpsiektebos[-sie]	Afrikaans	*Kalanchoe paniculata* Harv.
kroonplakkie	Afrikaans	*Kalanchoe crenata* (Andrews) Haw.
lanternplakkie	Afrikaans	*Kalanchoe prolifera* (Bowie ex Hook.) Raym.-Hamet
large orange kalanchoe	English	*Kalanchoe paniculata* Harv. *Kalanchoe rotundifolia* (Haw.) Haw.

Continued

TABLE 5.4 Alphabetical list of the common names of *Kalanchoe* species compiled for the *Flora of Southern Africa* region. For the sake of accuracy, the author citations of the *Kalanchoe* species names are given in the third column—cont'd

Common name	Language	Species
life plant	English	*Kalanchoe gastonis-bonnieri* Raym.-Hamet & H.Perrier
Madagascar widow's-thrill	English	*Kalanchoe blossfeldiana* Poelln.
mahogwe	isiZulu	*Kalanchoe crenata* (Andrews) Haw.
makplakkie	Afrikaans	*Kalanchoe pinnata* (Lam.) Pers.
maternity plant	English	*Kalanchoe daigremontiana* Raym.-Hamet & H.Perrier
mathongwe	isiZulu	*Kalanchoe crenata* (Andrews) Haw.
mbohlolololeshate	isiZulu	*Kalanchoe hirta* Harv.
meelblomplakkie	Afrikaans	*Kalanchoe pumila* Baker
meelplakkie	Afrikaans	*Kalanchoe thyrsiflora* Harv.
Mexican hat plant	English	*Kalanchoe daigremontiana* Raym.-Hamet & H.Perrier
mfayisele	isiXhosa	*Kalanchoe rotundifolia* (Haw.) Haw.
moêthimodisô/ moethimodiso	Setswana	*Kalanchoe rotundifolia* (Haw.) Haw. *Kalanchoe lanceolata* (Forssk.) Pers.
mother of millions	English	*Kalanchoe tubiflora* (Harv.) Raym.-Hamet
mother-of-thousands	English	*Kalanchoe daigremontiana* Raym.-Hamet & H.Perrier
mountain kalanchoe	English	*Kalanchoe montana* Compton
narrow-leaved kalanchoe	English	*Kalanchoe lanceolata* (Forssk.) Pers.
nenta(bos)	Afrikaans	*Kalanchoe rotundifolia* (Haw.) Haw.
nentebos	Afrikaans	*Kalanchoe rotundifolia* (Haw.) Haw.
nyêthi	Setswana	*Kalanchoe lanceolata* (Forssk.) Pers.
orange forest kalanchoe	English	*Kalanchoe crenata* (Andrews) Haw.
paddle plant	English	*Kalanchoe luciae* Raym.-Hamet
palm beachbells	English	*Kalanchoe gastonis-bonnieri* Raym.-Hamet & H.Perrier
panda plant	English	*Kalanchoe tomentosa* Baker
pennie-kalanchoe	Afrikaans	*Kalanchoe alticola* Compton
perskalanchoe	Afrikaans	*Kalanchoe fedtschenkoi* Raym.-Hamet & H.Perrier

TABLE 5.4 Alphabetical list of the common names of *Kalanchoe* species compiled for the *Flora of Southern Africa* region. For the sake of accuracy, the author citations of the *Kalanchoe* species names are given in the third column—cont'd

Common name	Language	Species
persplakkie	Afrikaans	*Kalanchoe fedtschenkoi* Raym.-Hamet & H.Perrier
pienkkalanchoe	Afrikaans	*Kalanchoe pumila* Baker
pig's ears	English	*Kalanchoe lanceolata* (Forssk.) Pers.
plakkie	Afrikaans	*Kalanchoe brachyloba* Welw. ex Britten *Kalanchoe crenata* (Andrews) Haw. *Kalanchoe hirta* Harv. *Kalanchoe rotundifolia* (Haw.) Haw. *Kalanchoe thyrsiflora* Harv.
plakkiesblom	Afrikaans	*Kalanchoe brachyloba* Welw. ex Britten
pregnant plant	English	*Kalanchoe tubiflora* (Harv.) Raym.-Hamet
purple scallops	English	*Kalanchoe fedtschenkoi* Raym.-Hamet & H.Perrier
rabbit's ear (kalanchoe)	English	*Kalanchoe paniculata* Harv.
red chandelier plant	English	*Kalanchoe* selection
red-cup kalanchoe	English	*Kalanchoe porphyrocalyx* (Baker) Baill.
red-leaved kalanchoe	English	*Kalanchoe sexangularis* N.E.Br.
rooiblaarplakkie	Afrikaans	*Kalanchoe sexangularis* N.E.Br.
sambreel-kalanchoe	Afrikaans	*Kalanchoe neglecta* Toelken
sand forest kalanchoe	English	*Kalanchoe leblanciae* Raym.-Hamet
sandwoud-kalanchoe	Afrikaans	*Kalanchoe leblanciae* Raym.-Hamet
scalloped kalanchoe	English	*Kalanchoe crenata* (Andrews) Haw.
sebraplakkie	Afrikaans	*Kalanchoe humilis* Britten
seeplakkie	Afrikaans	*Kalanchoe longiflora* Schltr. ex J.M.Wood
segolobe	Setswana	*Kalanchoe paniculata* Harv.
semonye	Setswana	*Kalanchoe lanceolata* (Forssk.) Pers.
serelile	Southern Sotho [Sesotho]	*Kalanchoe thyrsiflora* Harv.
serethe	Setswana	*Kalanchoe rotundifolia* (Haw.) Haw.

Continued

TABLE 5.4 Alphabetical list of the common names of *Kalanchoe* species compiled for the *Flora of Southern Africa* region. For the sake of accuracy, the author citations of the *Kalanchoe* species names are given in the third column—cont'd

Common name	Language	Species
sopbordplakkie	Afrikaans	*Kalanchoe luciae* Raym.-Hamet
spepejanie	Unknown	*Kalanchoe luciae* Raym.-Hamet
spookplakkie	Afrikaans	*Kalanchoe gastonis-bonnieri* Raym.-Hamet & H.Perrier
tshinyanyu	Tshivenḓa	*Kalanchoe brachyloba* Welw. ex Britten
tshirndidza	Tshivenḓa	*Kalanchoe paniculata* Harv.
turquoise kalanchoe	English	*Kalanchoe longiflora* Schltr. ex J.M.Wood
uchane	isiZulu	*Kalanchoe rotundifolia* (Haw.) Haw.
umadinsane	isiZulu	*Kalanchoe rotundifolia* (Haw.) Haw.
umahogwe	isiZulu	*Kalanchoe crenata* (Andrews) Haw.
umbrella kalanchoe	English	*Kalanchoe montana* Compton
umfayisele yasehlatini	isiXhosa	*Kalanchoe rotundifolia* (Haw.) Haw.
uquwe	isiXhosa	*Kalanchoe crenata* (Andrews) Haw.
utshwala benyoni	isiZulu and siSwati	*Kalanchoe thyrsiflora* Harv.
utywala bentaka	isiXhosa	*Kalanchoe thyrsiflora* Harv.
velvet kalanchoe	English	*Kalanchoe brachyloba* Welw. ex Britten
velvet leaf	English	*Kalanchoe beharensis* Drake
vlamplakkie	Afrikaans	*Kalanchoe sexangularis* N.E.Br.
voëlbrandewyn	Afrikaans	*Kalanchoe thyrsiflora* Harv.
vuurblaar	Afrikaans	*Kalanchoe sexangularis* N.E.Br.
white lady	English	*Kalanchoe thyrsiflora* Harv.
witplakkie	Afrikaans	*Kalanchoe gastonis-bonnieri* Raym.-Hamet & H.Perrier

TABLE 5.4 Alphabetical list of the common names of *Kalanchoe* species compiled for the *Flora of Southern Africa* region. For the sake of accuracy, the author citations of the *Kalanchoe* species names are given in the third column—cont'd

Common name	Language	Species
wolkalanchoe	Afrikaans	*Kalanchoe lanceolata* (Forssk.) Pers.
Wolkberg kalanchoe	English	*Kalanchoe winteri* Gideon F.Sm., N.R.Crouch & Mich.Walters
wolplakkie	Afrikaans	*Kalanchoe lanceolata* (Forssk.) Pers.
woolly kalanchoe	English	*Kalanchoe lanceolata* (Forssk.) Pers.
woudkalanchoe	Afrikaans	*Kalanchoe crenata* (Andrews) Haw.
woudplakkie	Afrikaans	*Kalanchoe crenata* (Andrews) Haw.
yasehlatini	isiXhosa	*Kalanchoe rotundifolia* (Haw.) Haw.
yellow hairy kalanchoe	English	*Kalanchoe hirta* Harv.
yellow kalanchoe	English	*Kalanchoe paniculata* Harv.
zebra kalanchoe	English	*Kalanchoe humilis* Britten

This is a direct reference to the thickened, strap-shaped leaves of several of these plants.

Kalanchoe rotundifolia is the species with the highest number of local common names (12 names in five languages), so reflecting the wide natural geographical distribution range of the species in the subcontinent (Smith & Figueiredo, 2017e).

Some southern African species that occur beyond the boundaries of the *FSA* region also have common names in the languages spoken in those countries. Where these could be traced, they are recorded and referenced in the species treatments. In this regard, Figueiredo & Smith (2012, 2017a) on the common names of Angolan plants and Kwembeya & Takawira (2002, n.d.) and Wild et al. (1972) on the common names of Zimbabwean plants were invaluable.

6

Geographical Distribution and Ecology

FIG. 6.1 *Kalanchoe sexangularis* var. *sexangularis* (bright red leaves and yellow-flowered inflorescences) growing strictly under the canopy of a large shrub. Plants do not occur beyond the protection provided by the canopy. Photograph: Gideon F. Smith.

Kalanchoe (Crassulaceae) in Southern Africa
https://doi.org/10.1016/B978-0-12-814007-9.00006-2

DISTRIBUTION

The majority of the southern African *Kalanchoe* species occur naturally in the eastern half of the subcontinent (Fig. 6.1). In this region, they occur predominantly in the arid to mesic, summer-rainfall, bushveld (savanna) areas. The centre of highest present-day diversity is located in South Africa's Mpumalanga and (northern) KwaZulu-Natal provinces, as well as in Swaziland (Fig. 6.2). In southern Africa, *Kalanchoe* species are therefore largely absent from the western and south-central, succulent-rich, winter-rainfall, and arid karroid parts (Smith et al., 2003). Within its southern African distribution range, kalanchoes are generally associated with predominantly dry habitats or at least locally dry sites.

Kalanchoes are found in a wide range of habitats and niches, occurring in sparse to dense grasslands (Retief & Meyer, 2017), on exposed rocky outcrops, on forest and savanna margins (Retief & Herman, 1997), and in thickets, but never in true deserts nor in fine-leaved shrublands (fynbos). Representatives of *Crassula* however abound in fynbos, which experiences a winter-rainfall, Mediterranean-type climate.

Kalanchoes occur from sea level (e.g. *Kalanchoe hirta*) to elevations of over 2500 m (e.g. *Kalanchoe paniculata* and *K. thyrsiflora*) on the KwaZulu-Natal and Lesotho Drakensberg. Certain species, such as *K. lanceolata* and *K. laciniata*, can have vast geographical distributions, with both spanning ranges of several thousand kilometres, from southern Africa to the Arabian Peninsula, and even beyond. One of these species, *K. laciniata*, has a more southwestern south-tropical African distribution range and barely crosses into the northern parts of Namibia and Botswana from southern Angola. This species therefore remains uncommon in southern Africa. The remarkable distribution range of this species lends some support for the existence of an arid corridor between western southern Africa and the Horn of Africa (see, e.g. Trimen, 1873: 8 on observations made in Angola by Friedrich Welwitsch in the 1860s).

Of the species of southern African *Kalanchoe* with a widespread distribution range, *K. rotundifolia* is the only one that occurs on continental Africa and on the Indian Ocean island of Socotra and islands in Maputo Bay, Mozambique. Southern Africa and Madagascar do not share any species of *K.* subg. *Kalanchoe* nor of *K.* subg. *Bryophyllum*.

In terms of geographical distribution ranges, the southern African kalanchoes can be broadly divided into two groups: (1) widespread species and (2) narrow-ranged species (Table 6.1).

Twelve of the 17 species of southern African *Kalanchoe* (Table 6.1, numbers 3, 6, 7, 9, 10, and 11–17) have geographical distribution ranges that are more or less restricted to the subcontinent. Seven of these (Table 6.1, B. Narrow-range, numbers 11–17) are restricted to specific, predominantly eastern and northeastern Regions or Centres of Endemism (Van Wyk & Smith, 2001; Steenkamp et al., 2004), while the other five have a typical subcontinental (southern African) distribution range but occur more widely across the subcontinent (Table 6.1, numbers 3, 6, 7, 9, and 10). The other five indigenous kalanchoes occur not only in but also well beyond southern Africa (Table 6.1, numbers 1, 2, 4, 5, and 8).

The occurrence of indigenous southern African kalanchoes in specific Regions or Centres of Endemism (Van Wyk & Smith, 2001) is indicated in Table 6.2. The Maputaland Centre of Endemism is host to the largest number of local endemics (three).

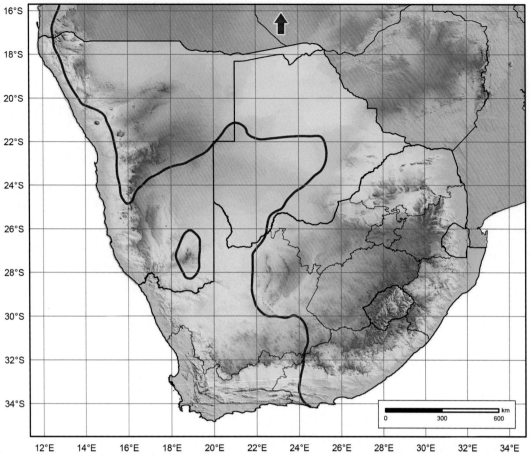

FIG. 6.2   Known geographical distribution range of the indigenous species of *Kalanchoe* in the *Flora of Southern Africa* region (Namibia, Botswana, Swaziland, Lesotho, and South Africa). The range of the genus as a whole is outlined in red and shown with brighter relief colours. The red *arrow* indicates that the genus is found further north in Africa and beyond especially in the Middle East, Madagascar and parts of Asia.

GENERAL MORPHOLOGY AND DEVELOPMENTAL STAGES

Kalanchoes are succulents that survive extended annual periods of environmental stress, such as long, usually predictable, dry seasons and sometimes significant, unpredictable droughts, by accumulating moisture predominantly in their leaves. The southern African *Kalanchoe* species generally do not have underground storage organs (Fig. 6.3); the rootstocks of most species are fibrous, somewhat woody, or only slightly tuberous, as in the case of *K. hirta*, rather than succulent. Plants such as geophytes (e.g. bulbous plants) that typically have substantial underground storage organs have the benefit that such structures provide them with significant tolerance for the impacts of drought, fire, freezing, and grazing and browsing; kalanchoes do not use this survival strategy. Rather, they are leaf succulents, with a number of leaf functional traits that represent adaptations enabling plants to survive, indeed thrive, under different,

sometimes adverse, environmental conditions. The leaves are thickened, very often quite large and pancake-like flattened, and borne at different angles, mostly ranging from horizontally spreading to erect (vertical). The leaves of many kalanchoes are almost invariably heterochromatic during different stages of their life cycle. For example, the leaves of several species, such as *K. sexangularis* and *K. thyrsiflora*, turn bright crimson red during winter. In general, different leaf colours, often on a single leafy stem or branch, displayed by many unrelated plant species is a further mechanism, along with succulence, that mediate the impacts of environmental stresses such as high levels of incident radiation (Wang et al., 2017). Red leaves, or at least red-infused leaves, are often larger than green ones, suggesting a significant investment to increase light capture and therefore photosynthetic efficiency because of their low leaf-chlorophyll concentration. With the southern Africa kalanchoes occurring exclusively in the summer-rainfall region, the dry season coincides with winter, which typically encompasses the months of May to August.

TABLE 6.1 Geographical distribution ranges of indigenous *Kalanchoe* species in the *Flora of Southern Africa* (FSA) region [Namibia, Botswana, Swaziland, Lesotho, South Africa]. *K. = Kalanchoe*

No.	Species	Distribution in the *FSA*
A.	Widespread	
1.	*K. brachyloba* Welw. ex Britten	Northwest, centre, north, and east*
2.	*K. crenata* (Andrews) Haw.	Predominantly in the east*
3.	*K. hirta* Harv.	Predominantly in the east and north*
4.	*K. laciniata* (L.) DC.	Extreme northwest*
5.	*K. lanceolata* (Forssk.) Pers.	Northwest, centre, north, and east*
6.	*K. luciae* Raym.-Hamet	North-central, northeast, and east*
7.	*K. paniculata* Harv.	Widespread in the east and centre*
8.	*K. rotundifolia* (Haw.) Haw.	Widespread in the east and centre*
9.	*K. sexangularis* N.E.Br.	Predominantly in the north and east*
10.	*K. thyrsiflora* Harv.	North-central, northeast, and east*
B.	Narrow-range	
11.	*K. alticola* Compton	East
12.	*K. crundallii* I.Verd.	North
13.	*K. leblanciae* Raym.-Hamet	East (KwaZulu-Natal)*
14.	*K. longiflora* Schltr. ex J.M.Wood	East (KwaZulu-Natal)
15.	*K. montana* Compton	East-central
16.	*K. neglecta* Toelken	East
17.	*K. winteri* Gideon F.Sm., N.R. Crouch & Mich.Walters	North

Distributions marked with an '*' indicate occurrence in and beyond southern Africa.

TABLE 6.2 Endemism of *Kalanchoe* species indigenous to the *Flora of Southern Africa* (FSA) region [Namibia, Botswana, Swaziland, Lesotho, South Africa] following Van Wyk & Smith (2001) and Steenkamp et al. (2004). *K. = Kalanchoe*

No.	Taxon	Distribution
A.	Narrow-range endemics restricted to specific Regions or Centres of Endemism	
1.	*K. alticola*	Barberton Centre of Endemism
2.	*K. crundallii*	Soutpansberg Centre of Endemism
3.	*K. leblanciae*	Maputaland Centre of Endemism
4.	*K. longiflora*	Maputaland-Pondoland Region of Endemism
5.	*K. montana*	Barberton Centre of Endemism
6.	*K. neglecta*	Maputaland Centre of Endemism
7.	*K. winteri*	Wolkberg Centre of Endemism
B.	Endemics or near-endemics in the *FSA*, that are not restricted to a specific Region or Centre of Endemism	
8.	*K. hirta*	*FSA* (also in Zimbabwe)*
9.	*K. luciae*	*FSA* (also in Mozambique and Zimbabwe)*
10.	*K. paniculata*	*FSA* (also in Mozambique and Zimbabwe)*
11.	*K. sexangularis*	*FSA* (also in Mozambique, Zambia, and Zimbabwe)*
12.	*K. thyrsiflora*	*FSA*
C.	Widespread species present in southern Africa	
13.	*K. brachyloba*	Southern, south-tropical, and central Africa
14.	*K. crenata*	Southern, south-tropical, central, west, and east-tropical Africa
15.	*K. laciniata*	Southern, eastern, northern Africa; Arabian Peninsula, India
16.	*K. lanceolata*	Southern, eastern, western Africa; Arabian Peninsula, India
17.	*K. rotundifolia*	Southern, south-tropical, and east-tropical Africa; some Indian Ocean Islands

Distributions marked with an '*' indicate occurrence slightly beyond southern Africa; these species are therefore near-endemic to the subcontinent.

The southern African kalanchoes essentially have two different growth forms: one group has soup plate-shaped and often soup plate-sized leaves and produces large, more or less club-shaped, densely flowered inflorescences (Fig. 6.4). This group consists of *Kalanchoe luciae*, *K. montana*, *K. thyrsiflora*, and *K. winteri*, with *K. crundallii* having a somewhat intermediate leaf morphology. The rest of the species are all small to medium-sized shrublets that remain low-growing until inflorescences that can reach a length of well over 1–2 m start developing.

In several species of *Kalanchoe*, specimens:

- are truly perennial (Fig. 6.5) and will grow and flower repeatedly, from season to season (e.g. *Kalanchoe sexangularis*);
- are biennial, typically with a one-season vegetative phase, after which they often rapidly enter a reproductive phase consisting of flower and fruit development, followed by seed set and dispersal the following year, after which they die (e.g. *K. paniculata*);

FIG. 6.3   In most instances, the roots of kalanchoes are rather short and grow close to the soil surface where they can access moisture derived from even low levels of precipitation. The roots of most kalanchoes are fibrous, rather than fleshy, and rarely are somewhat thickened, as here in the case of *Kalanchoe hirta*. Photograph: Gideon F. Smith.

FIG. 6.5   Most southern Africa *Kalanchoe* species are either perennial, weakly perennial, or sometimes biennial. This specimen of the low-growing *K. rotundifolia*, here nestling among the leaf litter produced by the deciduous trees under which it grows, is perennial and sprouts from the base season after season. Photograph: Gideon F. Smith.

FIG. 6.4   Especially, the soup plate-leaved species of *Kalanchoe* tend to be monocarpic and die after having flowered, such as this specimen of *K. thyrsiflora*. It can take several years before such a monocarpic species reaches flowering maturity, especially under abnormally dry periods that may last a few years. Photograph: Gideon F. Smith.

- are multiannual (also referred to as "short-lived perennial") and will grow vegetatively for a number of years (typically three to five), enter a reproductive phase, and then die.

Especially the second and third mechanisms have been observed in the same species and seem to be clone related. Further, observations thus far have not linked the lifespan of clones of some *Kalanchoe* species directly to climate or weather, such as prolonged dry or cold periods. Rather, lifespan seems to be under significant genetic control.

The flowers of the monocarpic *Kalanchoe* species are borne horizontally to suberect, while those of the shrubby species are invariably carried erectly, often in loose, few-flowered inflorescences. In contrast to representatives of *K.* subg. *Bryophyllum* and *K.* subg. *Kitchingia* from Madagascar, none of the southern African *Kalanchoe* species have truly pendent flowers, although those of *K. crundallii* are borne somewhat pendent at anthesis.

The developmental stages in the life cycle of kalanchoes that cover the flowering, fruiting, and seeding phases all occur while the plants are in leaf. This enables the plants to draw on the nutrients and moisture accumulated required to produce inflorescences and fruit from the often large leaves. Most kalanchoes have a life cycle that can be divided into:

- a spring and summer (warm, to hot, and rainy) growing phase during which resources are accumulated; and
- an autumn and winter (cold, but not freezing, and dry) flowering phase during which the plants do not show vegetative growth of any consequence.

All southern African kalanchoes produce few- or many-flowered inflorescences; no species with a very limited number of large, pendent flowers, such as found in, for example, *Kalanchoe uniflora* Raym.-Hamet from Madagascar, are known from southern Africa. Some forms of *K. rotundifolia* produce very few flowers in its inflorescences, but these are always small and carried erectly. Those *Kalanchoe* species that produce a cylindrical, many-flowered, terminal inflorescence are usually monocarpic, not unlike the majority of New World *Agave* L. species. In the case of the kalanchoes that have cylindrical, dense, many-flowered inflorescences, withdrawing the resources to sustain these structures almost invariably leads to the demise of the plants, with the leaves slowly withering to a papery, dry state. These flowers, as is the case with many representatives of the Crassulaceae, remain open for many days, so ensuring multiple visits by pollinators and therefore enhanced chances of successful pollination by especially flying and other insects. In addition, since kalanchoes sometimes grow in groups and produce flowers simultaneously, chances of outbreeding are heightened, so ensuring that the next generation will be genetically diverse.

Several kalanchoes occur sympatrically in the eastern half of southern Africa, with virtually all of them flowering in the cold, dry winter months. However, natural hybrids are less common than would be predicted by such closely overlapping biogeographical patterns.

Although the rule of winter flowering holds for by far the majority of the species of *Kalanchoe*, a few exceptions exist to this rule: *K. crundallii* can flower as early as midsummer, with its flowering season extending to autumn. Some forms of *K. rotundifolia*, especially from around Mbombela, Mpumalanga, an eastern province of South Africa abutting Mozambique, have been similarly observed as producing flowers in December in the southern hemisphere.

SOLAR RADIATION REQUIREMENTS

In their natural habitats, kalanchoes grow either in full sun or in dappled shade (Fig. 6.6) and only rarely in deep shade, as in the case of, for example, the forest-dwelling *Kalanchoe crenata* and *K. crundallii*. However, virtually, all the species display a preference for dappled shade (Fig. 6.7), rather than direct sunlight. When exposed to direct solar radiation, the leaves of most kalanchoes almost invariably take on a rich reddish hue (Fig. 6.8) that ranges from deep wine red to bright orange.

When growing in the shade of the canopies of bushveld trees, kalanchoes display very specific zonation, with some species thriving directly under the drip zone of a canopy (e.g. *Kalanchoe brachyloba*), while others grow

FIG. 6.6   In southern Africa, such as here in South Africa's North-West province, *Kalanchoe brachyloba* often occurs in bushveld (savanna) habitats where plants grow within the drip line of the canopies of bushveld trees with low-bending branches. Photograph: Gideon F. Smith.

FIG. 6.7   The leaves of *Kalanchoe paniculata* are lengthwise-folded upwards, unlike those of the similar-looking *K. luciae* and *K. thyrsiflora* that tend to remain rather flat, if rarely somewhat wavy. Here, in northeastern Swaziland, *K. paniculata* grows on a rocky outcrop, often a preferred habitat, on the margin of a dense stand of grass. Photograph: Gideon F. Smith.

FIG. 6.8   Plants of *Kalanchoe sexangularis* var. *sexangularis* will be greener, but still conspicuously red-infused, if they grow in the shade of surrounding vegetation. Photograph: Gideon F. Smith.

closer to the trunk of the tree (e.g. *K. rotundifolia*), so avoiding the drip zone (Figs. 6.1 and 6.7).

The vegetation directly under the tree canopies is often very sparse, and competition from other forb and graminoid species for resources is seemingly diminished.

TEMPERATURE REQUIREMENTS

In general terms, the present-day climate of southern Africa is comparatively mild. However, some areas do record very high or very low temperatures, and annual rainfall can be erratic over large parts, with droughts in both the winter- and summer-rainfall areas being a regular occurrence.

Heat tolerance

Temperature-wise, kalanchoes inhabit some of the hottest areas in southern Africa, for example, some river basins in the KwaZulu-Natal, Mpumalanga, and Limpopo provinces. In some of these areas, the air humidity can be very high though. The species are however virtually absent from the hot, dry Groot Karoo of South Africa.

Cold tolerance

Very few kalanchoes are able to tolerate sub-zero temperatures. The high mountain ranges of the Drakensberg Alpine Centre of Endemism therefore represent a significant and largely impassable barrier that prevents species that occur naturally along the eastern seaboard of southern Africa from crossing over into the climatically more severe South and southern African interior. Two exceptions are *Kalanchoe paniculata* and *K. thyrsiflora*, both of which have been recorded for the mountainous Lesotho and surrounding areas.

SOIL REQUIREMENTS

The distribution of kalanchoes in general does not seem to be limited by soil type. Especially, the widespread species (Table 6.1) have no pedological preferences. However, *Kalanchoe leblanciae* occurs mainly in aeolian sands that are present in the Maputaland Centre of Endemism. Some species with a more westerly geographical distribution range in southern Africa, for example, *K. brachyloba*, have been recorded from areas where reddish, windblown Kalahari sands have accumulated, but none of these species are restricted to this substrate.

While some species of *Kalanchoe* often grow in dense grassland vegetation, such as *Kalanchoe paniculata*, and at least in the seedling stage benefit from inter alia the shade provided by grasses and forbs, this is not a requirement for successful establishment, and seed of kalanchoes often germinates in soil in otherwise denuded areas.

PLANT–WATER RELATIONS

Both the large crassuloid genera *Crassula* and *Sedum* contain some representatives that qualify as true aquatics that grow exclusively and thrive in at least seasonally inundated habitats (Smith et al., 2017b; Wingate & Yeatts, 2003). These include *C. natans* Thunb., *C. peploides* Harv., and *C. vaillantii* Roth, as well as *S. rhodanthum* A.Gray [=*Clementsia rhodantha* (A.Gray) Rose], and *S. integrifolium* A.Nelson in J.M.Coult. & A.Nelson [=*Rhodiola integrifolia* Raf.]). However, no southern African kalanchoes grow directly in such often consistently wet, waterlogged habitats. In general, kalanchoes prefer soil substrates that can be very dry and largely do not occur in sodden soils.

As noted above, the leaves of kalanchoes are thick and succulent that enable them to survive the dry season, which for all the southern African species are the cold winter months. Kalanchoes make use of the Crassulacean Acid Metabolism (CAM) photosynthetic pathway, a mechanism that curtails more rapid water loss associated with standard photosynthesis (Lyndon, 1962; Milburn et al., 1968; Mooney et al., 1977; Heyes, 1989). The leaves of especially kalanchoes with soup plate-shaped leaves, for example *Kalanchoe winteri*, are often covered in a white, powdery, sometimes aromatic wax that further assists with preventing water loss. In other species, for

example *K. hirta*, the leaves are covered in short hairs that reflect sunlight and shade the leaf surfaces and also create a more substantial boundary layer. During times of drought, plants of *Kalanchoe* may shed the lower sets of leaves on the stems, with only the youngest ones remaining higher up on the stems. However, this is not a regular occurrence.

The root systems of kalanchoes are adventitious and fairly shallow, which enables them to benefit from comparatively little moisture. However, these plants generally do not survive in true desert conditions and need to have their water resources replenished from time to time through rainfall.

FUNCTIONAL ROLES OF KALANCHOES IN THE ECOSYSTEMS IN WHICH THEY OCCUR

The flowers of kalanchoes are nectariferous and offer considerable rewards for nectar-feeding insects and even birds. In many instances, the stamens and anthers protrude at the mouth of a flower, and pollen-collecting visitors have easy access to pollen. In southern Africa, the flowers of kalanchoes are visited by butterflies and other flying insects. Seed production is copious and seems to be in excess of the number of observed potential pollinator visits (Fig. 6.9). It is therefore likely that at least some of the species are self-compatible. Certainly in cultivation, seed is set in vast quantities even when no other species are in flower. Populations persist by annually reseeding in situ (Fig. 6.10).

At least some species are known to be poisonous, causing 'nenta' or 'krimpsiekte', a chronic form of heart glycoside poisoning, especially among small livestock

FIG. 6.9 In terms of pollination and fertilisation, the southern and south-tropical African species of *Kalanchoe* are invariably extraordinarily successful, and fruit and seed are produced in vast quantities, as here with *Kalanchoe salazarii*. Photograph: Gideon F. Smith.

FIG. 6.10 Seed of *Kalanchoe sexangularis* var. *sexangularis* is very small and fine and easily germinates in the thin soils and decomposed plant material that accumulate in the slightest of rock cracks in the often boulder-strewn natural habitats of the species. Photograph: Gideon F. Smith.

(Onderstall, 1984: 94; Smith & Van Wyk, 2008b: 152; Kirby, 2013: 231). Kalanchoes are generally not grazed by ruminants and others mammals.

POPULATION DYNAMICS

In their natural habitats, either kalanchoes can grow in very dense stands, especially in the case of smaller species, such as *Kalanchoe rotundifolia*, or the plants in a population can be sparsely dispersed, especially in the case of the larger-growing species, such as *K. thyrsiflora*.

While several species grow sympatrically, sometimes literally within only a few millimetres of each other, natural hybrids are hardly known, at least in southern Africa. In cultivation, plants do hybridise though, and a few such hybrids have been given cultivar names (see, e.g. Bischofberger, 2015 on *Kalanchoe* 'Vivien', of which the parents are *K. luciae* and *K. sexangularis*).

FIRE ECOLOGY

Along with very cold and frosty conditions, regular and at times quite severe fires are one of the main natural threats that impact on the in situ survival of succulents in general, including of kalanchoes. Succulents in general have several mechanisms by means of which they either avoid or tolerate fire in habitats that experience occasional or regular fires. The most common mechanism is to preferentially occupy sheltered sites, such as crevices under rocks or on top of rock sheets and rocky outcrops,

which are protected from fires because of a lack of combustible material. When growing in such refugia, the foliage of leaf succulents are often scorched by fires sweeping the area, but given the high leaf-moisture content, plants generally do not completely succumb to the flames (Smith, 2018).

In the case of root and caudiciform succulents, the underground perennating organs (thick, water-filled roots; tubers; and the stem-root continuum) are well protected from the heat, while the aboveground parts (stems and often flimsy or twining branches) usually easily succumb to fires. Such plants usually recover vegetatively from the effects of fires; they therefore resprout, rather than recruiting seedlings from the seed bank held in the soil.

When fires occur during the flowering season of succulents, which are the dry winter months for most kalanchoes, for example, inflorescences will be scorched or completely destroyed, which will result in decreased seed set and therefore less successful subsequent seedling recruitment.

Fire avoidance

Some kalanchoes tend to avoid fire by growing in places where fire will not reach them, such as on rocky outcrops or in gravel beds with either low-growing or minimal surrounding vegetation that can fuel fires. This type of habitat is favoured by numerous succulents, including aloes, mesembs, and portulacas.

The areas directly under bushveld trees are often largely devoid of kindling fuel, and fires sweeping through such patches tend to be of a lower intensity than where the grass and forb layer is very dense. Kalanchoes often occur in such microhabitats where the lower fuel load serves as an additional fire avoidance mechanism.

Fire tolerance

Kalanchoes in general do not have underground tubers. Under favourable conditions, plants possessing such, often deep-seated, organs usually sprout regenerative growth from the perennating tubers, after sacrificing annual or biennial aboveground growth to fire, frost, or extreme aridity (desiccation). This is also a very effective way of tolerating fire. Some species, for example, *Kalanchoe hirta*, which is a forest or savanna margin specialist, do develop a robust rootstock, which serves the same purpose as a tuber.

Under regular rainfall conditions, *Kalanchoe* species remain visible during the dry, cold winter months and do not lose all their leaves, with only the lower ones progressively drying and becoming detached. In their native habitats, kalanchoes are at times exposed to fires in the winter season, their persistent succulent leaves so providing a further mechanism that protects against fires. Note though that leaves are shed by the monocarpic species with soup plate-shaped leaves once flowering is complete. This group of species generally also has thickened, robust wax-covered stems that provide further protection against scorching fires.

RESEEDING AND RESPROUTING

Kalanchoes are essentially reseeders, that is, they produce copious amounts of small, very light seeds that are easily dispersed by wind or when a roving animal disturbs the inflorescences that are held aloft. While plants with subterranean perennating organs, for example, those that develop bulbs, often grow in clumps established through vegetative division, the especially large-growing southern African kalanchoes tend to rather grow in sometimes dense stands established from the successful recruitment of seedlings, so ensuring significant genetic diversity.

Seed of most species of *Kalanchoe* germinates easily, both in their natural habitats and in cultivation.

SEED PRODUCTION AND DISPERSAL

Reproduction of kalanchoes in nature in southern Africa is predominantly by seed. Unlike several Madagascan species of the genus, no southern African species have the trait of forming plantlets on their leaf margins. Several species do sprout vegetatively from near the base of the stem that has produced an inflorescence, so forming small clumps.

Reproduction from the rooting of a leaf with a wholly or partly intact petiole is known for some Madagascan *Kalanchoe* species. Some southern African species, for example, *K. neglecta*, will very slowly form callus-like tissue over the tip of the petiole wound and so remain green for a long time, but root formation is slow, if it happens at all. Therefore, while some species may possibly reproduce by means of this mechanism, it hardly seems like a viable reproductive method.

ECONOMIC BOTANY

Kalanchoes as medicinal plants and in African traditional medicine

Some species of *Kalanchoe* are traditionally used, including as medicine, in their countries of origin

(Riley, 1963: 157; Cole, 1995: 228). At present, there is no indication that such uses are not sustainable.

Horticultural trade in kalanchoes

Three species, the soup plate-leaved *Kalanchoe luciae*, and shrubby *K. longiflora* and *K. sexangularis*, have become popular in the nursery trade as waterwise species for growing either as groundcovers or as specimen plants. Trade in these species has not impacted natural populations as the shrubby ones are preferentially and very easily propagated on a massive scale using vegetative material, while seed of *Kalanchoe luciae* germinates readily.

CONSERVATION OF SOUTHERN AFRICAN *KALANCHOE* SPECIES

With very few exceptions, kalanchoes are known to occur in large numbers over extended areas in their natural habitats. They therefore tend to be locally common where present and in the case of the larger-growing species often comprise conspicuous and prominent components of the flora of a region. The smaller-growing species, such as *Kalanchoe rotundifolia* and *K. alticola*, while less conspicuous, often grow densely under bushes and in the dappled shade of trees, grasses, and forbs. As far as is known, very few species have small world populations, the recently described *K. winteri*, a habitat specialist, being a notable exception. The species is largely restricted to the Wolkberg, a part of the mountain chain that forms the northern Drakensberg escarpment in northern South Africa, and it therefore has a rather small known geographical distribution range; it is at present known from only three localities within a 50 km range. A conservation status of 'Rare' is therefore accorded to *Kalanchoe winteri*.

The comparatively little-known *Kalanchoe crundallii*, which was described in the mid-1940s, is the only other southern African kalanchoe that should be treated as 'Rare'. This species is found in the mist-belt regions of the western Soutpansberg where it grows in humus-enriched soils.

Kalanchoe longiflora, a local endemic from South Africa's KwaZulu-Natal province, has become very common in cultivation, likely because of the ease of propagation using vegetative material and its quick growth rate. Given the small natural geographical distribution range of the species, it could easily have qualified for inclusion on a Red List, but the species is not threatened.

In general, no threats to wild populations of the southern African kalanchoes have been recorded.

Characters, Pollination Biology and Life Cycle

FIG. 7.1   A dense indumentum of simple glandular hairs is diagnostic for *Kalanchoe lanceolata* when it enters the reproductive phase. Note the flowers with corolla lobes shortly tailed (ending in a pronounced apiculum). Photograph: Abraham E. van Wyk.

INTRODUCTION

Taxonomic evidence for the classification of plants is gathered from a variety of sources. Hitherto, both vegetative and reproductive morphologies have provided the most important characters in the infrageneric classification of *Kalanchoe*, but in recent times, molecular evidence has greatly assisted in the construction of phylogenetic hypotheses. A molecular phylogeny inferred from nucleotide sequences of the internal transcribed spacer 1 (ITS-1), and ITS-2 region suggests that within the Crassulaceae, the genus *Kalanchoe* forms a monophyletic clade (Gehrig et al., 2001). In addition, the ITS phylogenetic trees show that all African *Kalanchoe* species form a distinct group within the most derived of the three main clusters. This supports the widely held view that the centre of phylogenetic radiation of the genus is located in Madagascar, from where the species have spread into continental Africa and beyond to Southeast Asia.

According to Thiede and Eggli (2007), the genus *Kalanchoe* is morphologically highly derived and distinguished from other genera of the Crassulaceae by the combination of being essentially evergreen, with the leaves decussate, usually flat (rarely cylindrical); the flowers tetramerous (4-merous), with petals united to form a distinct tube; the stamens eight (obdiplostemonous), with each anther tipped by a ± spherical connective gland (appendage); and seeds with few (4–6) longitudinal ridges (costae) in side view, as well as a proliferation of the testa around the micropyle to form a so-called corona.

In this chapter, selected characters used in the formal description of the species (Chapter 12) are explained and discussed. The emphasis is on particularly diagnostic characters used in the key to the species and the possible link between their structure, adaptation, and function. The morphology of the reproductive structures should be viewed as part of a dynamic functioning system related to pollination biology. Hence, we provide some observations and speculations on this component of the species' interactions with their environment. Aspects/structures for which existing observations are deficient and in need of further study are also highlighted.

Variation relating to the life cycle (duration) of the various southern African species of *Kalanchoe* is provided as part of the taxon descriptions in Chapter 12.

Here, we provide more clarity on descriptive terms used in that chapter and in particular further elaborate on the question of whether some members of the group are monocarpic, a matter already referred to earlier (Chapter 6) where the ecology of the group was discussed. Although emphasis in the present chapter is on the indigenous southern African kalanchoes, occasional reference is made to some of those species introduced to the subcontinent mainly from Madagascar, but also from elsewhere in Africa. More details on the morphology of the Madagascan species are supplied by Boiteau & Allorge-Boiteau (1995). For ease of reference, the morphological characters of all the kalanchoes included in the present work are tabulated in Tables 7.1–7.4. Some of the technical terms are defined in the Glossary.

VEGETATIVE MORPHOLOGY

Growth form in *Kalanchoe* is diverse and covers the full spectrum from small subshrubs to larger shrubs and even small- to medium-sized trees. One of the largest members is the Madagascan *K. beharensis*, a robust shrub of up to 3 m tall with a well-developed woody stem and bark. It is widely cultivated throughout the world and has also become naturalised in southern Africa. Most of the indigenous southern African members are perennial, evergreen subshrubs. Some forms of *K. rotundifolia*, especially those from the eastern parts of its range, are relatively small plants in the vegetative phase, and only about 10 cm tall or smaller. A number of species, notably those with large paddle-shaped or roundish leaves, have stems with reduced internodes during the vegetative phase, resulting in a more or less basal rosette of leaves. In this group, the stems invariably lengthen during the reproductive phase, with the leaves remaining more or less of similar shape, but becoming smaller upwards along the emerging peduncle.

In all southern African species, the roots are quite shallow, fibrous, and only rarely slightly thickened (*Kalanchoe luciae*) to tuberous (*K. hirta*) (Fig. 6.3). When cultivated in sandy soils or naturally growing in a thin soil layer and decomposing plant debris that collected in rock cracks, the shallow root system may result in some species becoming top-heavy and toppling over during the

TABLE 7.1 Comparison of the vegetative and reproductive characters of *Kalanchoe* species indigenous to southern Africa. Species are arranged alphabetically according to specific epithet

Character	K. alticola	K. brachyloba	K. crenata	K. crundallii	K. hirta	K. laciniata	K. lanceolata	K. leblanciae	K. longiflora
Height (m) when in flower	0.1–0.25	0.6–2.0	0.3–2.0	Up to 1	Up to 1.5	0.3–1.2	Up to 1.5	0.3–1.6	0.2–0.4
Habit	Perennial, few- to many-leaved, multi-branched, glabrous throughout, small, haphazardly branched succulent	Perennial or biennial, few-leaved, solitary or, rarely, sparsely branched, glabrous but velvety, robust, subrosulate, succulent	Perennial or multiannual, solitary or sparsely branched, often clumped, glabrous or velvety, flimsy or robust succulent	Perennial or rarely multiannual, few-to many-leaved, rarely branched, glabrous, sparse succulent; roots thin, wiry	Perennial or, more rarely, multiannual, solitary or sparsely branched, sometimes clumped, conspicuously pubescent throughout, medium-sized to robust succulent; roots tuberous	Perennial or biennial, many-leaved, unbranched to sometimes once-branched, hairy to rarely glabrous, flimsy to robust succulent	Annual or biennial, many-leaved, unbranched, glabrous (vegetative phase) to hairy (reproductive organs), flimsy to robust succulent	Perennial, glabrous, succulent herb, with slight whitish bloom on preserved specimens	Perennial, few-leaved, sparsely branched, glabrous, medium-sized, succulent shrub
Stems	Thin, usually leaning and tangled, light green to distinctively pinkish green, several arising from a shallow rootstock near ground level, round in cross section	Erect to slightly leaning, terete, not angled, light bluishgreen, usually one arising from a rosette, unbranched, or rarely once-branched	Erect, usually simple, sometimes branched, round, fleshy, terete, light green to reddish green, glabrous or velvety towards base, glabrous or glandular-pubescent above, hairs short	Usually solitary, erect to leaning, rooting where they touch the ground	Erect from a curved base, usually simple, sometimes branched, round, somewhat fleshy, terete, light green, pubescent, hairs prominent, to 1mm long	One to few, erect, often quadrangular-angled lower down, green, glabrous lower down, minutely and densely hairy higher up	One, erect, often quadrangular-angled lower down, green, glabrous lower down, glandular hairy higher up	Simple or once-branched, erect to leaning, woody at the base, ± distinctly 4-angled, 5–9mm in diameter, tapering upwards	Few, unbranched or sparsely branched, brittle, erect to leaning to creeping, 4-angled, greenish pink to greenish orange, surface white-flaky
Leaves	Opposite-decussate, leaf pairs widely spaced, leaves clasping the stem, almost linking with the leaf opposite, succulent, erect to slightly spreading	Opposite-decussate, subrosulate, the lowest the largest, the uppermost becoming smaller and bract-like, succulent throughout, straight when young, becoming gracefully arched to recurved, lower, oldest ones becoming variously floppy, drying papery chartaceous, sometimes caducous	Opposite-decussate, subrosulate, succulent throughout, horizontal to variously deflexed, the lowest the largest, the uppermost becoming smaller and bract-like, not very crowded below, upper bract-like leaves shield-shaped attenuate, drying papery chartaceous, sometimes caducous	Spreading to patent, erect to downcurved, often arranged in four distinct vertical rows, succulent, ± flattened above to concave, slightly convex below	Opposite-decussate, subrosulate, succulent throughout, more or less horizontal, sessile higher up, the lowest the largest, the uppermost becoming smaller, crowded below when young, drying papery chartaceous, sometimes caducous	Succulent	Succulent, longitudinally folded, often floppy, caducous when inflorescence develops, drying papery	Narrowing basally to a subpetiolar base, not amplexicaul, flat to slightly lengthwise folded, fleshy, gracefully recurved to completely downcurved	Succulent, opposite, erect, subpetiolate to cuneate

Continued

TABLE 7.1 Comparison of the vegetative and reproductive characters of *Kalanchoe* species indigenous to southern Africa. Species are arranged alphabetically according to specific epithet—cont'd

Character	K. alticola	K. brachyloba	K. crenata	K. crundallii	K. hirta	K. laciniata	K. lanceolata	K. leblanciae	K. longiflora
Leaf colour and surface	Light bluish green to glaucous, shiny but with a dull powdery bloom	Mostly uniformly greyish green to dull light green, rarely midgreen, velvety-smooth	Mostly uniformly light green to dull greyish light green, rarely midgreen	Green to yellowish green, flaky-waxy with age	Mostly uniformly light green to yellowish light green	Green, basal ones sometimes glabrous, upper leaves minutely hairy	Green to yellowish green, basal ones glabrous, upper leaves glandular hairy	Midgreen, sometimes with a purplish hue during periods of environmental stress	Light sea green to turquoise, strongly infused with a orange or pink, never red
Petiole (mm, if present)	Usually absent	Absent	Up to 40, flattened, grooved above, slightly broadened at base, often clasping stem, sometimes connate with opposite one	5–20, Not stem clasping	Up to 40, flattened, grooved above, slightly broadened at base, often clasping stem, sometimes connate with opposite one	1–10, flattened, canaliculate, ± amplexicaul	Absent	Absent	2–10
Blade size (cm)	2–4 × 0.9–1.5	10–20 × 4–7	4.5–30 × 1.5–20	3–7 × 2–5	8–11 × 6–9	2–12 × 2–8	5–20 × 2–10	up to 10 × 1–3	4–7 × 4–7
Blade shape	Flat, ovate-oblong to oblong	Usually distinctly folded lengthwise along the midrib, narrowly oblong-lanceolate	Ovate or obovate-oblong to spatulate	Not folded lengthwise, suborbicular to broadly oblong	Flat, ovate, or obovate-oblong	Dissected, trifid-lobed, rarely lacking incisions, leaflets ovate, lanceolate or linear	Narrowly oblong, lanceolate, elliptic, ovate-spatulate	Oblong to oblong-spatulate	Slightly incurved along margins, somewhat spoon-shaped, obovate to nearly orbicular
Indumentum	Absent	Absent	Glabrous throughout or sometimes upper leaves sparsely glandular-pubescent	Glabrous throughout	Densely glandular hairy on both surfaces	Absent	Upper leaves glandular hairy	Absent	Absent
Bulbils	Not produced	Not produced	Not produced	Not produced	Not produced	Not produced	Not produced	Not produced	Not produced
Apex	Rounded, sometimes apiculate	Rounded obtuse, often curved downwards	Blunt, attenuate	Rounded, obtuse	Blunt, attenuate	Blunt, attenuate	Obtuse, often indented, more rarely acute	Apex rounded or obtuse	Apex rounded, obtuse, blunt
Base	Cuneate	Cuneate, subpetiolar, tapering towards midblade mostly clasping the stem, sometimes slightly auriculate basally	Cuneate	Cuneate	Cuneate	Attenuate towards the base	Attenuate to cuneate, stem clasping	Base, attenuate	Cuneate

Margins	Entire	Coarsely and irregularly serrate-lobed to crenate or undulate-crenate into irregular, rounded, harmless crenations, rarely almost smooth, light greenish white, sometimes lightly reddish tinted	Irregularly crenate to doubly crenate to sub-lobed, sometimes edged with red	Crenate-dentate, rarely entire, infused with a reddish tint	Irregularly dentate	Crenate or dentate, rarely entire	Variably crenate or dentate, rarely entire	Crenate, sinuate, crenate or serrate-lobed, reddish brown tinted	Scalloped into rounded, harmless, patelliform teeth, teeth becoming smaller to obsolescent towards leaf base
Inflorescence height (m)	0.1–0.2	0.3–1.5	Up to 0.3	Up to 0.8	Up to 0.4	0.4–0.8	0.3–0.5	0.3–1.6	Up to 0.5
Inflorescence morphology	Erect, apically closely packed and few-flowered, all parts with a distinctive waxy bloom, rounded to flat-topped thyrse with several dichasia	Erect to slightly leaning, apically subdense, few- to many-flowered, flat-topped thyrse with several subdense dichasia, round to ellipsoid in outline when viewed from above, branches opposite, projecting erectly away from main axis at 45°, subtended by small leaf-like bracts, leafy branchlets in axils mostly lacking, axis light green or covered in a dense, white, waxy substance that obscures the colour	Terminal; a flat-topped to rounded thyrse, usually with many dichasia ending in monochasia, glabrous or glandular-pubescent branches at ± 45° with axis	A slender, erect, densely flowered, shortly cylindrical to club-shaped, flat-topped thyrse consisting of several dichasia	Terminal; a flat-topped to rounded thyrse, usually with many dichasia ending in monochasia, glandular hairy branches at ± 45° with axis, gracefully curved upwards	Erect to slightly leaning, apically dense, flat-topped, few- to many-flowered, arranged in cymes that form corymbs, rather elliptic in outline when viewed from above branches opposite, growing upwards at a 45° angle, subtended by small, undivided, leaf-like bracts, with very small leafy shoots in axils, axis light green, all parts finely and densely glandular hairy	Erect, apically dense, flat-topped, many-flowered, arranged in an elongated thyrse consisting of a number of dichasia ending in elongated monochasia that can be recurved or coiled, which is especially evident in bud and fruit, branches opposite, growing upwards at a 30° angle, subtended by small leaf-like bracts, sometimes with very small leafy shoots in axils, axis light to midgreen, all parts finely and densely glandular hairy	Narrow, much elongated, apically dense, many-flowered corymbose cymes or panicles, branches opposite, erect to erect-patent, subtended by leaf-like bracts, with small, leafy shoots often produced in axils, greenish yellow	An erect, sparsely flowered, ± flat-topped thyrse consisting of several open dichasia, peduncle often densely covered in large, succulent bracts the same colour as the leaves
Flower disposition	Erect	Erect	Erect	Erect at bud stage to spreading or pendent at anthesis, dry flowers erect	Erect	Erect to slanted sideways	Erect to slanted sideways, rarely pendent	Erect	Erect

Continued

TABLE 7.1 Comparison of the vegetative and reproductive characters of *Kalanchoe* species indigenous to southern Africa. Species are arranged alphabetically according to specific epithet—cont'd

Character	K. alticola	K. brachyloba	K. crenata	K. crandallii	K. hirta	K. laciniata	K. lanceolata	K. leblanciae	K. longiflora
Flower colour	Light yellow to sulphur yellow to light orange	Greenish yellow (tube) and light yellow (lobes)	Basally greenish becoming light orange upwards (tube) and light orange longitudinally infused with yellow (lobes)	Buds yellowish to yellowish green, red at anthesis, drying dark reddish brown	Basally greenish becoming light yellow upwards (tube) and yellow (lobes)	Greenish yellow (tube) and light yellow upwards (lobes), sparsely glandular hairy	Yellowish orange (tube) and bright to yellowy orange (lobes), basal swollen part mid to light green, densely glandular hairy	Yellowish green and yellow; reportedly also varying from salmon to bright red, but these colours not observed	Greenish yellow (tube) and bright yellow (lobes)
Pedicel length (mm) and morphology; indumentum	2–3, robust	Absent	Up to 10; glabrous or glandular-pubescent	10–12	Up to 6, prominent, glandular hairy	2–10, slender	1–8, slender	± obsolete, 3.0–4.5 when present, slender	10–18
Calyx colour and indumentum	Bluish to purplish green	Light bluish green	Light green infused with red	Light green, remaining so postanthesis, contrasting against corolla tube	Light green	Light green, glandular hairy	Midgreen, glandular hairy	Shiny midgreen	Green, infused with orange towards tip
Calyx tube length (mm)	Basally fused for ± 0.5	–	–	–	Connate for up to 1.5 at base	–	Basally fused for 1 (–8)	–	–
Sepal size (length or size in mm)	± 5	2–5 × 1.5–3.0	4–5 × 0.7–1.5	2.0–2.5 × 1	3–6	± 4 × 2	± 3–12 × 2–5	1.2–1.8 × 0.8–1.0	3–4
Sepal fusion	± separate	Basally fused	Free to scarcely connate at base	Almost free to the base	± separate, basally adnate	± separate, basally adnate	Basally fused, above separate	± separate, basally adnate	Separate above, basally adnate
Sepal shape, colour, incumentum	Elongated-lanceolate, acute, contrasting against yellowish green corolla tube, clasping the corolla tube	Lanceolate to triangular-lanceolate, acute, distinctly contrasting against corolla tube	Lanceolate to linear-lanceolate, acute, hardly contrasting against corolla tube	Deltoid-lanceolate, contrasting against corolla tube especially at and after anthesis	Lanceolate-triangular, acute, hardly contrasting against corolla tube	Triangular-lanceolate, acute, tips recurved, hardly contrasting against corolla tube	Lobes curved away from corolla, triangular-lanceolate or ovate-lanceolate, acute, obscuring and hardly contrasting against basal, swollen part of corolla	Triangular-lanceolate, acute, shiny midgreen, contrasting darker green against corolla tube	Elongated-triangular, acute
Corolla length (mm)	9–13	13–17	8–16(–22)	11–15	11–17	9–12	6–13	8–12	16–20
Corolla shape; indumentum, if present	Somewhat inflated lower down and enlarged lower down, twisted apically after anthesis	Somewhat enlarged lower down, often twisted apically at and after anthesis	Distinctly enlarged lower down, often twisted apically at and after anthesis, glabrous or glandular-pubescent	More or less cylindrical, somewhat inflated and enlarged in the middle	Distinctly enlarged lower down, finely glandular-pubescent	Somewhat enlarged lower down and in the middle, very slightly twisted apically after anthesis	Distinctly enlarged lower down, very slightly twisted apically after anthesis, glandular hairy	Not twisted apically after anthesis	Elongated-urceolate, cylindrical-quadrangular, enlarged lower down, tapering to the tip, distinctly to slightly angled, hardly ever pyramidal

Corolla colour	Light green below, light yellow to sulphur yellow to light orange above	Yellowish green (tube) and light yellow (lobes)	Orange (in southern Africa), bright salmon to red, deep red, or brick-coloured elsewhere, lower part pale or greenish orange or whitish (in southern Africa), sulphur yellow to bright or deep yellow elsewhere (tube) and light orange longitudinally infused with yellow (lobes)		Light yellow (tube) and yellow (lobes)	Greenish yellow (tube) and light yellow to creamy yellow (lobes)	Yellowy orange (tube) and bright to yellowy orange (lobes)	Yellowish green and yellow [reportedly also varying from salmon to bright red, but these colours not observed]	Greenish yellow
Tube length (mm)	8–12	10–15	7–21	10–14	9–15	8–11	5–12	6–8	15–18
Tube shape	Very slightly 4-angled, rounded below and above	Quadrangular-cylindrical, distinctly 4-angled, obtusely rounded at bottom, box-shaped, square when viewed from below, longitudinally fluted above	Quadrangular-cylindrical, obscurely 4-angled, obtusely rounded at bottom, indistinctly box-shaped, square when viewed from below	More or less cylindrical, somewhat thickened in the middle, slightly 4-angled especially postanthesis	Quadrangular-cylindrical, lengthwise fluted to obscurely 4-angled, obtusely rounded at bottom, very indistinctly angled	Indistinctly 4-angled, rounded when viewed from below, narrowly cylindrical above	Distinctly to indistinctly 4-angled fluted, rounded when viewed from below, narrowly cylindrical above	Distinctly 4-angled, box-shaped, square when viewed from below	Elongated-urceolate, cylindrical-quadrangular, enlarged lower down, tapering to the tip, distinctly to slightly angled, hardly ever pyramidal
Tube colour	Light green below, turning light yellow to light orange above	Yellowish green, becoming creamy white to reddish brown and scarious when dry	Orange, becoming scarious when dry, papery and becoming scarious when dry and somewhat rigid in fruit	Yellowish to yellowish green especially lower down, uniformly red at anthesis but parts not exposed to direct sunlight remaining yellowish green	Yellowish, becoming scarious when dry, papery and somewhat rigid in fruit	Greenish yellow	Yellowish orange	Yellowish green	Shiny greenish yellow
Lobe size (mm)	5–6 × 2–4	2–5 × 2–4	4–8 × 2.5–5.0	3–4 × 3	4–8 × 2.5–3.0	3.0–3.5 × 2–3	2–6 × 1–4	2–3 × 1.8–2.0	4–5 × 4–5
Lobe shape	Lanceolate, spreading to recurved, tapering to an acute apex	Ovate to subcircular, acute, apiculate, usually connivent, slightly twisted with upper part of tube becoming slightly spreading, depending on time of day	Oblong-lanceolate to elliptic, acute or subacute, with apical mucro 0.5–1.0mm long, becoming fully spreading depending on time of day	Ovate to slightly square, straight to very slightly curved outwards, tapering to a point, minutely apiculate	Oblong-lanceolate to elliptic, acute, apiculate, erect, becoming fully spreading depending on time of day	Ovate to oblong, acuminate towards apex, apiculate	Obovate or lanceolate, acuminate towards apex, apiculate-mucronate	Suboblong to ovate, tip pointed, rounded, or obtuse	Broadly obovate to rounded, apiculate

Continued

TABLE 7.1 Comparison of the vegetative and reproductive characters of *Kalanchoe* species indigenous to southern Africa. Species are arranged alphabetically according to specific epithet—cont'd

Character	K. alticola	K. brachyloba	K. crenata	K. crundallii	K. hirta	K. laciniata	K. lanceolata	K. leblanciae	K. longiflora
Lobe colour	Light yellow to sulphur yellow to light orange	Light yellow, faintly white-margined	Light orange longitudinally infused with yellow	Yellow within, reddish without	Yellow	Light yellow to creamy yellow, faintly green-tipped	Bright to yellowish orange	Bright yellow, faintly brown-tipped	Yellow
Stamen inclusion	Included	Often included because of lobe connivence	Both whorls included	Included	Included	Well included	Well included	Included	Hardly exserted
Stamen insertion	—	In 2 obscure ranks ± above middle of corolla tube	In 2 distinct ranks ± at and just above middle of corolla tube	In 2 obscure ranks just above middle of corolla tube	Upper rank just above middle of corolla tube and lower rank just below middle	In two ranks well below the middle of corolla tube	In ± two ranks at about the middle of the corolla tube	In, or slightly above, the middle of the corolla tube	Three quarters of the way up corolla tube
Filament length (mm); morphology; colour	Short	1.0–1.5	1.0–2.5	3; thin; light yellow	± 1	2–4; thin; whitish green	2–4; thin; yellowish green	± 1.5; thin; light green	4–5; thin; greenish yellow
Anther length (mm); visibility; colour; shape	0.6–0.8; visible at the mouth of the corolla tube	0.5–1.0	0.5–1.0; broadly oblong	0.5–1.5; oblong to arrowhead-shaped	± 0.5	0.5–1.0; greenish yellow	0.5–1; yellowish grey; oblong	± 0.5; brownish yellow	0.4–0.5; yellowish brown
Carpel length (mm); colour	—	8–12	5–10	6–7; light green	4–6	4–5; light yellowish green	4–5; light yellowish green	5–6; light yellowish green	10–11; light green
Style length (mm); colour	—	0.5–1.5	1–5	3; light yellow	1–2	2–3	1–3	1.0–1.5	5–7
Stigma expression; colour	—	—	—	Slightly capitate	—	Very slightly capitate; whitish green	Very slightly capitate; whitish green	Very slightly capitate; light yellow	Very slightly capitate; yellowish green
Scale length (mm); shape; colour	3–4; Linear	Up to 4; linear-lanceolate	Up to 4; linear	± 2.5–3.0; linear-oblong; light yellow	Up to 2–3; linear	± 4; linear, sometimes apically bifid; whitish green	2–4; linear; yellowish green	± 1; linear	± 4; linear; yellowish green
Follicle length (mm); morphology; colour	Not seen	8–12; brittle, grass spikelet-like, enveloped in dry, creamy white remains of corolla, brown	5–7; brittle, grass spikelet-like, enveloped in whitish remains of the corolla, ultimately turning dull brown	5–6; brittle, grass spikelet-like, enveloped in shiny brown remains of the corolla, remaining shiny brown for a long time, ultimately turning dull brownish black	5–6; brittle, grass spikelet-like, enveloped in dry, light whitish grey remains of corolla, remaining dull light green for a long time, ultimately turning dull brownish black	Not seen	5–8; brittle, grass spikelet-like, enveloped in dry, straw-coloured remains of corolla	Not seen	10–11; brittle, grass spikelet-like, enveloped in dry, reddish brown remains of corolla, drying greyish black
Seed size (mm); shape; colour	Not seen	Not seen	Not seen	0.4–0.5; light to dark brownish black	0.4–0.6; dark brownish black	Not seen	0.25–0.5, Cigar-shaped to rectangular; light to dark brown	Not seen	0.50–0.75; Urceolate to banana shape curved, brown

TABLE 7.2 Comparison of the vegetative and reproductive characters of *Kalanchoe* species indigenous to southern Africa. Species are arranged alphabetically according to specific epithet

Character	K. luciae	K. montana	K. neglecta	K. paniculata	K. rotundifolia	K. sexangularis	K. thyrsiflora	K. winteri
Height (m) when in flower	Up to 2	Up to 1	Up to 1	Up to 1.2	0.2–1.5	0.2–1.0	Up to 1.5	0.5–0.9
Habit	Perennial or short-lived multiannual, few- to many-leaved, sparsely branched from near the base, glabrous or rarely minutely pubescent, robust succulent	Perennial or short-lived multiannual, few- to many-leaved, basally rosulate, solitary or sparsely branched from near the base, consistently minutely pubescent, robust succulent	Perennial, few-leaved, usually unbranched, glabrous, small to medium-sized succulent	Multiannual, few-leaved, usually solitary, sparsely branched, glabrous, robust succulent, sometimes perennial through basal offsets; rootstock somewhat fattened	Perennial, rarely annual or biennial, few-leaved, sparsely branched, glabrous, low-growing small to medium-sized, succulent	Perennial, few-leaved, sparsely branched, glabrous, robust succulent	Biennial or short-lived multiannual, few- to many-leaved, usually solitary, glabrous throughout, robust succulent	Perennial, many-leaved, 1–3 rosettes, sparsely to profusely branched from near the base and higher up, glabrous, waxy, robust succulent
Stems	Arising from a slightly swollen rootstock, erect to curved upwards, horizontally ridged where leaves abscised, greenish white	Fleshy, erect to leaning to creeping, curved upwards, greenish white	Somewhat brittle, few, unbranched, arising from herbaceous base, erect, green	Robust, usually unbranched, erect to leaning, often with one or more lengthwise running ridges, round to slightly angled, light green to yellowish green, drying reddish brown, same colour as leaves	Few, usually unbranched, thin, erect to leaning to creeping, rooting along the way, sometimes with sterile, small-leaved branchlets at the base, usually simple or sometimes branching near the base, rarely branched higher up, green with a slight bloom	Few, unbranched, arising from a brittle, corky base, erect to leaning to creeping, often with several lengthwise running ridges, 4-angled at least on sterile shoots, green to deep wine-red	Usually solitary, arising from a slightly swollen rootstock, erect, slightly horizontally ridged, greenish white	Erect to leaning and curved upwards, glabrous, waxy especially at internodes, light green
Leaves	Erect to patent-erect, succulent, ± flattened above, slightly convex below	Simple, erect to patent-erect, succulent, ± flattened above, slightly convex below	Opposite-decussate, succulent, spreading, papery on drying	Opposite-decussate, sessile lower down, petiolate higher up, succulent, spreading, smooth, coriaceous and papery on drying	Opposite-decussate, never peltate, succulent, erect to spreading, dense low down on stem and branches, sparse higher up	Opposite-decussate, succulent, spreading or recurved, coriaceous and rigid on drying	Erect to patent-erect, succulent, ± flattened above, slightly convex below	Opposite, erect to mostly spreading to variously floppy, succulent, flattened above and below

Continued

TABLE 7.2 Comparison of the vegetative and reproductive characters of *Kalanchoe* species indigenous to southern Africa. Species are arranged alphabetically according to specific epithet—cont'd

Character	K. luciae	K. montana	K. neglecta	K. paniculata	K. rotundifolia	K. sexangularis	K. thyrsiflora	K. winteri
Leaf colour and surface	Light yellowish green to bluish green, infused with ruby red especially apically and along margin	Light yellowish green, rarely infused with red especially apically and along margin, very minutely pubescent	Light green	Light green to yellowish green, sometimes lightly infused with red especially along margin	Midgreen to dull-green to pale greyish green, covered in a slight bloom	Green infused with red to a deep crimson red	Light greyish green to bluish green, infused with red, pink, or orange, especially along margin	Waxy, light green to bluish green, glabrous
Petiole (length in cm, if present)	Absent	Absent	2–8; slightly channelled above, not stem clasping	Absent or up to 5; channelled above, not or scarcely clasping the stem	Up to 1; slightly channelled above, subcylindrical in cross section	0.5–5.0; Channelled above, lower leaves not clasping the stem but upper leaves distinctly so	Absent	Absent
Blade size (cm)	4–16 × 2–9	4–18 × 2–7	4–12 × 3–8	6–20 × 4–10	1–8 × 1.0–5.5	6–14 × 4–10	6–14 × 2–9	14–16 × 8–14
Blade shape	Not folded lengthwise, round to obovate to oblong	Not folded lengthwise, obovate to oblanceolate	Broadly elliptic, ovate, cordate, or peltate, saucer-like	Often folded lengthwise, oblong to ovate to, more rarely, almost round, sometimes gracefully recurved in upper half	Not folded lengthwise, oblong, lanceolate, spatulate, narrowly to broadly cymbiform, sometimes staghorn-like divided, not recurved in upper half	Often ± folded lengthwise, broadly elliptic or obovate to oblong, recurved in upper half	Not folded lengthwise, obovate to oblanceolate	Not folded lengthwise, obovate to ± oblong
Indumentum	Absent	Very minutely pubescent	Absent	Absent	Absent	Absent	Absent	Absent
Bulbils	Not produced	Not produced	Not produced	Not produced	Not produced	Not produced	Not produced	Axils often carrying small leafy shoots and short branches that produce flowers in season
Apex	Rounded, obtuse, or truncate	Rounded, obtuse, or truncate	Rounded, obtuse	Rounded, obtuse, or slightly indented	Rounded, obtuse	Rounded, obtuse	Rounded, obtuse	Rounded, obtuse, or truncate, usually indented
Base	Narrow	Narrow	Deeply cordate in lower leaves, often cuneate higher up	± cuneate to abruptly tapering into petiole	Tapering, narrowly triangular	Obcordate in lower leaves, narrowly triangular to cuneate in upper leaves	Narrow	Narrow, sometimes distinctly auriculate

Margins	Entire, with a substantial red tint	Entire, sometimes undulate	Entire or coarsely crenate or undulate, crenate into rounded, harmless, crenations, patelliform	Entire	Entire, coarsely dentate to crenate	Coarsely crenate or undulate-crenate into rounded, harmless crenations	Entire, with a reddish tint	Glabrous, slightly lighter green than blade, sometimes infused with red
Inflorescence height (m)	Up to 1.6	Up to 0.75	Up to 1	Up to 1.2	0.01–0.40	0.25–0.80	Up to 1.5	0.5–0.9
Inflorescence morphology	A slender, erect, densely flowered, cylindrical to club-shaped thyrse consisting of several dichasia terminating in monochasia	An erect, densely flowered, cylindrical to club-shaped thyrse consisting of several dichasia terminating in monochasia	A short to long, mostly erect, apically dense to sparse, many-flowered, flat-topped thyrse with several dense dichasia, rather round in outline when viewed from above, branches opposite, without leafy branchlets in axils, axis green to bluish green	An erect to leaning, apically dense, many-flowered, flat-topped thyrse with several dichasia, more or less round in outline when viewed from above, branches opposite, erect, straight, slanted upwards at 45° or less, subtended by succulent, boat-shaped, leaf-like bracts, without, or very rarely with, leafy branchlets in axils, axis light green to yellowish green	Floriferous only at the top, erect to leaning, apically sparsely branched, few-flowered, corymbose cyme, rather round in outline when viewed from above, branches opposite, subtended by small leaf-like bracts that dry soon, without leafy branchlets in axils, axis green to bluish green	Erect to leaning, apically dense, few- to many-flowered, flat-topped thyrse with several dichasia, rather ellipsoid in outline when viewed from above, branches opposite, erect to gracefully curved upwards, subtended by leaf-like bracts, without or very rarely with, leafy branchlets in axils, axis reddish green to bright crimson red	A slender, erect, densely flowered, cylindrical to club-shaped thyrse consisting of several dichasia terminating in monochasia	A slender, erect, densely flowered, cylindrical thyrse consisting of several dichasia terminating in monochasia
Flower disposition	Erect to slanted horizontally	Erect to slanted horizontally	Erect	Erect	Erect	Erect	Erect to slanted horizontally	Erect to slanted horizontally
Flower colour	Usually white, pale greenish white, sometimes cream, pinkish, or yellowish, all parts haphazardly covered with a thin to mostly substantial white waxy bloom	Greenish yellow, all parts minutely pubescent	Upper ½ of tube and lobes orange to yellowish orange, lower ½ of tube often gradually becoming light green infused, lobes same colour as upper ½ of tube	Yellowish orange, gradually green lower down	Usually bicoloured, upper ½ of tube and lobes crimson red, orange, or yellowish orange, lower ½ of tube usually gradually becoming green, lobes same colour as upper ½ of tube	Greenish yellow (tube) to bright yellow (lobes)	Usually greyish green, all parts covered with a substantial white waxy bloom	Pale yellowish green to greenish white (tube) and yellow (lobes), all parts excepting tepal lobes above covered with a substantial white waxy bloom, highly scented, resinous to the touch

Continued

TABLE 7.2 Comparison of the vegetative and reproductive characters of *Kalanchoe* species indigenous to southern Africa. Species are arranged alphabetically according to specific epithet—cont'd

Character	K. luciae	K. montana	K. neglecta	K. paniculata	K. rotundifolia	K. sexangularis	K. thyrsiflora	K. winteri
Pedicel length (mm) and morphology; indumentum	4–6	4.0–6.0(–10.4)	4–7; slender	4–5; slender	1–8; slender	4–8; slender	Short	9–10
Calyx colour and indumentum	Light greenish white	Light yellowish green, reddish brown-tipped, conspicuously so when flowers in bud, minutely pubescent	Light green	Dull yellowish green	Bluish green, covered with a slight bloom	Shiny reddish green, strongly infused with small red spots especially towards base	Greyish green, hardly contrasting against corolla tube	Midgreen, contrasting against lighter green corolla tube
Calyx tube length (mm)	–	–	–	1	0.5	–	–	1
Sepal size (length or size in mm)	3–6	5–7	2–3	± 2 × 1	± 1.0–2.0 × 1.0–1.5	± 2 × 1	3.0–4.5	3–4
Sepal fusion	Almost free to the base	Almost free to the base	Separate, basally slightly fused	Basally fused for 1mm	± separate, basally fused for ± 0.5mm	± separate, basally adnate	Almost free to the base	Basally fused for 1mm
Sepal shape, colour, indumentum	Short-triangular, acute	Lanceolate-elongated, acute	Elongated-triangular, acute, hardly contrasting against green part of corolla tube, curved away from base of corolla tube	Deltoid-triangular, acute, hardly contrasting against corolla tube	Elongated-triangular, acute, hardly contrasting against green part of corolla tube	Triangular-lanceolate, acute, contrasting reddish green against corolla tube	Short-triangular to ovate, acute	Elongated-triangular, acute
Corolla length (mm)	8–14	15	6–8	10–12	8–15	14–17	13–18	13–15
Corolla shape; indumentum, if present	More or less quadrangular-urceolate, enlarged in the middle	More or less pyramidal, broadest and somewhat cigar-shaped enlarged slightly below the middle	Distinctly enlarged lower down around carpels, distinctly and tightly twisted apically after anthesis	Somewhat enlarged lower down, very slightly once twisted apically after anthesis	Distinctly enlarged lower down around carpels, distinctly and tightly twisted apically after anthesis	Somewhat enlarged lower down, not twisted apically after anthesis	More or less quadrangular-urceolate, slightly 4-angled	More or less quadrangular, ellipsoid, cigar-shaped enlarged in the middle, distinctly 4-angled
Corolla colour	Light greenish yellow	Light greenish yellow	Yellowish orange, gradually green lower down	Yellow to yellowish green	Crimson red, orange or yellowish orange, usually gradually becoming green lower down	Yellowishgreen and yellow	Greyish green, densely covered in a waxy bloom	Light greenish yellow

	7–12	13	5–7	9–10	6–15	13–16	12–18	11–12
Tube length (mm)	7–12	13	5–7	9–10	6–15	13–16	12–18	11–12
Tube shape	More or less quadrangular-urceolate, enlarged in the middle, slightly 4-angled	More or less pyramidal, broadest and somewhat cigar-shaped enlarged slightly below the middle, slightly 4-angled	Rounded, 4-angled, box-shaped, square lower down when viewed from below, longitudinally narrowing above beyond carpels	Distinctly 4-angled, box-shaped, square when viewed from below, longitudinally fluted above	Rounded, 4-angled, prominently urceolate, strongly globose lower down to somewhat box-shaped, square when viewed from below, longitudinally sharply narrowing beyond carpels above	Distinctly 4-angled, box-shaped, square when viewed from below, longitudinally fluted above	More or less quadrangular-urceolate, slightly 4-angled	More or less quadrangular, ellipsoid, cigar-shaped-enlarged in the middle, distinctly 4-angled
Tube colour	Light greenish yellow	Light greenish yellow	Yellowish orange, green lower down	Yellow to yellowish green	Crimson red, orange or yellowish orange	Yellow to yellowish green	Greyish green, densely covered in a waxy bloom	Light greenish yellow
Lobe size (mm)	2.5–4.0 × 3–4	5 × 2	2.5–3.0 × 1.5–2.0	2.0–2.5 × 2.0–2.5	2.0–4.5 × 2–3	2.0–3.5 × 2–3	2–3 × 2–3	6–8 × 3.5–4.0
Lobe shape	Broadly ovate, obtuse to nearly acute, sometimes apiculate	Elongated-triangular, erecto patent to slightly spreading, margins slightly folded in, apiculate	Lanceolate, distinctly acute apically, spreading	Deltoid to rather narrowly triangular, pointed at apex, apiculate	Lanceolate to elliptic, distinctly acute apically, subfalcate, spreading but showing diurnal movement (closing at night)	Ovate to subcircular to somewhat pyriform, rounded at apex, apiculate	Somewhat square, recurved, tapering to a point	Triangular, margins slightly to distinctly enrolled, truncated
Lobe colour	Lobes mostly white, fading pinkish brown when spent	Lobes yellow, fading light brown when spent	Yellowish orange, orange more intense towards the lobe margins	Bright yellow	Crimson red, orange or yellowish orange	Bright yellow, faintly brown-tipped	Bright yellow	Bright yellow
Stamen inclusion	1–2 mm exserted	1–2 mm exserted	Included	Included	Included	Slightly exserted	Included to hardly exserted, visible at mouth of corolla	1–2 mm exserted

Continued

TABLE 7.2 Comparison of the vegetative and reproductive characters of *Kalanchoe* species indigenous to southern Africa. Species are arranged alphabetically according to specific epithet—cont'd

Character	K. luciae	K. montana	K. neglecta	K. paniculata	K. rotundifolia	K. sexangularis	K. thyrsiflora	K. winteri
Stamen insertion	Middle of the corolla tube	–	–	In two ranks at about middle of corolla tube	In several ranks at or just above middle of corolla tube	Well above middle of corolla tube	In two ranks well above middle of corolla tube	Just below or in middle of corolla tube
Filament length (mm); morphology; colour	±6; thin, light greenish white	Thin; light greenish yellow	Thin	3–5; thin; yellow	3–5; thin; yellow-infused to reddish	3–5; thin; yellow	Short	3.0–5.5; thin; light greenish white
Anther length (mm); visibility; colour; shape	± 0.5; brownish orange	± 0.5; purplish brown	0.5–0.7	0.3–0.5; yellowish brown	0.3–0.5; yellow	0.3–0.5; greenish yellow	1.5–2.0; yellow	1.4–1.6; yellow
Carpel length (mm); colour	6–8; light yellowish green	Not seen	6–7; midgreen	5–6; light yellowish green, red spotted	4–6; mid green	7–8; light yellowish green	10–14	9–10; light green
Style length (mm); colour	2–4	Slender	Short	2–3	2–3	3–7	1.5–3.5	±4
Stigma expression; colour	Shortly exserted, very slightly capitate; light yellow	Inserted, minutely capitate; light yellowish white	Very slightly capitate	Very slightly capitate; whitish yellow	Very slightly capitate; yellowish green	Very slightly capitate; whitish yellow	Very slightly capitate	Very slightly capitate, exserted as far as or slightly less than anthers; light yellow
Scale length (mm); shape, colour	± 2; square to transversely oblong	Oblong	± 2; narrowly tapering, linear	1.5–2.0; linear, light green	± 2; linear, light yellowish	± 2; reddish brown in upper half	± 2.5; transversely oblong, truncate	2.3–2.5 × 1.8–2.1; narrowing at the base, truncate, repand
Follicle length (mm); morphology; colour	9–10; with the appearance of a grass spikelet, surrounded by dark brown dry corolla tube; dull light green	Not seen	6–7; enveloped in dry, dark brownish purple remains of corolla, sharply recurved like a peeled banana at tips; drying dark brown	8–9; brittle, grass spikelet-like, enveloped in dry, light to dark brown remains of corolla; dull whitish green	4–6; brittle, grass spikelet-like, enveloped in dry, light brownish purple remains of corolla; dull brownish purple	6–7; brittle, grass spikelet-like, enveloped in dry, light brown remains of corolla; dull whitish green	Not seen	6–7; brittle, grass spikelet-like, enveloped in dry, light whitish brown remains of corolla; light to dark brown when dry
Seed size (mm); shape; colour	1.0–1.5; oblong; light brown	Not seen	0.8–1.0; dark brown	0.75–1.00; cylindrical to slightly banana shape curved; reddish brown to dark brown	0.50–0.75; brown	0.50–0.75; brown	Not seen	0.75–1.25; dark brown

TABLE 7.3 Comparison of the vegetative and reproductive characters of Madagascan *Kalanchoe* species naturalised in southern Africa. Species are arranged alphabetically according to specific epithet

Character	K. beharensis	K. daigremontiana	K. fedstchenkoi	K. gastonis-bonnieri	K. pinnata	K. prolifera	K. tubiflora
Height (m) when in flower	Up to 3	Up to 0.8	Up to 0.75	Up to 0.7	Up to 2	Up to 2–3	Up to 2
Habit	Perennial shrubs or small trees, sparsely branched, with haphazardly rounded canopies, robust succulent	Biennial shrubs, unbranched, glabrous, robust succulent	Perennial, spreading, robust, tuf-like, annually increasing in size, glabrous, brittle shrubs, sparsely to densely branched succulent, with haphazard, untidy canopies	Perennial, sometimes biennial and monocarpic, few-leaved, few-branched, glabrous, small- to medium-sized, terrestrial, succulent, forming dense groups	Perennial, sometimes monocarpic, many-leaved, unbranched but sprouting from the base, glabrous throughout, medium-sized to large succulent	Perennial, many-leaved, unbranched but sprouting from the base, glabrous throughout, large succulent	Biennial or semiperennial, erect to leaning to procumbent, glabrous, brittle shrubs, sparsely branched, or suckering near the base, with canopies terminating in an inflorescence, succulent
Stems	Simple lower down, branched from ± 1m above the ground, thick, to 15cm in diameter, ± straight or variously curved, surface prominently covered by sharp projections left by expanding, prominent leaf scars, bark flimsy, light brownish grey	Simple, erect, or leaning and then curved upwards, brownish	Thin, green to greenish purple when young, becoming brown, variously erect or prostrate, rooting along the way, leaning branches developing long, near-woody stilt-like roots, bark peeling to flaking with age, flimsy, brown	Short, sometimes branched low down, erect	Robust, erect to leaning-erect, somewhat woody below	Robust, erect to leaning, somewhat woody below, ± 4-angled, raised scars of abscised leaves conspicuous	Thin to medium-sized in diameter, light yellowish brown to yellowish grey, usually erect or toppling over under the weight of the inflorescence
Leaves	Few to many towards terminal ½ of branches, shed lower down to yield leaf scars, erect to variously curved, ± succulent, ± flattened above and below	Sometimes peltate with blade 'wings' stretching beyond point of petiole attachment, ± flattened above and below	Many, densely packed in young plants, more widely dispersed on older stems, erect to variously slanted away from branches, succulent, flattened above and below, often slightly thickened below in line with petiole	Basally laxly rosulate to subrosulate, opposite-decussate, succulent, spreading to erectly spreading	To 20cm long, opposite-decussate, leaf pairs widely spaced, softly leathery-succulent to succulent, spreading to slightly down curved	To 30cm long, opposite-decussate, leaf pairs widely spaced, succulent, spreading to downcurved	Many, sparsely arranged and evenly spaced throughout, slanted away from branches at ± 45°, succulent, terete, cylindrical to narrowly oblong

Continued

TABLE 7.3   Comparison of the vegetative and reproductive characters of Madagascan *Kalanchoe* species naturalised in southern Africa. Species are arranged alphabetically according to specific epithet—cont'd

Character	K. beharensis	K. daigremontiana	K. fedstchenkoi	K. gastonis-bonnieri	K. pinnata	K. prolifera	K. tubiflora
Leaf colour and surface	Dull yellowish green to bluish green, densely pubescent	Green and variously spotted with purplish, pinkish or brownish blotches, glabrous	Bluish to purplish to brownish green, glabrous, distinctly waxy, wax easily rubs off	White-pruinose especially above, green or bluish green with irregular brownish green markings	Green, sometimes with purplish lines, somewhat shiny	Green, somewhat shiny	Bluish to purplish to brownish green to grey-green with irregular dark green or bluish purple spots, glabrous, somewhat waxy
Petiole (length in cm, if present)	4–10, same colour as leaf blade	1–4, amplexicaul	0.1–1.0, thin, same colour as leaf blade	Short or prominent, broad, often not very distinct	Up to 10, subterete, amplexicaul to half-clasping the stem	Up to 16, broadened at base, clasping the stem, purplish red	Absent
Blade size (cm)	6–40 × 8–30	5–20 × 2–5	2–7 × 1.5–4.0	12–55 × 5–10	5–20 × 4–10	7–15 × 1–5 [segments]	3–12 × 0.2–0.5
Blade shape	Irregularly folded lengthwise and in width; ± elongated-triangular, deltoid to peltate	Often irregularly folded lengthwise; ovate, obovate, elongated-triangular	Obovate to nearly circular	Folded lengthwise; ovate-lanceolate	Often lengthwise slightly boat shape-folded upwards; first leaves simple, 3- to 5-foliolate higher up, sometimes reduced to terminal leaflet, pinnate blade segments sometimes slightly asymmetrical, circular to oblong to oblong-oval to ovate-elongate, terminal leaflet the largest	Often lengthwise boat shape-folded upwards; oblong to ovate-elongate, pinnatisect or pinnate, rarely undivided	Cylindrical to narrowly oblong
Indumentum	Leaves densely pubescent	Absent	Absent	Absent	Absent	Absent	Absent
Bulbils	Not produced	Formed along leaf margins	Formed along leaf margins, especially once leaves are detached	Leaf apex forming plantlets, these developing roots while still attached	Produced in the notches between the crenations	Formed on inflorescence	Apex with teeth and usually bulbils borne in their axils, also once leaves are detached
Apex	Rounded, obtuse, or truncate	Acute	Rounded obtuse	Harmlessly acute	Obtuse rounded	Rounded	With 2–9 small, conical teeth
Base	Flat, flared downwards beyond point of attachment of petiole, rarely somewhat cuneate	Broadened, rounded	Narrow, cuneate	Cuneate	—	Sometimes asymmetrical, decurrent at the base	Narrow

Margins	Irregularly toothed	Irregularly crenate with plantlets developing on the marginal crenations	Irregularly toothed with coarse, acute or subacute crenations in the upper half, depressions between teeth sometimes dark brownish purple, somewhat angled, forming bulbils (adventitious buds) in the crenations especially once leaves are detached, lower half of leaves sometimes with a few small, irregular, hardly noticeable crenations only	Sinuate to coarsely crenate	Broadly crenate to harmlessly dentate, reddish purple, with bulbils produced in the notches between the crenations	Crenate to harmlessly dentate, reddish purple	–
Inflorescence height (m)	Up to 0.4	Up to 0.2	Up to 0.2	0.40–0.55 × 0.3–0.6	Up to 0.8	0.35–0.90 × 0.20–0.45	Up to 0.2
Inflorescence morphology	An axillary, densely branched, many-flowered panicle, rounded, erect to gracefully curved sideways and upwards	A terminal, apically branched, many-flowered, head-shaped corymb, erect, peduncle straight, light greenish purple	Terminal, apically branched, many-flowered head-shaped corymb, erect, some corymb branches terminating in sterile, curved, pedicel-like branchlets peduncle straight, light brownish purple	Erect, apically dense, many-flowered, lax cyme, corymbose	A very large, erect to leaning, branched, paniculate cyme	A very large, erect to leaning panicle, often with aborted flowers and bulbils	A terminal, apically branched, many-flowered, head-shaped to rounded thyrse, erect, peduncle long, straight, light yellowish grey
Flower disposition	Erect or slanted in various directions, subtended by prominent bracts that are soon shed	Pendulous, subtended by small bracts that soon shrivel	Pendulous, subtended by small bracts that soon shrivel	Pendulous to slightly spreading	Pendulous	Pendulous	Pendulous, subtended by small bracts that soon shrivel

Continued

TABLE 7.3   Comparison of the vegetative and reproductive characters of Madagascan *Kalanchoe* species naturalised in southern Africa. Species are arranged alphabetically according to specific epithet—cont'd

Character	K. beharensis	K. daigremontiana	K. fedstchenkoi	K. gastonis-bonnieri	K. pinnata	K. prolifera	K. tubiflora
Flower colour and indumentum	Pale yellow to yellowish green with longitudinal reddish-purple stripes prominent on corolla lobes, less so on corolla tube, waxy bloom absent, ± densely pubescent	Dull light pinkish purple, longitudinally infused with yellow in centre of petal, waxy bloom absent, papery when dry, glabrous	Orange, longitudinally infused with yellow in centre of petal, waxy bloom absent, papery when dry, dries same purplish colour as calyx, glabrous	Yellowish green infused with reddish-purple veins, glabrous	Greenish white below where enveloped by calyx, red, purplish or greenish purple above where exposed beyond calyx	Green below where enveloped by calyx, red above where exposed beyond calyx, sometimes light green to yellowish green throughout	Buds yellowish, mature flowers various shades of red, from crimson to deep orange, basally longitudinally infused with yellow along tube angles, waxy bloom ± absent, papery when dry, drying purplish brown, glabrous
Pedicel length (mm) and morphology; indumentum	5–12; densely pubescent	5–10; glabrous	6–10; glabrous	5–15; slender	10–25; slender	8–15; slender	6–12; glabrous
Calyx shape	Shortly tubular	Tubular for ± two-thirds	Tubular for ± three-quarters	Tubular, urceolate	Calyx inflated, broadly cylindrical to campanulate-tubular, thinly succulent, hardly depressed at the base	Calyx inflated, tubular	Campanulate, tubular for ± one-third
Calyx colour and indumentum	Light greenish white, very lightly infused with purple, especially towards tip, with or without feint longitudinal reddish-purple stripes, ± densely pubescent	Shiny light green infused with purple, purple more prominent towards calyx base and sepal tips, purple arranged in feint longitudinal lines	Light greenish purple, purple more prominent towards sepal tip, purple arranged in feint longitudinal lines	Prominent, pale green with reddish-purple veins	Green with purple markings to purple with green markings, smooth, shiny, ± cylindrical	Green, campanulate, 4-angled, slightly scabrid	Light greenish red, red-infusion more prominent towards sepal tip, purplish when young, red arranged in feint longitudinal lines
Calyx tube length (mm)	2	4–6	10–15	15–19	20–30	12–16	3–4
Sepal size (length or size in mm)	5–7	7–9	15–20	20–25; [5–6 × 4–5 (lobes)]	25–40; [5–10 × 5–11 (lobes)]	15–20; [2.5–4.0 × 2.5–4.0(–7) (lobes)]	10–11
Sepal fusion (mm)	Below for ± 2	Below for ± 4–6	Below for ± 10–15	Below for 15–19	Below for 20–30	Below for 12–16	Below for ± 3–4
Sepal shape, colour, indumentum	Elongated-triangular, acute, obscuring the corolla tube, ± densely pubescent	Short-triangular to deltoid, acute, obscuring one-quarter of corolla tube	Elongated-triangular, acute, obscuring two-thirds to three-quarters of corolla tube	± deltoid, acute, greenish white-pruinose, glabrous	Apically acuminate-cuspidate, softly acute, contrasting against corolla tube, clasping the corolla tube	Apically acuminate-cuspidate, softly acute, green contrasting against red corolla tube, clasping the corolla tube	Elongated-triangular, acute, obscuring lower ± one-third of the corolla tube, purplish when young, red arranged in feint longitudinal lines

Corolla length (mm)	7–8	16–20	19–22	40–50	27–44	18–25	25–30
Corolla shape; indumentum, if present	More or less quadrangular-urceolate; ± densely pubescent	More or less cylindrical-tubular to somewhat campanulate	More or less cylindrical-tubular to somewhat campanulate	Cylindrical	Somewhat balloon-like inflated, longer than calyx tube	Somewhat inflated and enlarged lower down	Cylindrical-tubular to distinctly campanulate, bulging in the middle
Corolla colour	Pale yellow to yellowish green with longitudinal reddish-purple stripes	Dull light pinkish purple, longitudinally infused with yellow in centre of petal	Orange to pinkish orange, longitudinally infused with yellow in centre of petal	Yellowish green infused with reddish-purple veins	Greenish white below where enveloped by calyx, red, purplish, or greenish purple above	Light green below where obscured by calyx, red above where exposed	Various shades of red, from crimson to deep orange, basally longitudinally infused with yellow along tube angles
Tube length (mm)	6–7	15–19	18–21	30–40	25–40	15–25	25–30
Tube shape	More or less quadrangular-urceolate, tapering to the mouth, 4-angled	More or less cylindrical-tubular to somewhat campanulate, flared at the mouth	More or less cylindrical-tubular to somewhat campanulate, flared at the mouth	Cylindrical	Tubular	Tubular, slightly constricted above carpels, suburceolate above constriction	Cylindrical-tubular to distinctly campanulate, bulging in the middle, slightly flared at the mouth
Tube colour	Pale yellow to yellowish green with longitudinal reddish-purple stripes	Dull light pinkish purple, longitudinally infused with yellow in centre of petal	Orange to pinkish orange, longitudinally infused with yellow in centre of petal	Yellowish green infused with reddish-purple veins	Greenish white below, merging into red, purplish or greenish purple	Light green below, merging into red above, sometimes light green to yellowish green throughout	Various shades of red, from crimson to deep orange, basally longitudinally infused with yellow along tube angles
Lobe size (mm)	3 × 4	± 7 × 4–5	± 5 × 5	± 10 × ± 6	9–14 × 4–7	2–3	± 15 × 5
Lobe shape	Strongly recurved, deltoid, apiculate	Obovate, apically rounded and acute-tipped	Spatulate to obovate, apically obtuse-flattened with a slight indentation	Spreading to slightly recurved, triangular-ovate, apically acuminate	Deltoid-cuspidate, tapering to an acute apex, somewhat spreading	Acuminate-cuspidate, tapering to an acute apex	Spatulate to obovate, apically obtuse-rounded, minutely apiculate
Lobe colour	Pale yellow to yellowish green with prominent longitudinal reddish-purple stripes	Dull light pinkish purple, longitudinally infused with yellow in centre of petal	Orangey within, orangey pink without, longitudinally infused with yellow in centre of petal	Yellowish green infused with reddish-purple veins	Greenish white below, merging into red, purplish or greenish purple	Green merging into red above, sometimes light green to yellowish green throughout	Various shades of red, from crimson to deep orange
Stamen inclusion	1–2mm exserted	Included or hardly exserted	Included or hardly exserted	Included	Very slightly exserted	Exserted	Included to hardly exserted
Stamen insertion	In the middle of the corolla tube or higher up	Below middle of corolla tube at ± upper level of carpels	Very low down in corolla tube at ± upper end of carpels	Below middle of the corolla tube	Below middle of corolla tube	Below middle of corolla tube	Very low down in corolla tube at ± same level as carpels

Continued

TABLE 7.3 Comparison of the vegetative and reproductive characters of Madagascan *Kalanchoe* species naturalised in southern Africa. Species are arranged alphabetically according to specific epithet—cont'd

Character	*K. beharensis*	*K. daigremontiana*	*K. fedstchenkoi*	*K. gastonis-bonnieri*	*K. pinnata*	*K. prolifera*	*K. tubiflora*
Filament length (mm); morphology; colour	± 5; thin; light greenish white	± 12; thin; light purplish red	± 20; thin; light purplish red	Short	Long	Long	25–28; thin; light greenish red, basally light green
Anther length (mm); visibility; colour; shape	± 0.5–1.0; exserted; yellow	± 1.0–1.5; hardly exserted; black	± 0.5–1.0; hardly exserted; black, pollen greyish yellow	± 3; reniform	± 3.0[× 1.5–2.0]; ovate	2.0[× 1.5]; visible beyond mouth of corolla tube; ovate	± 1; included to hardly exserted; black; pollen greyish yellow
Carpel length (mm); colour	3–4; light yellowish green	± 6; shiny midgreen	6–8; light shiny green	8–11	12–14	7–8	5–8
Style length (mm); colour	7–9	± 5	18–20	Prominent	20–30	1.5–2.0	20–25
Stigma expression; colour	Very slightly capitate; light brownish yellow	Very slightly capitate; green	Very slightly capitate; green; later slightly exserted	Capitate	–	–	Stigmas very slightly capitate; green; inserted, later hardly exserted
Scale length (mm); shape; colour	± 3[× 1]; rectangular, connate, truncate above, slightly apiculate	± 1.5[× 1.5]; ± square, free, slightly indented above	± 1[× 1]; rectangular to square, free, slightly indented above	Half-round to square	1.6[–2.6 × 1.5–2.0]; rectangular	1.4[–1.6 × 2.0–2.5]; ± orbicular	± 0.5 ×1.0; rectangular, free, flat or slightly indented above
Follicle length (mm); morphology; colour	6–7; brittle, grass spikelet-like, enveloped in dry, remains of corolla, greyish black dark brownish black	Not seen	Not seen	Not seen	Not seen	Not seen	Not seen
Seed size (mm); shape; colour	0.75–1.00; rectangular to slightly banana shape curved, tapering at both ends; dark brown to black	Not seen	Not seen	Not seen	Not seen	Not seen	Not seen

TABLE 7.4 Comparison of the vegetative and reproductive characters of Madagascan and south-tropical African *Kalanchoe* taxa widely cultivated in southern Africa. Taxa are arranged alphabetically according to specific epithet and cultivar name

Character	K. blossfeldiana	K. humilis	K. 'Margrit's Magic'	K. porphyrocalyx	K. pumila	K. tomentosa
Height (m) when in flower	0.1–0.4	Up to 0.8	Up to 0.6	Up to 0.4	Up to 0.30	Up to 1
Habit	Perennial, usually many-leaved, densely branched, glabrous throughout, small- to medium-sized, rounded, succulent shrublet	Perennial, low-growing, glabrous, sparsely branched, tuft-like rosettes, with small, rounded canopies, succulent	Perennial, few- to many-leaved, multi-branched, glabrous or finely pubescent, tuft-forming succulent	Perennial, tuft-like, smooth, sparsely to much-branched shrublets, with small, untidily rounded canopies, succulent	Perennial, many-leaved, branched, glabrous, small, epiphytic, lithophytic or terrestrial, succulent	Perennial, many-leaved, sparsely branched, entirely hairy, medium-sized, terrestrial, succulent
Stems	A few arising from a sometimes brittle, corky base, erect to slightly leaning, round in cross section, rarely with one or two lengthwise running ridges, green to reddish green	Thin, erect, leaning, creeping or trailing, weak, purplish green, arising from a thickened, slightly rhizomatous base	Few, unbranched or sparsely branched, erect to leaning, sometimes creeping, rooting along the way, leaning branches developing short, near-woody stilt-like roots, woody, somewhat brittle, nodes thickened, round, brown to reddish brown, older internodes with longitudinal light brownish or greenish stripes, sterile and reproductive stems glabrous to finely pubescent	Thin, ± erect, sturdy, leaf scars obvious when young, light green to light brownish green	Branched, erect to arched to leaning, ± round in cross section, raised scars of abscised leaves conspicuous, white-pruinose	Branched, basally ± woody and covered in longitudinally flaking, yellowish bark, erect to arched to leaning, ± round in cross section, scars of abscised leaves obscure
Leaves	Opposite-decussate, spreading to slightly erect, somewhat succulent, papery on drying	Few to many, sparsely to densely arranged along branches, erect to patent-erect, succulent	Opposite-decussate, lower older ones spreading to horizontal to decurved, upper younger ones ± vertical, succulent, papery on drying	Many, variable in shape, sparsely to densely arranged towards upper ends of branches, erect to slanted away from branches at a 45° degree angle, succulent, lengthwise folded along midrib	Opposite-decussate, erect, succulent	Alternate, often rosulate, erect, distinctly succulent
Leaf colour and surface	Midgreen to dark green, reddish infused if grown in exposed positions, shiny	Light green to bluish green, densely and irregularly reddish-purple-spotted depending on age and exposure to insolation, bloom not waxy, glabrous	Green to variously infused with red, not waxy	Light green, not waxy, glabrous	Purplish pink, covered with a very fine dusty, white, waxy layer	Dull light green to dull bluish green

Continued

TABLE 7.4 Comparison of the vegetative and reproductive characters of Madagascan and south-tropical African *Kalanchoe* taxa widely cultivated in southern Africa. Taxa are arranged alphabetically according to specific epithet and cultivar name—cont'd

Character	K. blossfeldiana	K. humilis	K. 'Margrit's Magic'	K. porphyrocalyx	K. pumila	K. tomentosa
Petiole (length in cm, if present)	0.5–3.0, channelled above, leaves not clasping the stem	Absent or up to 0.5, same colour as leaf blade	Up to 0.2, channelled above, not clasping the stem	0.1–1.0, same colour as leaf	Petiole absent or very short, not clasping the stem	Absent, leaves not clasping the stem
Blade size (cm)	4–7 × 2.5–4.0	2–10 × 1–3	1–3 × 1–2	2–5 × 0.5–2	2.5–4.5 × 1.5–2.5	2.5–8.0 × 1.0–2.5
Blade shape	Flat, ovate-oblong to oblong	Lower surface convex, upper surface flat or convex, obovate to spatulate	Flat, curved upwards towards margins, obovate, obovate-orbicular, orbicular, to somewhat oblong	From near-linear to near-orbicular, mostly oblong-ovate	Flat, obovate	Concave above, convex below, obovate, ovate, oblong, or subcylindrical
Indumentum	Absent	Absent	Very finely pubescent	Absent	Absent	All plant parts, especially the leaves, covered with hairs; hairs fine to coarse, erectly spreading, white, brown or reddish brown
Bulbils	Not produced	Not produced but severed leaves root easily	Not produced	Not produced	Not produced but severed leaves root easily	Not produced but severed leaves root easily
Apex	Rounded obtuse	Ending in a rounded tooth	Rounded obtuse	Rounded obtuse	Rounded obtuse	Rounded obtuse
Base	Rounded to somewhat cuneate	Narrow, cuneate	Cuneate	Narrow, attenuate	Cuneate	Cuneate
Margins	Coarsely crenate or undulate-crenate into rounded, harmless teeth especially above the middle, often curved upwards, red	Upper half irregularly toothed with small crenations, reddish purple infused	Entire to weakly crenate especially in upper two-thirds	Irregularly toothed with coarse crenations, concolourous or faintly red-infused	From midleaf upwards coarsely crenate or undulate-crenate into rounded, harmless teeth, purplish red	Often ± upcurved, entire, from ± midleaf upwards coarsely crenate with rounded, harmless light to dark brown teeth
Inflorescence height (m)	0.15–0.20	0.1–0.4	0.18–0.20	0.02–0.10	0.1–0.2	± 0.7
Inflorescence morphology	An erect, apically dense, many-flowered, flat-topped thyrse, branches opposite, erect, subtended by leaf-like bracts, without leafy branchlets in axils, axis green to reddish green to brown	A diffuse, terminal, branched, few- to many-flowered head-shaped cyme, dichasial branches subtended by fleshy, carpel-shaped bracts, erect to leaning to one side, peduncle ± straight, light green infused with purple	A terminal, branched, erect, apically sparse to dense, few- to many-flowered, flat-topped cyme with several dichasia, rounded when viewed from above, branches opposite, erect, subtended by very small leaf-like bracts, without leafy branchlets in axils, axis bright crimson red, minutely white-hairy	A terminal, apically branched, often one- to few-flowered head-shaped corymb, erect peduncle ± straight, deep red-purple	Short, erect, few-flowered, apically dense, corymbose, greyish whitewaxy	Erect to leaning, few-flowered, apically dense, corymbose

Flower disposition	Erect	Erect to horizontal, subtended by small succulent bracts that are soon shed	Erect	Pendent, subtended by small succulent bracts	Erect	Erect to spreading
Flower colour	Crimson red to scarlet (selected colour forms include virtually all colours, except black and blue), light green lower down, scarlet higher up (tube) to bright scarlet (lobes)	Light green (base of tube) to very light purple (upper part of tube) with a network of distinct, purple veins, waxy bloom absent, glabrous or very finely pubescent	Bright crimson red (tube and lobes), light green at level of calyx, obsolescently pubescent	Red to deep purple-red, lobes or margins only of corolla lobes light yellowish green, insides of lobes yellowish green, waxy bloom absent, glabrous or obsolescently pubescent	violet-pink, purple veined	Yellowish brown to greenish, often ± purplish
Pedicel length (mm) and morphology; indumentum	3–4	5–15	8–10	5–15	6–10, slender	5–10, slender
Calyx colour and indumentum	Shiny reddish green, strongly infused with small red spots especially towards tips and margins	Purple	Light reddish green, strongly infused with small red spots especially towards sepal margins	Red to deep purple-red, occasionally light greenish purple	Green to strongly reddish purple infused	Yellowish brown to greenish, often ± purplish, hairy
Calyx tube length (mm)	0.5	<1	± 1	± 2–3	± 1	± 1
Sepal size (length or size in mm)	± 4–6	1.0–1.5	± 4–6 × 3–4	5–6	3–5 × 1.0–2.5	3–5 × 1.0–2.5
Sepal fusion (mm)	± separate, basally hardly adnate for ± 0.5	Basally fused for <1	± separate, basally fused for ± 1	Fused for ± 2–3	± separate, basally fused for ± 1	± separate, basally fused for ± 1
Sepal shape, colour, indumentum	Triangular-lanceolate, acute, hardly contrasting against corolla tube, flared away from tube	Deltoid- to elongated-triangular, acute, apically slightly curved away from the corolla tube, glabrous or rarely very finely pubescent	Triangular-lanceolate, acute, hardly contrasting against light green basal part of corolla tube, minutely white-hairy	Lanceolate- to deltoid-triangular, acute, curved away from the corolla tube at a 90° angle, glabrous or rarely very finely pubescent	Abruptly triangular to lanceolate, acute, white-pruinose	± deltoid, obtuse

Continued

TABLE 7.4 Comparison of the vegetative and reproductive characters of Madagascan and south-tropical African *Kalanchoe* taxa widely cultivated in southern Africa. Taxa are arranged alphabetically according to specific epithet and cultivar name—cont'd

Character	K. blossfeldiana	K. humilis	K. 'Margrit's Magic'	K. porphyrocalyx	K. pumila	K. tomentosa
Corolla length (mm)	12–13	5–7	18–20	10–25	6–11	8–13
Corolla shape; indumentum, if present	4-angled, rounded to slightly box-shaped, square when viewed from below somewhat enlarged lower down, not twisted apically after anthesis	4-angled, basally slightly broadened, box kite-shaped above, very slightly flared at the mouth	Distinctly 4-angled, box-shaped, square when viewed from below, slightly enlarged above the middle, not twisted apically after anthesis	More or less campanulate to cylindrical-urceolate, flared at the mouth	Campanulate to suburceolate	Campanulate to very slightly urceolate, hairy
Corolla colour and indumentum	Light green and crimson red to scarlet	Light green (base of tube) to very light purple (upper part of tube) with a network of distinct, purple veins, waxy bloom absent	Bright crimson red, light green lower down, minutely white-hairy, drying purple-red	Red to deep purple-red (tube), lobes or lobe margins only light yellowish green, waxy bloom absent	Violet-pink, purple veined	Yellowish brown to greenish, sometimes ± purplish, especially on the inside of the flower mouth, hairy
Tube length (mm)	7–10	4–6	16–18	18–21	5–10	9–12
Tube shape	4-angled, rounded to slightly box-shaped, square when viewed from below, slightly longitudinally fluted above	4-angled, basally slightly broadened, box kite-shaped above, very slightly flared at the mouth	Distinctly 4-angled, box-shaped, square when viewed from below	More or less campanulate to cylindrical-urceolate, flared at the mouth	Campanulate to suburceolate	Campanulate to very slightly urceolate
Tube colour	Light green and crimson red to scarlet	Light green (base of tube) to very light purple (upper part of tube) with a network of distinct, purple veins	Bright crimson red, light green lower down, minutely white-hairy	Red to deep purple-red (tube), lobes or lobe margins only light yellowish green	Violet-pink, purple veined	Yellowish brown to greenish, sometimes ± purplish, especially on the inside of the tube mouth
Lobe size (mm)	5–6 × 3–4	± 1.0–2.5 1–2	4.5–5.0 × 4.5–5.0	± 5–6 × 6–7	5–10 × 3–4	3–4 × 4–6
Lobe shape	Ovate to oblong-ovate to oblong-deltoid, rounded-acute at apex, mucronate	Erect to hardly flared, oblong, apiculate	Ovate to subcircular, rounded at apex, apiculate	Ovate, apically obtuse-flattened with a slight indentation	Prominent, spreading to strongly recurved, partly obscuring the corolla tube, ovate to oblong-ovate, mucronate	Prominent, deltoid to round, erectly spreading, mucronate
Lobe colour	Bright crimson to scarlet	Very light purple	Bright crimson red	Yellowish green inside	Violet-pink, purple veined	Purplish, especially on the inside of the tube mouth
Stamen inclusion	Included	Included	Included	Included	Included to slightly exserted	Included

	Well above the middle of the corolla tube	At ± middle of corolla tube	At about the middle of the corolla tube	Very low down in corolla tube at ± lower end of carpels	Well above the middle of the corolla tube	± the middle of the corolla tube
Stamen insertion	Well above the middle of the corolla tube	At ± middle of corolla tube	At about the middle of the corolla tube	Very low down in corolla tube at ± lower end of carpels	Well above the middle of the corolla tube	± the middle of the corolla tube
Filament length (mm); morphology; colour	1.0–1.5; thin; yellow	± 3–4; thin; light greenish yellow	6–7; thin; yellow	± 20; thin; light yellowish green	Short	Short, filiform
Anther length (mm); visibility; colour; shape	0.30–0.75; rounded to slightly elongated, greenish yellow turning brown	± 0.75–1.00; included; yellowish brown	0.5; purplish brown	± 1; included; black	0.40–0.75; conspicuously yellow; reniform	± 1; visible inside the mouth; yellow; ovate
Carpel length (mm); colour	5–6mm long, bright light green	2.0–2.5mm long, light shiny green	6–7mm long, light green	7–8mm long, light shiny green	8–11mm long	7–11mm long
Style length (mm); colour	2–3	2.5–3.0	8–9	15–17	Short	Short
Stigma expression; colour	Capitate, whitish yellow, visible at mouth of corolla tube	Stigmas very slightly capitate, brownish, included	Capitate, whitish yellow	Very slightly capitate, brownish, included	Capitate	–
Scale length (mm); shape; colour	± 2; linear, light yellowish green	± 2; linear, obscurely bifid, free	± 2; narrowly columnar to slightly linear, light yellowish green	± 2 × 1; rectangular, free, distinctly indented above	Oblong to rectangular	Rectangular
Follicle length (mm); morphology; colour	5–6; grass spikelet-like, enveloped in dry, light brown remains of corolla; dull whitish green	Not seen	6–7; brittle, grass spikelet-like, enveloped in dry, purplish remains of corolla; dull whitish green	Not seen	Not seen	Not seen
Seed size (mm); shape; colour	0.3–0.4; light brown	Not seen	0.50–0.75; light brown	Not seen	Not seen	Not seen

reproductive phase. This has often been seen in cultivated specimens of *K. thyrsiflora*, a species in which under favourable conditions the club-shaped and quite heavy inflorescence can reach a height of about 2 m.

All kalanchoes are leaf succulents, with the leaves mostly decussately arranged (always so in *Kalanchoe* subg. *Kalanchoe*). Leaf character variation in all the kalanchoes treated in this book is summarised in Tables 7.1–7.4. Whereas some of the Madagascan species of *K.* subg. *Bryophyllum* have cylindrical to narrowly oblong leaf blades (e.g. *K. tubiflora*), those of the indigenous species are all flattened, although characteristically folded upwards along the midrib in some species (e.g. *K. brachyloba*, *K. paniculata*, and *K. sexangularis*).

Vegetative characters are generally very useful for distinguishing among the indigenous southern African species and have therefore been preferentially used in the identification key to the species (see Chapter 12). Particularly diagnostic are the size, shape, and disposition of the leaves, whether the leaf margin is entire, crenate, toothed, or variously lobed, the degree to which the leaves are covered by a white waxy bloom or flushed red by anthocyanins (see Chapter 8), length of the petiole, shape of the leaf base, and the presence or absence of an indumentum. A single leaf character is often sufficient to identify some of the species; for example, *Kalanchoe laciniata* can readily be identified by its deeply lobed, almost feather-like leaves, *K. neglecta* by its peltate leaves, and *K. sexangularis* by its distinctly petiolate, reddish leaves. On the other hand, some vegetative characters can be quite variable in some of the more widespread species, for example, those of the leaves in *K. rotundifolia*. However, in this species, it is possible that some of its forms eventually may be best treated as infraspecific taxa or even as distinct species.

Leaf colour is quite useful to distinguish among otherwise superficially rather similar-looking species. For example, the leaves of *Kalanchoe luciae* tend to be more readily flushed red than those of *K. thyrsiflora*; such red leaves are particularly striking in some horticultural forms of *K. luciae*. The turquoise-orange leaves of *K. longiflora* are a useful diagnostic character to distinguish it from *K. sexangularis*, the species with which it has until recently often been confused (Smith and Figueiredo, 2017b). The leaves of *K. sexangularis* are green, and most often strongly red-infused to bright wine-red but never turquoise (Smith and Figueiredo, 2017a).

An indumentum of simple glandular hairs is diagnostic for the indigenous *Kalanchoe lanceolata* when it enters the reproductive phase (Fig. 7.1), but such hairs have also been recorded in *K. hirta* and some forms of *K. montana*. It is noteworthy that the leaves of all these species are essentially glabrous in young plants but become pubescent higher up, as are the inflorescences and flowers, when plants enter the reproductive phase. The possible functional significance of this increase in hairiness on the inflorescences and flowers requires further study as it contradicts the usual pattern in plants where older growth in otherwise hairy plants tends to be glabrescent. Interestingly, hairy forms of *Cotyledon barbeyi* Schweinf. ex Baker, a variable crassuloid species from southern Africa, are considerably more pest-resistant than the glabrous forms of this species. It is therefore possible that the hairiness of kalanchoes in the reproductive (mature) phase has significance as a structural deterrent for nectar- and pollen-thieving microfauna or even inattentive browsing animals (although kalanchoes are known to be poisonous). Among the introduced and cultivated Madagascan species, a dense indumentum is characteristic for *K. beharensis* and *K. tomentosa*. However, the hairs in these two species are nonglandular and stellate.

A diagnostic character of the members of the Madagascan *Kalanchoe* subg. *Bryophyllum* is the formation of bulbils (also called pseudobulbils) in the indentations (see, e.g. Fig. 10.10, *K. laxiflora*) or on the tips of crenations (see, e.g. Fig. 12.19.1, *K. daigremontiana*) along the leaf edges and/or on the inflorescence. These bulbils are highly specialised adventitious buds (Johnson, 1934; Bell, 2008). They are sometimes borne upon a grooved peg that facilitates mechanical detachment. Among the species introduced into southern African, bulbils are produced in *K. daigremontiana*, *K. fedtschenkoi*, *K. gastonisbonnieri*, *K. pinnata*, *K. prolifera*, and *K. tubiflora*. In *K. fedtschenkoi* and *K. tubiflora*, bulbil formation is promoted in detached leaves. These plantlets may sprout roots while still attached to the mother leaf. The bulbils eventually drop to the ground, develop their own roots, and become whole new plants, which are genetically identical clones of the parent. The production of bulbils no doubt contributes towards the invasiveness of some of these species in southern Africa (see Chapter 10).

Although bulbils of the kind found in members of *Kalanchoe* subg. *Bryophyllum* are absent in the indigenous southern and south-tropical African kalanchoes, some of the species (*K. rotundifolia* and *K. humilis*) may sprout adventitious plantlets from the broken base of detached leaves (also see Chapter 11). This is known as gemmipary and is a phenomenon widely encountered in the Crassulaceae (Stoudt, 1938). However, among the introduced species, this mode of vegetative reproduction has only been noticed in *K. beharensis*, a species that does not form bulbils.

REPRODUCTIVE MORPHOLOGY

One of the most useful characters for distinguishing among the kalanchoes indigenous to southern African can be found in the morphology of the inflorescences. Two easily recognisable inflorescence types may be distinguished, namely, a sturdy-peduncled, erect, densely flowered, cylindrical- or club-shaped type, and a comparatively thinly peduncled, spreading, sparsely flowered, flimsy type. For wanting of a better word, we have adopted the term 'flimsy' for the latter type to signify that these inflorescences often appear top-heavy, with the relatively weak peduncles occasionally resulting in them leaning sideways, in some instances even coming to lie on the ground. Possible alternative terms to describe the 'flimsy' type of inflorescence would be 'slender' or, considering the spreading nature of the inflorescence axes and more sparsely arranged flowers, 'diffuse'.

All indications are that at least four of the species with club-shaped inflorescences form a closely knit natural group (*Kalanchoe luciae*, *K. montana*, *K. thyrsiflora*, and *K. winteri*), all sharing a vegetative phase of comparatively large, soup plate- or paddle-shaped leaves with entire margins that are often covered with a waxy bloom. There is also a tendency in the vegetative phase for the leaves to be arranged in a basal rosette, although this is not so marked in *K. winteri*. Although *K. crundallii* shares a similar inflorescence type with these species, it stands somewhat apart by having smaller, well-spaced leaves with crenate-dentate (rarely entire) margins and only a faint waxy bloom.

The kalanchoes with club-shaped inflorescences are thus far only known to occur in the *Flora of Southern Africa* (*FSA*) region and slightly further north. *Kalanchoe montana*, *K. thyrsiflora*, and *K. winteri* are endemic to the five-country (*FSA*) region (Namibia, Botswana, Swaziland, Lesotho, and South Africa), while *Kalanchoe luciae*, a predominantly southern African species, also occurs, to a lesser extent, further north in south-tropical Africa (Zimbabwe and Mozambique; see Fig. 12.10.2). *Kalanchoe wildii* Raym.-Hamet ex R.Fern. is the only species in this group (soup plate-shaped leaves and dense, club-shaped inflorescences) that does not occur in the *FSA* region and has thus far only been recorded from Zimbabwe (Lebrun & Stork, 2003: 221). The relationship between *K. luciae* and *K. wildii*, which occur sympatrically in Zimbabwe (Lebrun & Stork, 2003: 217, 221), is not well understood.

The uniqueness of this growth that form in *Kalanchoe* prompted the indefatigable Berger (1930: 404, 407) to propose the inclusion of *K. luciae* and *K. thyrsiflora*, the only two species with soup plate-shaped leaves and club-shaped inflorescences known at that time, in his 'Transvaalenses', which was equivalent to 'Group 14' proposed earlier by Hamet (1907: 877–879) (see Table 4.2).

The species with flimsy inflorescences tend to have smaller leaves, the margins of which are more often crenate, dentate, or lobed. They seem to comprise a heterogeneous assemblage of species that are not necessarily closely related, although at least *Kalanchoe alticola*, *K. neglecta*, and *K. rotundifolia* are to some extent united by having their flowers conspicuously twisted postanthesis (see later). While this character is also present in several other species, it is most pronounced in the three listed above. A further character peculiar to some of the southern African *Kalanchoe* species with flimsy inflorescences is that their flowers display distinct diurnal movement, closing at night and opening during the day. Diurnality in these (and other flimsy-inflorescenced) species is broken at different times of day—the corolla lobes of some species and those of some forms of others remain connivent until nearly midday, whole in other species; they become fully reflexed much earlier during the day.

The possible functional significance of the two inflorescence types (club-shaped vs flimsy) is further considered in the present chapter under 'Pollination biology'.

Whereas the definition and recognition of several standard types of inflorescences are straightforward, the interpretation of others can be quite difficult considering that many combinations and integrations between types exist (Bell, 2008). In southern African indigenous kalanchoes, the growth in length of a vegetative shoot is terminated by an inflorescence, a state described as hapaxanthic (see also in this chapter under 'Pollination biology'). These sometimes intricately branched inflorescences are essentially made up of partial inflorescences in the form of mainly not only dichasial but also monochasial cymes (Fig. 7.2).

A cyme is a sympodial inflorescence in which the central axis is terminated by a flower that opens first, followed by the growth being continued by buds in the axil(s) of the bract(s) below this central flower. If there are two opposite bracts (as in the case of kalanchoes with their decussate leaf arrangement) and a bud from each gives rise to either an axis or stalked flower, the result is a dichotomously branched unit called a dichasium. In its simplest form, a dichasium is 3-flowered (also called a triad), that is, an older central flower (that opens first), supported lower down by two younger ones. Repeated branching of the inflorescence axis in such a regular dichotomous way results in a compound dichasium. However, in some cymes of kalanchoes, the axillary bud of only one of the two opposite bracts below the terminal flower gives rise to a stalked flower. The latter is then called a monochasium. Likewise, the sprouting bud not only may give rise to a terminal flower but also may have a new pair of lower bracts from which a single axis may sprout to continue the one-sided branching. The resultant inflorescence is called a compound monochasium.

The flimsy inflorescence type of indigenous kalanchoes can be broadly classified as a determinate thyrse (Weberling, 1989) (Fig. 7.3). The latter author defines a thyrse as 'an indefinite complex inflorescence with cymose branches on a dominating main axis'. In this case,

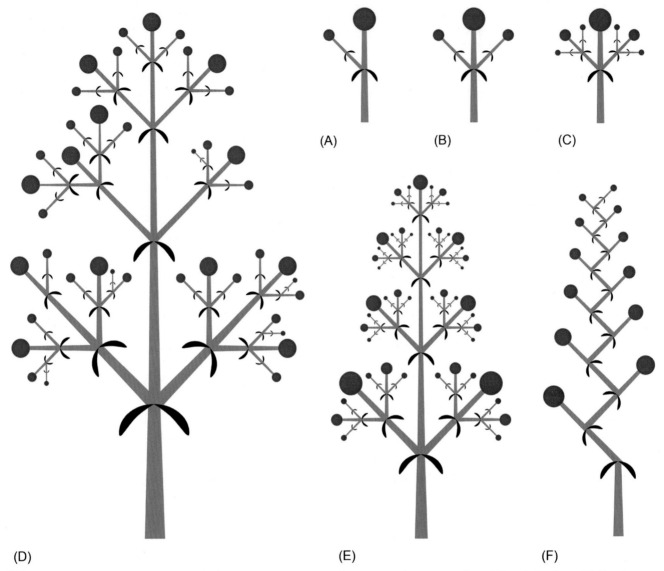

FIG. 7.2 Diagrammatic representation of inflorescence types and their elements in indigenous southern African kalanchoes. (A) Simple monochasium. (B) Simple dichasium (3-flowered cyme; triad). (C) Compound dichasium. (D) Dichasial partial inflorescence terminating in mainly 3-flowered cymes (triads), often with the two lateral flowers flanked by 1- to few-flowered monochasiums (representing area enclosed by rectangle in Fig. 7.3). (E) Determinate thyrse. (F) Monochasial cicinnus-like partial inflorescence (representing area enclosed by rectangle in Fig. 7.4). Flowers shown as red circles; size indicates sequence of opening. Bracts are shown in black.

the *ultimate* partial inflorescences of each inflorescence axis can be quite variable but basically are either 3-flowered cymes (triads), often with the two lateral flowers flanked by 1- to few-flowered monochasiums (e.g. *Kalanchoe sexangularis*) or dichasiums ending in multiflowered monochasiums approaching a cincinnus (e.g. *K. lanceolata*) (Fig. 7.4). The thyrse may be conical in outline or more or less flat-topped or dome-shaped. In the case of the latter two states, also described as a corymb-like thyrse, the lateral axes of different orders may elongate so that collectively all the flowers are elevated to lie at, or approach, the same level.

The club-shaped inflorescence type is essentially a determinate thyrse with a sturdy main axis but with short to extremely reduced lateral axes (Fig. 7.5). This reduction in the length of the lateral axes gives rise to the relatively tight clustering of flowers along the main axis. Moreover, the decussately arranged reduced leaves (bracts) with their congested axillary cymes are often well spaced along the main axis. The result is a sturdy main axis with often interrupted clusters of inflorescences. Each of these clusters may approach a verticillaster in appearance, notably in some forms of *Kalanchoe luciae*. According to Bell (2008), in the case of a

FIG. 7.3 The complex inflorescence of *Kalanchoe sexangularis* var. *sexangularis* is a determinate thyrse with cymose partial inflorescences ending mainly in dichasia. The general structure of one of the partial inflorescences (black rectangle) is depicted diagrammatically in Fig. 7.2. Photograph: Abraham E. van Wyk.

FIG. 7.5 One of the flower clusters along the main inflorescence axis of *Kalanchoe luciae* show the two much reduced lateral axes, each sprouting from the axil of the two more or less opposite bracts. Photograph: Abraham E. van Wyk.

FIG. 7.4 *Kalanchoe lanceolata* differs from the other indigenous southern African species in having thyrsoid inflorescences ending in multiflowered monochasiums approaching a cicinnus. The structure of such a monochasium (white rectangle) is depicted schematically in Fig. 7.2. Photograph: Abraham E. van Wyk.

verticillaster, the reduced axes bearing the cymes should be arranged in whorls. But this is not an absolute requirement as the arrangement at a node should merely be in a *seeming* whorl. The verticillaster is, for example, common in Lamiaceae (the Mint family) where it is usually associated with decussately arranged leaves/bracts similar to the pattern in kalanchoes. The club-shaped inflorescences in kalanchoes can therefore at times also be described as a raceme of verticillasters. However, we have largely steered clear from using the terms 'raceme' and 'panicle' for the inflorescences found in southern African kalanchoes.

Floral construction in all the indigenous southern African kalanchoes is similar in that they are actinomorphic, 4-merous, with four variously fused sepals, four petals fused into a distinct tube but with free lobes, eight stamens inserted on the corolla tube, four free carpels, and four free nectaries (nectary scales or squamae). This type of floral construction is in line with the general pattern found in the Crassulaceae (Thiede & Eggli, 2007). However, many species-specific modifications exist, and these are summarised in Tables 7.1–7.4. A peculiarity of *Kalanchoe* (and the three other genera of Crassulaceae subfam. Kalanchooideae, viz., *Adromischus*, *Tylecodon*, and *Cotyledon*) is the presence on the anthers with a more or less spherical connective gland (appendage) (Raadts, 1979). Also note that basic floral characters are enlightening similar in the *Kalanchoe* species introduced into southern Africa, with, additionally, sufficient floral differences existing to differentiate with reasonable accuracy among the three subgenera of *Kalanchoe*, although some representatives of especially *K.* subg. *Kitchingia* show considerable similarities with those of *K.* subg. *Bryophyllum* (see Chapter 12).

Characters useful to distinguish at species level among indigenous members of the genus include the following: size and orientation of the flowers; length, colour, and indumentum of the sepals; length, shape, colour, and indumentum of the corolla tube, as well as its degree of distal twisting postflowering; colour, shape, size, and orientation of the petal lobes; level of insertion on the corolla tube of the stamens; relative length of the two stamen whorls and degree to which they are included or exserted from the corolla tube; shape of the carpels; and size and shape of the nectar glands.

The taxonomic significance of flower morphology is well illustrated in the superficially very similar-looking *Kalanchoe luciae* and *K. thyrsiflora* (Figs. 7.6, 7.7 and 7.8). The former has the corolla 8–14 mm long, the tube is urceolate and constricted below the throat with the lobes mostly white, the stamens exserted 1–2 mm, and the undehisced anthers brownish orange, whereas in the latter, the corolla is 13–18 mm long and the tube is urceolate but less swollen, therefore appearing less constricted below the throat with the lobes invariably bright yellow, the stamens included or hardly exserted, and the undehisched anthers yellow. In addition, the flowers of *K. luciae* have no detectable odour, whereas those of *K. thyrsiflora* are clearly scented, the latter usually perceived as having a sweet undertone.

FIG. 7.7　In *Kalanchoe*, each anther is tipped by a ± spherical connective gland, here shown just prior to anthesis in the still undehisced brownish orange anthers of *Kalanchoe luciae*. Photograph: Abraham E. van Wyk.

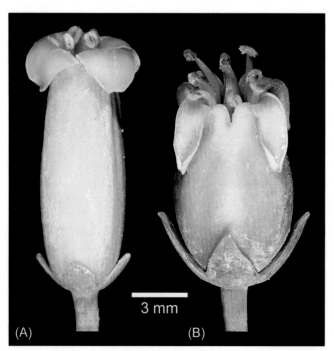

FIG. 7.6　Plants of *Kalanchoe thyrsiflora* (A) and *K. luciae* (B) are superficially very similar-looking, and the two species are often confused. The flowers, however, are quite different. Photograph: Abraham E. van Wyk.

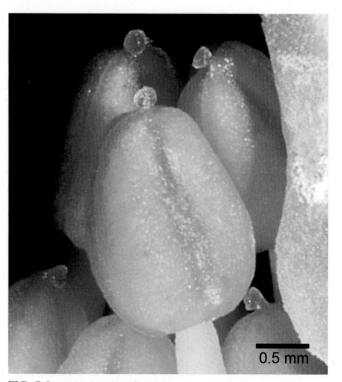

FIG. 7.8　*Kalanchoe thyrsiflora* differs from *K. luciae* in that the mature but still undehisced anthers are yellow and not brownish orange. As is diagnostic for the genus, each anther is tipped by a ± spherical connective gland. Photograph: Abraham E. van Wyk.

FIG. 7.9   Age sequence of *Kalanchoe rotundifolia* flowers, from preanthesis (left) to postanthesis (right). Note the anti-clockwise twisting of the distal region of the fading corolla tube in postanthesis flowers. Photograph: Abraham E. van Wyk.

Flower colour is fairly constant in those species with yellow/light orange flowers. Bright magenta flowers have only been recorded in *Kalanchoe rotundifolia*, but flower colour in this widespread and heterogeneous species can be quite variable with all gradations between yellow, orange and red, although the colour is usually similar in plants from the same geographical area. In *K. crundallii*, the colour of the corolla tube is influenced by the amount of direct sunlight it receives, uniformly red at anthesis when exposed to direct sunlight but otherwise remaining yellowish as when in bud. This is the only southern African species in which we have noticed this particular phenomenon.

A peculiarity of the flowers in some indigenous species is the anti-clockwise twisting of the distal region of the fading corolla tube in postanthesis flowers (Fig. 7.9). This action effectively encloses the ripening fruit within the shrivelling tube. Twisting of the corolla is consistent and particularly tight and distinct in *Kalanchoe rotundifolia* and its close relatives *K. alticola* and *K. neglecta* but was also recorded to some degree in specimens of *K. brachyloba*, *K. crenata*, and *K. hirta*. This slowly desiccating, tight 'wrapper' effectively protects the developing fruit and additionally signifies to potential pollinators that no further rewards are to be had from flowers that have reached this stage of development, so forcing visits to flowers that are still to be pollinated (and ovules of which to be fertilised). Apart from these functions, the additional significance of this floral twisting is unknown.

All the kalanchoes with large leaves and dense, club-shaped inflorescences have flowers with prominent nectary scales that are relatively short (about 2 mm) and square to transversely oblong (up to about 2.5 mm wide) and produce nectar in volume (Fig. 7.10). All the other

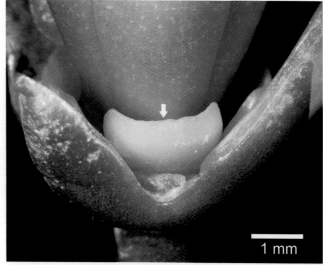

1 mm

FIG. 7.10   All the kalanchoes with large leaves and dense, club-shaped inflorescences have flowers with nectary scales that are relatively short and square to transversely oblong shown here (arrowed) in a flower of *Kalanchoe luciae*. Photograph: Abraham E. van Wyk.

indigenous species have much narrower linear to linear-oblong nectary scales of up to about 4 mm long (Fig. 7.11). Hence, nectary morphology supports the conclusion that the species with club-shaped inflorescences comprise a close-knit group.

In kalanchoes, the four free carpels develop into many-seeded follicles, each dehiscing in the dry state and under dry conditions, along the ventral suture. In the indigenous southern African species, the dehisced fruits are quite persistent on old dry inflorescences, in some cases still present during the following flowering season.

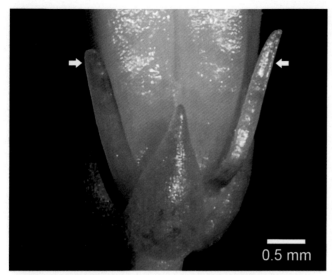

FIG. 7.11 Indigenous southern African kalanchoes with the flimsy inflorescence type have relatively narrow linear to linear-oblong nectary scales of up to about 4 mm long, here depicted (arrowed) in a flower of *Kalanchoe sexangularis* var. *sexangularis*. Photograph: Abraham E. van Wyk.

At this stage, no comparative study of the fruit has been conducted, and its taxonomic significance still needs to be elucidated. Likewise, a comprehensive comparative study of the seeds in *Kalanchoe* remains lacking. Seeds of kalanchoes are usually ellipsoid, very small, and variously striated. A broad survey of seed morphology in Crassulaceae by Knapp (1994) has identified a fair degree of taxonomically informative variation in the family. Seeds of indigenous species have not yet been studied in detail and are still unknown in several species. Dispersal in the group is also poorly known but appears unspecialised and dependant mainly on gravity, wind, and rainwater.

POLLINATION BIOLOGY

Flowers of Crassulaceae show a wide range of floral-biological syndromes, most of which seem to have evolved more than once in different groups (Vogel, 1954). In the case of southern African members of *Kalanchoe*, Vogel (1954) classified the flowers of *K. rotundifolia* as butterfly-pollinated (psychrophilous). He observed butterflies of the Pieridae visiting flowers of a 'related species' (probably a form of *K. rotundifolia* or *K. crenata*) at the mouth of the Umhlali River in KwaZulu-Natal. In this rather limited study, Vogel ascribed bee-pollinated (melittophilous) flowers to members of *Kalanchoe* in Madagascar, whereas flowers of the African *K. marmorata* were considered moth-pollinated (phalenophilous) and that of the Madagascan *K. tubiflora* as bird-pollinated (ornithophilous).

We are not aware of any systematic studies on the pollination biology of any southern African kalanchoes in their natural habitat. Vogel's, 1954 claim that *Kalanchoe rotundifolia* is butterfly-pollinated has subsequently been supported by observations that these flowers are frequently visited by butterflies (N.R. Crouch, personal communication) (Fig. 7.12). Butterflies were also noted as visitors to flowers of *K. longiflora* (N.R. Crouch, personal communication). Presumably based on casual observations mainly in gardens, flowers of *K. sexangularis* are said to be visited by bees, butterflies, and other insects, as well as insect-eating birds (Christoffels, 2016); we also have observations (C. de Wet, personal communication) of an Amethyst Sunbird (*Chalcomitra amethystina*) feeding on nectar in the flowers of this species (Fig. 7.13). Flowers of *K. luciae* are also visited by sunbirds (Nectariniidae; see further on), which are probably one of the pollinators of this species (M. Heigan, personal communication). There are also observations of sunbirds feeding on the flowers of *K. thyrsiflora* (see further on). We have one observation of a day-flying hawk moth (probably *Macroglossum trochilus*) taking nectar from the flowers of *K. thyrsiflora* (A. du Preez, personal communication). In a Pretoria garden, the flowers of *K. winteri* were visited by carpenter bees, *Xylocopa* cf. *caffra*, and in a Durban garden occasionally by honey bees (*Apis mellifera*) (Crouch et al., 2016b).

Of the two groups of kalanchoes indigenous to southern Africa, that is, those with densely flowered, club-shaped inflorescences and paddle-shaped leaves versus those with comparatively sparsely flowered, flimsy inflorescences and most often small(-er) leaves, it is the first group that has been reported as having members with fragrant flowers. However, *Kalanchoe luciae*, a member of the first group, is a notable exception in that its flowers

FIG. 7.12 African monarch butterfly (*Danaus chrysippus aegyptius*) sucking nectar of *Kalanchoe rotundifolia*. Photograph: Steve Woodhall.

FIG. 7.13 Female Amethyst Sunbird (*Chalcomitra amethystina*) feeding on nectar in the flowers of *Kalanchoe sexangularis* var. *sexangularis*. The yellowish pollen on the bird's head is that of *Tecomaria capensis* (Bignoniaceae). Photograph: Callie de Wet.

species of *Kalanchoe* secretions accumulate between the glandular tissue and the cuticle. This pattern where the secretory product accumulates in a subcuticular space is commonly encountered in glands secreting volatile compounds such as essential oils (Evert, 2006). It is noteworthy that in some of the investigated species, Raadts (1979) suspected the glands to be rudimentary or vestigial (nonsecreting).

Flowers of *Kalanchoe luciae* are very popular with sunbirds (Nectariniidae) with the birds deliberately probing the flowers to feed on the nectar. Observations by Martin Heigan (personal communication) have shown frequent visits by several sunbird species, including the Greater Double-collared Sunbird (*Cinnyris afer*) (Fig. 7.14) and White-bellied Sunbird (*Cinnyris talatala*). Although sunbirds (Africa and Asia) can feed while hovering in front of flowers (the preferred feeding behaviour in the Neotropical hummingbirds, Trochilidae), they prefer to perch when doing so (Skead, 1967). The sturdy inflorescence axis in *Kalanchoe luciae* is ideally suited as a 'bird perch'. The dense clustering of the flowers along the inflorescence axis also places them within easy reach of perching birds. It is hypothesised that the characteristic club-shaped inflorescences of this and the other southern African species with similar inflorescences (*K. montana*, *K. thyrsiflora*, and *K. winteri*) may well have evolved to

have no detectable scent, as is also the case in the group of species with flimsy inflorescences.

The flowers of some of the species in the scented group (at least *Kalanchoe thyrsiflora* and *K. winteri*) can perhaps best be described as having a musty, rather than pleasantly fragrant scent. Subtle differences can be detected between the scents given off by the flowers of these species. For example, in the case of *K. thyrsiflora*, the scent is strong, quite sweet, and reminiscent of that which one encounters when opening a sealed sugar tin after it was kept closed for some days. In contrast, the flowers of *K. winteri* lack that sugary sweetness to their scent, and at least to some humans, their scent is noted as unpleasant.

The precise origin of the floral scent is unknown, but one potential source is the thick, waxy bloom that the flowers of these species have in common. It has been claimed that the fragrance lingers on one's fingers when the bloom is rubbed off, but the merits of this observation require further study as not everybody perceives this effect. The scent also seems to originate, at least in *K. thyrsiflora*, from the anthers/pollen. We hypothesise that the anther gland (Figs. 7.7 and 7.8) may contribute towards scent production, one of the potential functions suggested in Myrtaceae for these little known appendages (Landrum and Bonilla, 1996). In an anatomical study, Raadts (1979) has shown that in the anther glands of some East African

FIG. 7.14 Male Greater Double-collared Sunbird (*Cinnyris afer*) using the sturdy inflorescence axis as a perch when visiting the flowers of *Kalanchoe luciae*. Photograph: Martin Heigan.

facilitate visitation by potential bird pollinators, notably sunbirds.

It is noteworthy that Skead (1967) regards the role of sunbirds as pollinators to be incidental. He argues that since both flowers and insects are more numerous than sunbirds, no flowers are dependent on sunbirds for pollination, although a few individual flowers may be pollinated by these birds. In a study in the USA, it was, for example, shown that bumble bees, often feed on nectar in flowers that are adapted for pollination primarily by hummingbirds (Irwin and Brody, 1999). Except for the relative concealment of nectar in the indigenous kalanchoes, their flowers are rather unspecialised with both the stigmas and anthers, the latter often on shortly exserted filaments, positioned close to and usually visible in the mouth of the corolla tube. Hence, at least pollen would be readily available to a broad range of insects that might visit the flowers.

It should be noted that the nectar in flowers of indigenous kalanchoes is well hidden deep within the corolla tube. Not only is the mouth of the corolla tube relatively narrow, but also it cannot stretch and is partly blocked by the anthers and the stigmas. Only insects with long, sucking mouthparts such as butterflies, moths, or long-tonged flies (family Nemestrinidae) would be able to reach the nectar in a non-destructive way. Sunbirds with their long, thin bills also seem well equipped for obtaining the nectar in some of the species with larger flowers.

Whereas the flowers of *Kalanchoe luciae* have no detectable scent (thus complying with the general syndrome in bird-pollinated flowers), those of *K. thyrsiflora* are sweet-scented, the latter being a feature generally associated with bee-pollinated flowers (Proctor et al., 1996). Our observations in the vicinity of Pretoria, Gauteng, South Africa, have shown that the flowers of *K. thyrsiflora* are popular with honeybees (*Apis mellifera scutellata*). The bees alight on the mouth of open flowers and actively collect pollen from the readily accessible anthers (Fig. 7.15). They cannot, however, enter the flowers to reach the nectar. No doubt this pollen-collecting may result in own or other pollen being deposited on the stigmatic surfaces that are positioned close to or just below the level of the anthers. Although the honeybees almost certainly contribute to pollination of the species, the question is for what purpose the flowers produce copious nectar deep in the corolla tube. We suggest that the nectar reward is most probably meant for sunbirds. Although we have no photographic records of sunbirds feeding on the flowers of this species, there are many anecdotal reports on the Internet that the flowers are visited by sunbirds. We suspect the species may well have a dual system of pollination, not only by sunbirds to which the flowers are best adapted but also by honeybees. This would tie in with the above-mentioned comment by Skead (1967) that bird-pollinated flowers are not

FIG. 7.15   African honeybee (*Apis mellifera scutellata*) collecting pollen of *Kalanchoe thyrsiflora* but because of its size cannot enter the flower to reach the nectar. This pollen-collecting may result in own or other pollen being deposited on the stigmatic surfaces that are positioned close to or just below the level of the anthers, as can be seen in the open flower. Photograph: Abraham E. van Wyk.

necessarily dependent on sunbirds for pollination. The possible involvement of day-flying hawk moths in the pollination of *K. thyrsiflora* (mentioned above) deserves further investigation.

Studies of nectar sugar composition have shown that floral nectars have only three major sugars (sucrose, fructose, and glucose) and that the ratios between them are of ecological significance (Baker & Baker, 1983). In an early general survey of nectar in angiosperms, Percival (1961) classified the nectar of three Madagascan species of *Kalanchoe* as being dominated by fructose and glucose with little or no sucrose (*Kalanchoe daigremontiana* and *K. blossfeldiana*) or sucrose-dominant with lots of fructose and glucose (*K. uniflora*). In a study of the invasive potential of *K. daigremontiana* in the Neotropics, Herrera & Nassar (2009) recorded that in this species, one flower produces on average 113.27 μL (±8.44 SE) of nectar, most of it (87%) during the first day after anthesis. Average nectar sugar concentration was 44.61% (±0.13 s.e.) and did not vary much along the day. They noted that the flowers were rarely visited by hummingbirds, most probably because the viscosity due to the high sugar concentrations makes it difficult for the birds to extract the nectar (see also Heyneman, 1983). They also suspected that birds may avoid the nectar because it may well contain toxins, seeing that the plant produce toxic bufadienolides in all its tissues, especially in the flowers, as was earlier shown by McKenzie et al. (1987). However, in southern Africa species of *Melianthus* (Melianthaceae), a group also rich in bufadienolides (see e.g. Kellerman et al., 2005), the unusual black nectar (and honey) has been claimed by some as toxic but edible by others (references in Watt and Breyer-Brandwijk, 1962). Whatever the case may be, sunbirds feed on the floral nectar without any apparent ill effect (Scott-Elliot, 1890; Linder et al., 2006).

One of the defining characteristics of bird-pollinated flowers is that they produce copious and dilute nectar, often in the range 20%–25% sugars w/w (Nicolson & Fleming, 2003). These nectars also tend to contain a high proportion of sucrose, whereas insect-pollinated flowers typically produce more concentrated nectars dominated by the hexose sugars fructose and glucose (Nicolson & Fleming, 2003). According to these authors, the majority of sunbird-pollinated flowers analysed (259 species) produce nectar that is hexose dominant. About half (47%) of sunbird plants produce nectar with <10% of the sugar present as sucrose. We could find little published information on the sugar composition of nectar in southern African kalanchoes. Based on unpublished data, it is briefly mentioned by Nicolson and Van Wyk (1998) that sucrose levels decreases with age in species of *Kalanchoe*.

Among the introduced species, *Kalanchoe daigremontiana* is autogamous and does not require flower visitation to produce seed (Herrera & Nassar, 2009), whereas *K. pinnata* is partially self-compatible (González de León et al., 2016). Hybridisation experiments on Madagascan species suggest that kalanchoes lack self-incompatibility mechanisms, at least to a degree, and emasculation is usually needed to avoid self-pollination (Kato & Mii, 2012; Kuligowska et al., 2015). The same may well apply to the southern African species, but this requires confirmation. Fruit-set is exceedingly prolific in the indigenous kalanchoes with essentially all flowers developing fruit, even in species grown under greenhouse conditions with little or no access to pollinators. However, this does not mean that seed production is similarly prolific or that the seeds are necessarily viable.

In any one inflorescence of the indigenous species, only relatively few flowers open every day. As old flowers fade, new ones open, and in some species, especially those with larger inflorescences, for example, *Kalanchoe thyrsiflora*, *K. sexangularis*, and *K. laciniata*, flowering in an inflorescence can extend over a couple of weeks. This sequential opening of flowers is obviously a consequence of the growth and proliferation of the cymose inflorescences, which results in flower buds having a range of ages. At the inflorescence level, these kalanchoes

can be described as having extended blooming (in contrast to mass blooming) (Bawa, 1983; De Jong et al., 1992). Possible advantages of such prolonged flowering in outcrossing plants is that it should increase an individual's chances of fertilising other plants and also improve chances to receive pollen donors from a larger range of genotypes. In addition, extended blooming should minimise the risk of reproductive failure resulting from a shortage of pollinators or bad weather (Bawa, 1983; De Jong et al., 1992).

LIFE CYCLE (DURATION)

Nearly all the southern African species of *Kalanchoe* flower mainly in autumn and winter; a similar timing of flowering seems to prevail in most members of the genus. This pattern is in line with that of so-called short-day plants. *Kalanchoe crundallii* is a notable exception among indigenous kalanchoes in flowering mainly in midsummer and early autumn. It has been shown experimentally that flowering in at least some members of *Kalanchoe* is readily induced by exposure to short days (Engelmann, 1960; Wareing & Phillips, 1970). More information on photoperiodism in the group is supplied in Chapter 8.

However, in kalanchoes, the age of the plant is also important for the transition from the vegetative to the flowering phase. This is most noticeable in those southern African species with club-shaped inflorescences. The latter plants do not flower before they have reached a certain age, usually at least 2 years but often up to 5 years; this may be referred to as the vegetative, juvenile phase. During this phase, the leaves are more or less congested in a basal rosette. Once the rosettes have gathered enough resources and when day-length conditions are favourable, the tall inflorescence develops quite quickly. However, once the rosette (or shoot in other species) has developed a terminal inflorescence, it has entered the mature (or adult) phase. Following seed set, the inflorescence and the particular rosette (often most of the particular shoot in other species) die.

Plants in which all resources are utilised for one episode of reproduction, followed by the death of the entire plant, are referred to as semelparous or monocarpic. Both short- and long-lived semelparous plants are known in the Crassulaceae (Young & Augspurger, 1991; Thiede & Eggli, 2007). Technically, monocarpy is defined as a life history type in which a single large sexual reproductive effort is directly associated with programmed *whole plant* death (Young and Augspurger, 1991). Note that for this purpose, 'whole plant' is meant to refer to each genet, that is, individuals originating from a specific single zygote, irrespective of its ramet (vegetative units of a specific genet) number (Lev-Yadun, 2017).

In the botanical literature, the term 'hapaxanthic' is often used as a synonym for 'monocarpic', but this is a cause for confusion. We prefer to use 'hapaxanthic' to describe the mode of flowering in a single vegetative shoot rather than a whole (often multi-stemmed) plant (Bell, 2008; Lev-Yadun, 2017). In their famous book on tree architecture, Hallé et al. (1978) define 'hapaxanthy' as follows: 'when a shoot apical meristem becomes wholly transformed into a flowering axis after a period of vegetative growth, i.e. the hapaxanthic shoot is determinate and ends in an inflorescence'. The alternative state where a vegetative axis bears lateral flowers and remains indeterminate is described as 'pleonanthy'.

Following the definition adopted above, the mode of flowering in all southern African species of *Kalanchoe* is clearly hapaxanthic. Vegetative shoots are invariably terminated by flowering. This pattern of shoots once flowering and then apparently dying has led to the description of some southern African species as being monocarpic. However, in such species, it is usually not the entire plant that dies because new shoots (plantlets) often sprout from the base of the dying rosette or the more persistent rootstock. Such new shoots are often initiated just prior to, or at the time of, flowering and have been seen in both species with flimsy (e.g. *K. rotundifolia*) (Figs. 6.5 and 7.16) and club-shaped (e.g. *K. thyrsiflora*, *K. montana*) inflorescences. In the description of *K. rotundifolia* (Chapter 12), for example, we have described such newly produced plantlets as 'sterile, small-leaved branchlets' (Fig. 7.16).

Southern African kalanchoes are best described as essentially perennial, polycarpic (=iteroparous), clonal plants. This means that the whole plant (a particular genet) is capable of reproducing sexually more than once. Whole plants often comprise a hierarchy of older and larger shoots (rosettes) that flower once then die and younger shoots/plantlets belonging to the same genet that may grow, flower, and die much later. An annual life cycle is rare and only occurs as an exception in some taxa (e.g. *Kalanchoe hirta*), often as a phenotypic plasticity response reflecting limiting growth conditions. Annuals have, for example, also been described in *K. lanceolata*. Because in practice, most people look at kalanchoe plants, which often grow in dense stands, not in terms of genets and ramets, we have in Chapter 12 described duration as it perhaps best applies to a specific sexually reproducing unit (shoot). Hence, descriptive terms such as 'perennial', 'multiannual', and 'biennial' must be seen as applying not necessarily to a genet as a whole. True monocarpy is also relatively rare but is encountered more often than annuals, especially in the group of species with club-shaped inflorescences. In this group, it may happen that a flowering ramet representing the whole genet for some or other reason either fails to produce new plantlets from the base or the plantlets may not survive for long, thus resulting in the death of the whole plant. For example,

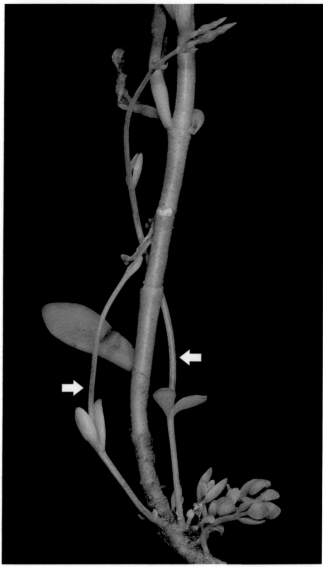

FIG. 7.16 Basal portion of a flowering stem of *Kalanchoe rotundifolia* (thick axis) with several replacement small-leaved plantlets already sprouting from the base, two of which have elongated (arrowed). The one on the right bears a few flowers and will probably die with the main flowering shoot. Photograph: Abraham E. van Wyk.

K. thyrsiflora is more prone to be monocarpic than the rather similar-looking *K. luciae*; consequently, the latter is considered a more desirable horticultural subject.

It needs to be pointed out that in members of *Kalanchoe* subg. *Bryophyllum*, bulbils produced on the leaves and/or inflorescences should be seen as highly specialised ramets. Plantlets derived from the bulbils are genetically identical to the mother plant and strictly speaking do not represent different genets. By analogy, an informative discussion on how bulbils from the inflorescence in *Agave* should be seen as belonging to the same genet is supplied by Lev-Yadun (2017).

CHROMOSOME NUMBER

Crassulaceae display a great diversity of chromosome numbers, and it has been considered as karyologically perhaps the most diverse family of the angiosperms (Thiede & Eggli, 2007). The base number for the family is $x=8$ and for the Kalanchooideae, to which the kalanchoes belong, $x=9$ [$n=9$, 17 (18)] (Stevens, 2001 onwards). The following chromosome numbers have been supplied for *Kalanchoe*: $n=17$, 18, 20, and 51 (for references, see the species treatments in Chapter 12). Edith Raadts [1914–2004] has made, among others, significant contributions towards our knowledge of chromosome numbers in African members of *Kalanchoe*; for an obituary, see Leuenberger (2004). Chromosome numbers are available for only five of the indigenous southern African species, but these are mostly in agreement with counts in kalanchoes from elsewhere, the most frequently reported number being $2n=34$. Other numbers for species from mainly Madagascar include not only $2n=36$, 40, and 68 (tetraploids) but also the odd $n=30$.

Recent advances in our understanding of polyploidy (genome doubling) and associated modifications of chromosome number have shown it to be a major force in the evolution of many plant lineages (Stuessy et al., 2014). Despite comparatively little practical value for the species-level classification of indigenous kalanchoes, knowledge of chromosome numbers in species is obviously also of considerable practical importance when it comes to the use of interspecific hybridisation for the breeding of new cultivars of succulents in general (e.g. Rowley, 2017) and *Kalanchoe* in particular (Mii, 2009).

Although cultivars mainly based on *Kalanchoe blossfeldiana* have been introduced with great commercial success globally, selection emphasis has hitherto been mainly on variation in floral features. Considering the great diversity in both vegetative and reproductive features in *Kalanchoe* as a whole, a vast but largely untapped genetic resource is available to plant breeders. Note though that at least one hybrid kalanchoe, *K. ×houghtonii*, has become a problematic invader in some parts of the world (see Chapter 10). By applying special breeding techniques, it is even possible to get viable progeny from hybrids between species belonging to *K.* subg. *Kalanchoe* and *K.* subg. *Bryophyllum* (Mii, 2009). Of special interest is the observation that it is possible to induce genome doubling of hybrids through shoot regeneration from leaf explants (Aida & Shibata, 2002; Mii, 2009). One inevitably wonders to what extent gemmipary could lead to a similar doubling of the genome of a species in nature.

Physiology and Anatomy

FIG. 8.1 Leaves of *Kalanchoe luciae* often display both a pronounced white powdery waxy bloom on the surface and a reddish flush due to the accumulation of anthocyanins in the leaf itself. Photograph: Abraham E. van Wyk.

Kalanchoe (Crassulaceae) in Southern Africa
https://doi.org/10.1016/B978-0-12-814007-9.00008-6

INTRODUCTION

Plant physiology is the study of plant function and behaviour, encompassing all the dynamic processes of growth, metabolism, reproduction, defence, and communication that account for plants being alive (Salisbury & Ross, 1992; Baluška et al., 2006; Scott, 2008). Considering that most of these processes take place at the level of cells, tissues, and organs, there is, because of the close association between structure and function in plants, also a close association between plant physiology and plant anatomy. Moreover, within the living cell, much of the metabolic activity is at the molecular level; therefore, a full understanding of a plant's physiology requires an essential background in chemistry and physics.

Many plant physiological insights into basic processes were gained from research based on a relatively small number of convenient experimental or model plants (e.g. beans, lettuce, maize, wheat, and in more recent times, *Arabidopsis thaliana* (L.) Heynh., the thale cress) (Pitzschke, 2013). Knowledge thus gained is then extrapolated to other plants as it is assumed that such processes operate similarly. For example, the vast majority of green plants have never been chemically analysed for the presence of chlorophyll, yet we assume that the green colour of their leaves is due to the presence of this pigment.

Although few indigenous southern African kalanchoes have been the subjects of physiological studies, a lot can be inferred about their physiology from work conducted on other members of the genus and other representatives of the Crassulaceae. Of particular significance is the fact that a few mainly Madagascan species of *Kalanchoe*, especially *K. blossfeldiana*, *K. daigremontiana*, *K. fedtschenkoi*, and *K. tubiflora*, have served as model experimental plants in many plant physiological studies, especially ones aimed at elucidating the mechanism of photosynthesis in succulents and how flowering is initiated. *Kalanchoe fedtschenkoi* is an emerging molecular genetic model for the study of crassulacean acid metabolism (see below) in the eudicots (Yang et al., 2017).

In this chapter, we will be highlighting crassulacean acid metabolism (CAM), a specific type of photosynthesis first discovered in a species of *Kalanchoe* and subsequently found to operate in most Crassulaceae and many other groups of succulents. Kalanchoes are known for their flowering that peaks during the winter months. We will explain this reproductive behaviour in some detail and show how knowledge of the underlying processes is being used to manipulate a selection of kalanchoes, especially *K. blossfeldiana* and hybrids of which it is one parent, for lucrative commercial purposes.

The leaves and stems of indigenous kalanchoes are not always green; sometimes, they are variously infused with red (anthocyanins) with the southern African *Kalanchoe sexangularis* (see Chapter 12.15) being a prime example, and/or their outer surfaces are often substantially covered in a powdery, whitish deposit (epicuticular wax) as in, for example, the group of southern African kalanchoes with club-shaped inflorescences and large, paddle-shaped leaves (Fig. 8.1). We will briefly explain the nature of these two modifications and their possible functions. Finally, some noteworthy anatomical features of the kalanchoes will be briefly mentioned. Information on the toxicity and chemistry of kalanchoes is provided in Chapter 9.

PHOTOSYNTHESIS

Many succulents grow in habitats where water is a scarce commodity, resulting in these plants having had to make evolutionary adaptations to survive—indeed thrive—in conditions where rainfall is at worst low and erratic or, at best, seasonal. Some succulents, like epiphytic Madagascan kalanchoes, such as *Kalanchoe uniflora* (Fig. 12.9), may occur in high-rainfall forests, but they almost invariably grow in, or on, substrates such as the debris collected in the forks of tree branches, or against vertical cliff faces, where water drains away very quickly. In fact, by far the most relevant abiotic constraint for growth and vegetative function in vascular epiphytes is water shortage (Zotz & Hietz, 2001). This means that succulents in general—even those occurring naturally in rainforests—need to be careful with their water consumption and have mechanisms to minimise water use yet maintain their physiological, life-sustaining processes.

Photosynthesis is the complex process by which plants convert the energy of sunlight into the energy of organic molecules (sugars), the latter based on carbon atoms derived from atmospheric carbon dioxide (CO_2) (Lawlor, 2001). Many succulents have developed strategies to assist them to save water. One such physiological strategy is a type of photosynthesis called crassulacean acid metabolism, or CAM for short. Reports of CAM as a phenomenon dated from Roman times through persons who noted that some common house plants have an acidic taste in the morning (Black & Osmond, 2003). The first record of the phenomenon in some detail was based on observations published in 1815 by the Scottish botanist, naturalist, and physician, Benjamin Heyne [1770–1819], on the Madagascan succulent, *Kalanchoe pinnata* (Bonner & Bonner, 1948; Kluge & Ting, 1978; Osmond, 1978)—at the time merely recorded as the observation that leaves of this species have a more acid taste in the morning than late in the afternoon. Elucidation of the complete pathway responsible for this diurnal fluctuation in acidity took nearly 150 years and encompassed many fundamental discoveries in plant biochemistry (Winter & Smith, 1996; Black & Osmond, 2003). CAM is one of three metabolic pathways found in the photosynthetic tissues of vascular plants for assimilation of atmospheric carbon dioxide, the other two being C3 and C4. CAM photosynthesis has evolved to overcome water stress and is advantageous to plants that grow in arid environments or in locally dry sites in moist habitats and is most often associated with succulence. However, CAM species inhabiting tropical rainforests far outnumber typical desert species, but many of these species are epiphytic and subject to the particular challenges of water supply in this type of habitat (Zotz & Hietz, 2001; Lüttge, 2004).

The term "crassulacean acid metabolism" was first used by plant physiologist Meirion Thomas [1894–1977] during a public talk presented in January 1947 and was subsequently published in the 4th edition of his textbook "Plant Physiology" in 1949. The name is derived from the family Crassulaceae, members of which were widely used in early research on the phenomenon (Black & Osmond, 2003). Essentially, CAM is a CO_2 acquisition, CO_2 temporary storage, and CO_2 concentrating mechanism based on organic acid synthesis (Lüttge, 2015). It needs to be pointed out that CAM photosynthesis is merely a specialised modification of regular C3 photosynthesis and most plants with CAM photosynthesis can at times switch to C3 photosynthesis. In fact, no particular biochemical reactions or enzymes are specific for CAM only. Although special variant forms of enzymes may be involved, basically, all reactions have well-known housekeeping functions in all green plants. It is the special linkage of metabolic elements that makes up the CAM network at times (Lüttge, 2015).

The movement of CO_2 and other gasses, including water vapour, into and out of leaves takes place through pores in the epidermis of leaves and stems. Known as stomata, these pores are surrounded by so-called guard cells that regulate their opening or closure. In regular C3 photosynthesis, plants need to keep their stomatal pores open during the day to allow the exchange of gasses into and out of the leaf. Opening these pores results in large quantities of water being lost from the leaf. It has been estimated that for every 1 g of CO_2 fixed by a C3 plant during photosynthesis, up to 1 litre of water is lost (Scott, 2008). In contrast to C3 and C4 plants, CAM plants (Fig. 8.2) close their stomata during the day and open them to take up CO_2 from the atmosphere predominantly at night, when temperatures are generally lower, so limiting water loss. But seeing that no light is available at night for photosynthesis, the CO_2 is temporarily stored as an organic acid (usually malic acid). Subsequently, the stored CO_2 is released during the following light period (the next day) and assimilated to the level of sugars.

One consequence of plants having CAM photosynthesis is that to humans, they taste more acidic or at least acrid, in the mornings, after the stomata had been open during the night. This is because of the accumulated malic acid in the vacuoles of the plants' cells. Malic acid contributes, among others, to the pleasantly sour taste of fruits and is widely present in fruit juices and wines. Although CAM photosynthesis has been studied scientifically only for the past 200 years, the higher level of acidity in the leaves of some succulents, such as the appropriately named *Sedum acre* (Fig. 8.3), a species of Crassulaceae, from continental Europe, in the mornings has been known since Roman times (see, e.g. Rowley, 1978: 12; Black & Osmond, 2003; Smith et al., 2015a). Conversely, in the afternoons and later in the day, the leaves taste much less acidic, given that the malic acid is metabolised during the day and the CO_2 is released into the photosynthetic pathway. Take care not to swallow any plant sap when doing this taste-test because many Crassulaceae are poisonous, including the kalanchoes (see Chapter 9).

This daily change in acidity can also be tasted in the edible leaves of the southern African spekboom (Afrikaans) or pork bush, *Portulacaria afra* (Didiereaceae or, until recently, Portulacaceae), but only if the particular plant is experiencing drought stress. In this species, CAM is a continuous trait that operates at low levels even during predominantly C3 photosynthesis (Guralnick & Gladsky, 2017). This variability illustrates the fact that many succulents are facultative CAM plants that can switch according to circumstances between CAM and C3 photosynthesis. Depending upon species, ontogeny, and environment, the contribution of nocturnal CO_2 fixation to total daily carbon gain can range continuously from close to 0% to 100% (Winter et al., 2015). Moreover, in members of the

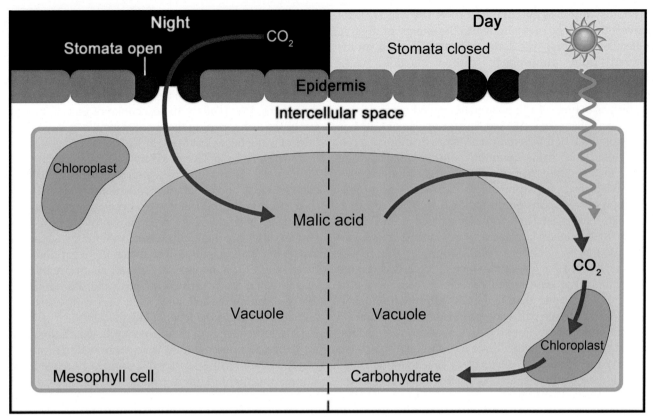

FIG. 8.2 Simplified schematic outline of the crassulacean acid metabolism (CAM) cycle showing temporal organisation of the fixing and release of carbon dioxide. At night, CAM plants open their stomata and fix carbon dioxide, which is then stored as malic acid in the vacuole. When the day begins, the stomata close, and the malic acid is broken down to release the carbon dioxide, which is then, in the presence of light, fixed by the C3 photosynthetic pathway into carbohydrates.

New World succulent genus *Dudleya* Britton & Rose, also of the Crassulaceae, it was found that most of the leaf biomass was the result of uptake during the day through C3, while the inflorescences indicated growth through the CAM pathway. This would suggest that not only a species may differ in pathways during seasonal shifts in moisture availability but also apparently because different organs may have a different history in terms of carbon fixation (Teeri, 1984, and references therein).

In past research aimed at unravelling the mechanism and ecological significance of CAM photosynthesis, members of the genus *Kalanchoe* occupied a prominent position. Today, many generalisations concerning CAM were derived from studies involving species such as *Kalanchoe daigremontiana*, *K. fedtschenkoi*, and *K. tubiflora*. An informative review of the physiological and ecological aspects of CAM in the genus *Kalanchoe* was provided by Kluge & Brulfert (1996). *Kalanchoe* is the first eudicot CAM lineage with a genome sequence to date and serves as an important reference for understanding the evolution of CAM (Yang et al., 2017). The available genome sequence is that for *K. fedtschenkoi*, an obligate CAM species. *Kalanchoe fedtschenkoi* is a Madagascan species that has been recorded as naturalised in southern Africa

and beyond (Smith & Figueiredo, 2017c). For more information on CAM in *Kalanchoe* compared with that in some other CAM plants, with both sequence convergence and in particular a number of changes in the temporal expression of genes, see Yang et al. (2017). The hypothesis that convergent evolution in protein sequence and/or temporal diel gene expression underpins the multiple and independent emergences of CAM from C3 photosynthesis is supported by these molecular studies.

A survey based on carbon isotope composition of the mode of photosynthesis in many Madagascan species of *Kalanchoe* showed the whole range from predominantly CAM to predominantly C3 (Kluge et al., 1991). This was followed by a study of carbon isotopes in some African members of the genus (Kluge et al., 1993). Indigenous southern African species surveyed include *K. brachyloba*, *K. hirta*, *K. lanceolata*, *K. laciniata*, *K. luciae*, *K. paniculata*, and *K. rotundifolia*. In contrast to the Madagascan species that cover the whole spectrum of C3 to CAM and great inter- and intraspecific diversity of CAM performance, nearly all the isotope values for African species showed a strong CAM behaviour with CO_2 uptake taking place mainly during the night. A subsequent molecular phylogenetic study of *Kalanchoe* (Gehrig et al., 2001) supports

FIG. 8.3 The leaves of *Sedum acre*, a species of Crassulaceae, from continental Europe, taste more acidic in the mornings than later in the day. This has been known since ancient times and is due to the fixation at night of carbon dioxide as malic acid, a feature of plants with crassulacean acid metabolism (CAM). The specific epithet *"acre"* refers to this acrid taste. Photograph: Gideon F. Smith.

the hypothesis that CAM has evolved from the C3-pathway. Species with mainly C3 photosynthesis turned out to be ancestral to those that are moderate to strong CAM performers. The phylogeny, combined with information on the mode of photosynthesis, also suggests that the centre of adaptive radiation of the genus is located in the more humid eastern regions of Madagascar. From there, the genus has spread into the arid areas of Madagascar and from there to the eastern parts of continental Africa and beyond.

To accommodate the accumulation of malic acid during the night, the parenchyma cells in the leaves of CAM plants possess large vacuoles, which can take up around 90% of the volume of the cell (Scott, 2008). Moreover, the accumulation of malic acid also increases the osmotic concentration of the sap in the vacuoles, allowing the plants to absorb and store water, thus leading to the observed succulence of the photosynthetic tissues (Ingram et al., 2016). Other adaptations found in CAM plants include having thick, succulent, or very much reduced leaves (with a low surface area:volume ratio), or no leaves at all, a thick, waxy cuticle (see section "Epicuticular wax" further on), and sunken stomata with prominent guard cells. The ability of succulents to be very frugal with water has also a number of consequences for the architecture of the plants themselves. One result of their ability to conserve water is a lesser need for a massive root system that can "fetch" water from very deep and far beyond where the plant becomes established. Succulents therefore often do not have elaborate root systems that run deep into the soil. The roots are often rather fine and radiate from the stem or plant body, in a dense network that spreads close to the soil surface, often horizontally rather than vertically. Apart from the dense network of mostly fine, fibrous roots of most kalanchoes (see Chapter 7), conspicuous "stilt" roots develop on the rather thin stems of some Madagascan species, notably *Kalanchoe fedtschenkoi* (see, e.g. Fig. 12.20.4) and *K. laxiflora*. The development of such roots that provide structural support and keep plants more or less erect is extremely rare in southern African *Kalanchoe* species and has only been noted in the case of a few specimens of the small-growing *K. alticola* from Swaziland.

PHOTOPERIODISM

Photoperiodism refers to a response to the length of day that enables an organism to adapt to seasonal changes in the environment. With one exception, all the southern African indigenous kalanchoes flower mainly in winter, provided the vegetative plants have reached a certain age (also see Chapter 7, under section "Life cycle). *Kalanchoe crundallii*, the odd one out, flowers in summer. Winter is also the prevailing flowering time in most other kalanchoes, making them so-called short-day plants (Thiede & Eggli, 2007; Akulova-Barlow, 2009; Currey & Erwin, 2010, 2011). Note, however, that the timing of darkness during the daily cycle is just as important as that of light. Besides flowering, day length may also have other responses in kalanchoes. For example, bulbil formation on leaves in *K. daigremontiana* and *K. tubiflora* is promoted by long days (Schwabe, 1971; Ingram et al., 2016), whereas short days, in addition to flowering, were also shown to promote anthocyanin formation in the vegetative growth of *K. blossfeldiana* (Neyland et al., 1963).

Indigenous kalanchoes are mainly confined to the summer-rainfall region of southern Africa. In this part of the subcontinent, winter is the dry season with many plants in a state of dormancy. However, leaf succulence and their short-day requirement enable the kalanchoes to flower, thus providing an abundant supply of nectar and pollen to potential insect and avian feeders during a season when alternative sources are scarce. For this reason, the ecological importance of the kalanchoes being short-day plants cannot be underestimated. It is noteworthy that in southern Africa, a similar flowering period is shared with many species of the succulent genus *Aloe*, a group of which the ecological significance of providing a source of nectar and pollen in winter is well established (Human & Nicolson, 2008; Symes et al., 2008).

From a phenological perspective, kalanchoes therefore flower during the dry season and produce, almost invariably, copious amounts of dustlike seed (see Chapter 7) towards the end of this season. Seed dispersal takes place towards the end of the dry season, with the seed, once having been deposited in suitable places, germinating with the onset of the rainy season. This is again a reproductive mechanism that kalanchoes share with many of the aloes (Asphodelaceae) from southern Africa's summer-rainfall region that flower during the dry, otherwise drab, winter season (Smith & Van Wyk, 2008a).

The Madagascan *Kalanchoe blossfeldiana* has been the subject of many classical studies on photoperiodism (e.g. Wareing & Phillips, 1970; Salisbury & Ross, 1992). It is an obligate phototrophic plant showing a well-marked critical day length below which flowering will occur. Such species are said to have an absolute or qualitative short-day requirement (Wareing & Phillips, 1970). Day length may also have flower-inhibiting effects. For example, if a single leaf of *K. blossfeldiana* is exposed to short days and the remainder of the plant is maintained under long days, the presence of long-day leaves between the short-day leaf and the shoot apex prevents flowering. However, if all leaves are removed above the short-day leaf and only long-day leaves below the short-day one are allowed to remain, the plant will flower (Wareing & Phillips, 1970).

Although the critical role of day length as the trigger for the induction of flowering in certain plants is without question, the exact mechanism involved remains elusive. In 1936, Mikhail Khristoforovich Chailakhyan [1902–1991], an Armenian-Russian plant physiologist, proposed the involvement of a hypothetical flowering hormone, florigen, produced in leaves, after the stimulus of inductive day length. Based on numerous experiments over many years, researchers have concluded that florigen must be a relatively stable mobile molecule that is synthesised in leaves in response to the favourable photoperiod and then migrates through the vascular system, most probably the phloem, to the apical meristem to promote floral initiation. After more than 70 years since florigen was first proposed, molecular studies on *Arabidopsis thaliana* (Brassicaceae), a long-day plant, have provided strong evidence that such a flowering hormone indeed exists. The current view is that the FT protein (produced by the gene *FLOWERING LOCUS T*) in *Arabidopsis* and corresponding proteins in other species may well be an important part of florigen (Notaguchi et al., 2008; Turck et al., 2008).

By knowing a species' day-length requirements, artificial day lengths can be used to manipulate flowering under greenhouse conditions. Providing short days (less than 12 lighted hours) over a period of two to three weeks will trigger the formation of flower buds in kalanchoes, provided plants have reached reproductive age. To obtain summer flowering of some short-day kalanchoes is also relatively easy. Plants just need to be covered with a light-excluding material over a period of two to three weeks to ensure light exposure is shorter than 12 lighted hours (the so-called critical day length). To prevent flowering, the short days of winter can be extended by using artificial light, but this is not really necessary. As pointed out above, the length of darkness is as important. In these short-day plants, flowering depends on whether or not a critical duration of darkness is exceeded. Hence, flowering in winter of these plants can be prevented by a relatively short exposure to light (using lamps with a high red/far-red ratio) at a particular time during the night (usually near midnight). In fact, such night-break lighting treatments are used by commercial growers to regulate the flowering time in a number of species, including cultivars and hybrids of *Kalanchoe blossfeldiana* (Ingram et al., 2016).

The flowering of cultivars selected from *Kalanchoe blossfeldiana*, popularly known as "Flaming Katy" or "Florist's kalanchoe", can be easily influenced as the plants respond to day length. This means that plants can be brought to flowering if placed in a darkened vessel, for a regulated period, thereby deliberately curtailing the number of sunshine hours they receive. When such cultivars are produced commercially, this is of course done on a massive scale. The ease with which these kalanchoes can be enticed into flowering out of season is the reason why showy specimens of these cultivars can be purchased from nurseries and florist shops at any time of the year. Of course, for most gardeners and succulent plant collectors, manipulating day length takes considerable effort, and acquiring fresh, shop-bought material whenever flowering "Florist's kalanchoes" are desired is preferable. In fact, this is often the best as the general quality of *Kalanchoe blossfeldiana* cultivars tends to deteriorate over time, with the flowering event becoming less substantial in subsequent years.

In addition to day length, temperature can also influence flowering in some kalanchoes. Even though short days are enough to induce flowering in easily induced species (e.g. *Kalanchoe blossfeldiana*), cold night temperatures are required to flower-induce other species (e.g. the East African *K. marmorata*) (Lopes Coelho et al., 2015). On the other hand, photoperiodic flower induction in *K. blossfeldiana* is inhibited when night temperatures during short days exceed 27°C—a phenomenon referred to as "heat delay" (Pertuit, 1977). For more details on the cultivation and manipulation of flowering in this commercially important species and other species of *Kalanchoe*, see among others Jacobsen (1986), Pertuit (1992), and Currey & Erwin (2011).

ANTHOCYANINS

A conspicuous feature of many kalanchoes is the reddish or purplish tint of various aboveground parts. The degree of reddish pigmentation can be of taxonomic significance. For example, one of the diagnostic characters of *Kalanchoe sexangularis* (see Chapter 12.15) is the reddish appearance of the plants, one of the features that not only helps to separate it from the closely related *K. leblanciae* (see Chapter 12.8) and *K. longiflora* (see Chapter 12.9) but also contributes towards its popularity as a garden subject. An infusion with various shades of red, pink, and orange can be particularly attractive in the large, roundish leaves of the group of indigenous southern African species with club-shaped inflorescences, with the red frequently very pronounced in *K. luciae*, especially towards the margins (Fig. 8.1). Reddish pigments are sometimes restricted to the leaf margins only. In addition to the infusion of leaves with reddish pigments, those in some mainly Madagascan and East African species are conspicuously mottled with well-demarcated dark purplish spots, flecks, and stripes (e.g. *K. diagremontiana*, *K. humilis*, *K. marmorata* (Fig. 4.2), *K. tubiflora*). Not only are reddish pigments present in the leaves, but also the flowers. In *K. montana*, for example, the carpels, sepals, and sometimes corolla lobes may be conspicuously reddish infused, whereas obvious reddish or purplish stripes and veins are present in the corolla lobes of some of the Madagascan (e.g. *K. beharensis*, *K. gastonis-bonnieri*) and African (*K. salazarii*) species. The corolla lobes are also reddish or pinkish in some of the indigenous but particularly the Madagascan kalanchoes.

Based on studies on other groups of plants, we have assumed that the reddish to purplish pigmentation in kalanchoes is caused by anthocyanins. These molecules are water-soluble vacuolar flavonoids and are the most important pigments of vascular plants. Chemically, the basic structure of anthocyanins is made up of the anthocyanidins (or aglycons) (Fig. 8.4). When the anthocyanidins are found in their glycoside form (bonded to a sugar), they are known as anthocyanins (Castaneda-Ovando et al., 2009; Landi et al., 2015). Studies have shown that anthocyanins are metabolically expensive to manufacture by the plant (Chalker-Scott, 1999). Another potential cost of anthocyanin accumulation is their interference with the light reactions of photosynthesis. These compounds absorb blue light and reflect red wavelengths; thus, their presence in the upper epidermis or mesophyll of leaves could theoretically compete with light harvesting by chlorophyll and carotenoids. It is unlikely that plants will expend energy they could otherwise use for growth and development on the synthesis of anthocyanins if these compounds were not benefitting the plant. Whereas some potential benefits of these pigments to plants are well established, many others are still the subject of debate and further study (Archetti et al., 2009; Landi et al., 2015). Here, we will only consider some of the potential benefits of anthocyanin accumulation in the leaves of kalanchoes.

In the leaves of kalanchoes, the reddish anthocyanins are mainly confined to the vacuoles of the epidermal cells (Figs. 8.5 and 8.6), although sometimes also visibly present in a narrow zone of the mesophyll immediately below the epidermis (Fig. 8.7). This superficial location suggests an adaptive role as a type of light filter. Most popular is the so-called photoprotection hypothesis, which suggests the red pigments protect against the harmful effects of light (Archetti et al., 2009). During photosynthesis, continuous damage occurs to photosystem II in the chloroplast, and this requires constant repair. If the light intensity is very high and especially under certain stressful environmental conditions, such as drought and cold, the rate of damage creation exceeds the rate of repair. This leads to long-term damage to the photosynthetic apparatus in the chloroplast, termed photoinhibition.

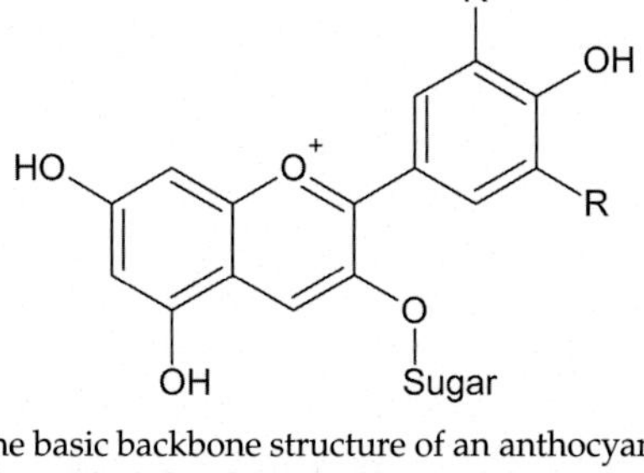

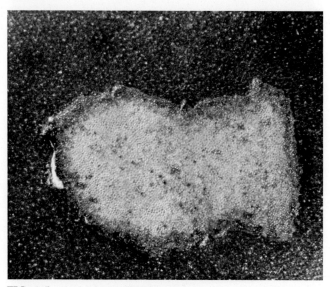

FIG. 8.5 Lower (abaxial) surface of the reddish purple leaf of *Kalanchoe sexangularis* var. *sexangularis* with a thin slice (mainly epidermis) cut away to show the underlying green photosynthetic tissue. This demonstrates the superficial location of the reddish anthocyanins. Photograph: Abraham E. van Wyk.

FIG. 8.4 The basic backbone structure of an anthocyanin. Anthocyanins are the glucosides of anthocyanidins, and they may have different sugars bonded to their ringed structure.

FIG. 8.6   Leaf of *Kalanchoe sexangularis* var. *sexangularis* seen in transverse section and showing the restriction of reddish anthocyanins to the epidermis. This superficial location suggests an adaptive role as a type of light filter. Photograph: Abraham E. van Wyk.

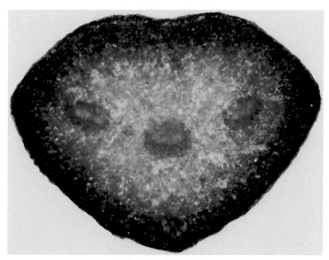

FIG. 8.7   Petiole of *Kalanchoe sexangularis* var. *sexangularis* seen in transverse section and showing the confinement of reddish anthocyanins to the epidermis and some of the adjacent but still superficially located spongy parenchyma. Photograph: Abraham E. van Wyk.

To avoid photoinhibition, plants possess various so-called photoprotective mechanisms, one of which is to reduce light absorbance in the chloroplast by filtering sunlight through nonphotosynthetic but light-reducing pigments, such as anthocyanins. There is substantial experimental evidence that foliar anthocyanins serve as a sunblock protecting chloroplasts from the adverse

effects of excess light (Landi et al., 2015 and references therein). It has also frequently been suggested that anthocyanin accumulation may be a possible protection mechanism in plants against DNA damage caused by UV radiation. Although there is some experimental evidence suggesting such a role for anthocyanins, the current consensus is that their UV-absorbing functions are unlikely to be a universal primary role for these pigments (Landi et al., 2015)—see under "Epicuticular wax (waxy bloom)" elsewhere in this chapter. It might, however, be significant that in a recent study on the New World *Sedum wrightii* A.Gray (Crassulaceae), it was found that anthocyanins accumulated significantly more when the adventitious root tips of plantlets sprouting from detached leaves were exposed to high amounts of UV-B radiation (Kerr et al., 2018). The accumulation of red anthocyanins in root tips is a peculiarity of members of the Crassulaceae (and a few other families), the function of which is still a mystery (Molisch, 1928).

A possible light-protective function for the anthocyanins in the leaves of indigenous kalanchoes is supported by casual observations that in nature, many of the species prefer to grow in semishaded habitats, such as under shrubs and trees (Fig. 6.1). This may indicate a sensitivity to direct sunlight. Plants of *Kalanchoe sexangularis*, for example, are much more red when grown in full sun than when in the shade. Leaves in the commonly cultivated *K. blossfeldiana* also become more reddish infused if grown in exposed positions, although pigment formation in this species is also promoted by short days (Neyland et al., 1963). An exception is those indigenous species with club-shaped inflorescences and fairly large, paddle-shaped leaves, which are usually found in more open sun-exposed habitats on rocky outcrops in dense or in open, sparse grasslands. But these species are initially low growing during the vegetative phase and may receive some light protection from the associated grasses and forbs. They also seem to quite readily accumulate anthocyanins in their leaves. It is possible that the anthocyanins may protect against mainly high-intensity visible light, whereas epicuticular wax provides more the UV protection, in addition to other functions (see section "Epicuticular wax (waxy bloom)"). The Madagascan *K. orgyalis* is seemingly an exception to the common pattern in that its leaves are reddish brown to coppery in dappled shade but more whitish green when in direct sunlight (Figs. 12.4, 12.5 and 12.11). However, the reddish brown colouration is not attributable to anthocyanins but rather to the presence of a dense indumentum of brownish hairs on the leaf surfaces of this species.

Thorny, spiny, and prickly plants also often have conspicuous colours. It is possible that their colours are aposematic and associated with poisonous or distasteful effects. This has led to some authors speculating that reddish colours imparted by anthocyanins may also warn potential herbivores that a particular plant is toxic

(Lev-Yadun, 2001 and references therein). Kalanchoes do in fact contain toxins (see Chapter 9), but the possibility that these pigments may primarily accumulate as a warning signal to potential herbivores is still viewed with scepticism (Archetti et al., 2009).

But what would be the adaptive value of the reddish or dark purple mottled leaves in mainly not only some of the Madagascan but also a few African kalanchoes? Again, this phenomenon, which has evolved independently in many groups of plants, is not well understood. It has been suggested that the dark leaf flecks in the African *Kalanchoe marmorata* (Fig. 4.2) may serve as warning signals (Farmer, 2014). According to the latter author, many poisonous plants have dark flecks on their leaves, especially to protect them against herbivores that feed during daylight hours; dark colours on green leaves could be seen in daylight by herbivores as tonal changes. Observations on the feeding of domestic livestock in grassland where *K. marmorata* grows on the Liban Plateau in Ethiopia have shown that these plants with their spotted leaves are avoided. In this case, the spots may indeed serve as a warning colouration that they are distasteful/poisonous (Farmer, 2014).

EPICUTICULAR WAX (WAXY BLOOM)

In most land plants, the outer surface of the cuticle is coated by a layer of wax, varying in thickness from barely detectable to visually quite noticeable. In the latter case, the wax layer manifests as a thin layer of white waxy powder, called a "bloom", with such plants sometimes described as "glaucous". Plant organs with a well-developed bloom have a white, bluish or whitish grey colour and some of the wax can easily be rubbed off with a finger. A powdery waxy bloom is widely encountered in Crassulaceae and is particularly pronounced in some of the indigenous southern African kalanchoes, especially *Kalanchoe luciae* (Fig. 8.1), *K. montana*, *K. thyrsiflora*, and *K. winteri*. However, the thickness of the bloom may vary among individuals of a species, and in the case of leaves, it sometimes diminishes with age. A particularly thick powdery white bloom may cover the inflorescences of *K. thyrsiflora*, hence the common names "white lady" and "*meelplakkie*" (Afrikaans). A waxy bloom is also quite pronounced in some of the Madagascan species, e.g., *K. pumila*, with the common names "flower dust kalanchoe" and "*meelblomplakkie*" (Afrikaans) also alluding to this powdery coating.

Functionally, epicuticular waxes are closely linked with the underlying plant cuticle from which they are derived. Together, both layers exhibit a multitude of important functions involving interfacial interactions with both the biotic and abiotic environments that enable plants to survive in many different terrestrial habitats (Bargel et al., 2006). Among the functions attributed to

FIG. 8.8 Water droplets on the leaf of *Kalanchoe thyrsiflora*, a species that may show a pronounced waxy bloom. These epicuticular waxes render the leaves hydrophobic (water-repellent), causing water to roll of the leaf as spherical droplets without wetting the leaf surface. Photograph: Abraham E. van Wyk.

a waxy bloom are protection against nonstomatal water loss and the shielding of plant cells from UV radiation and high light intensities by mechanisms such as reflection, absorption, and scattering (Koch & Ensikat, 2008). Moreover, epicuticular waxes are hydrophobic (water-repellent) (Fig. 8.8) and particularly in so-called superhydrophobic surfaces, which is responsible for self-cleaning of the leaf surface, a phenomenon also known as the "lotus effect" (Barthlott & Neinhuis, 1997). Such self-cleaning is believed to protect the plant against pathogen spores and dust particles. An often overlooked function is that in some plants, a waxy surface may provide a slippery surface, thus restricting access by certain insects or, as in pitcher plants, assisting in trapping others (Federle et al., 1997; Riedel et al., 2003).

Although epicuticular waxes show considerable ultrastructural and chemical diversity and are often of taxonomic significance (Barthlott et al., 1998), no broadscale comparative studies of this layer have been conducted on the kalanchoes. Chemically, these waxes are composed of a large mixture of different very-long-chain fatty acids and their derivatives. In addition, triterpenoids, tocopherols, or aromatic compounds can be present in various species of *Kalanchoe*, including the indigenous *K. thyrsiflora* (Siems et al., 1995). For a detailed study on the development and chemical composition of the epicuticular waxes of *Kalanchoe daigremontiana*, and scanning electron micrographs of the epicuticular wax

FIG. 8.9 The powdery white bloom on the leaves of the Mexican *Dudleya brittonii* was shown to have the highest ultraviolet light reflectivity (up to 83% in UV-B) of any known naturally occurring biological substance. Note the development of reddish anthocyanins (rather than a whitish bloom) in the inflorescences. Photograph: Stickpen [Public domain], from Wikimedia Commons.

FIG. 8.10 In some Crassulaceae, maximum anthocyanin and wax accumulation tend to be mutually exclusive in any one plant. The waxy layer on the peripheral parts of older leaves seems to wear away over time, and these more exposed surfaces then assume a reddish colour. This is clearly shown here in a plant of *Dudleya farinosa*, photographed along the shoreline in Salt Point State Park, northern California, USA. Photograph: Adam W. Braziel.

layer on its leaf surfaces, see Van Maarseveen & Jetter (2009) and Van Maarseveen et al. (2009).

Although the exact functions of the waxy bloom in the kalanchoes remain to be elucidated, one can safely assume that several of the known functions may, to some extent, be involved. As leaf succulents with water-storing tissues, the bloom almost certainly helps to prevent excessive water loss. The bloom is also likely to protect the plants against damage by UV radiation. Towards this end, studies on the spectral properties of the visually very similar powdery white bloom of *Dudleya brittonii* Johansen (Fig. 8.9), a succulent member of the Crassulaceae indigenous to Mexico, have shown it to have the highest ultraviolet light reflectivity (up to 83% in UV-B) of any known naturally occurring biological substance (Mulroy, 1979). In addition, the bloom is also responsible for high reflectances of visible and near-infrared radiation. As is the case in many species of *Kalanchoe*, *D. brittonii* also accumulates red anthocyanins in the epidermal cells of some sun-exposed tissues, notably those of the inflorescences (Fig. 8.9). Anthocyanins are also known to have radiation protection properties (see section "Anthocyanins" in the present chapter). It might be of significance that in some species of *Dudleya*, notably

D. farinosa Britton & Rose (see Fig. 8.10), and some of the indigenous kalanchoes with a dense waxy bloom (e.g. *K. luciae*, *K. thyrsiflora*), maximum anthocyanin and wax accumulation *tend* to be mutually exclusive in any one plant. Although reddish surfaces may still be covered by a bloom, the latter tend to be thinner on these surfaces than elsewhere on the leaf, so making the red-infused parts of the plants (e.g. the leaf margins) quite conspicuous. Additionally, the waxy layer on the peripheral parts of older leaves seems to wear away over time, and these more exposed surfaces then assume a reddish colour. This suggest that the one (anthocyanins) may to some degree compensate for the other (bloom) as far as radiation protection is concerned.

The bloom may also provide protection against attack by herbivorous insects in native kalanchoes. It has, for example, been shown that in *Dudleya brittonii* the larva of the moth *Rhagea stigmella* (Pyralidae), a major leaf- and stem-boring predator, less frequently attacks plants with a well-developed bloom (Mulroy, 1979). Kalanchoes have been recorded as larval food plants for a number of butterflies (Lycaenidae) (Fig. 8.11) in southern Africa (e.g. Kroon, 1999). The possibility that the waxy bloom in indigenous kalanchoes provides protection against these butterfly larvae deserves investigation. Another possible function of the waxy bloom is to serve as a "slippery" surface preventing ants from

FIG. 8.11   The common hairtail butterfly, *Anthene definita definita*, on *Kalanchoe rotundifolia*. This species not only feeds on the floral nectar, like some of the other butterflies of family Lycaenidae, but also has members of *Kalanchoe* as one of its larval food plants. The pronounced waxy bloom on some southern African kalanchoes may provide such plants with some protection against excessive breeding, on a single plant, by these butterflies. Photograph: Neil R. Crouch.

reaching the flowers. Despite copious nectar production, we have never seen ants visiting the flowers of indigenous kalanchoes. Federle et al. (1997) have shown that many ant species are unable to walk on the waxy bloom covering the stems of trees in the genus *Macaranga* Thouars (Euphorbiaceae). It is also known that a waxy bloom can in this way prevent nectar-robbing ants from reaching the nectar in flowers (Harley, 1991). This may well also be the case in some or all of the indigenous southern African species of *Kalanchoe*, especially some of those with club-shaped inflorescences covered with a thick layer of epicuticular wax.

ANATOMY

In general, succulents are not popular subjects for comparative anatomical studies. The vast amount of highly vacuolated water-storage parenchyma is often seen as unspecialised and the variation rather uninformative for taxonomic purposes. Instructive overviews of known anatomical variation in Crassulaceae as a family are provided by Metcalfe & Chalk (1950) and Gregory (1998). Some of the salient anatomical feature based mainly on the latter source is summarised by Thiede & Eggli (2007). Whereas crystals of calcium oxalate are present as prisms or druses in many Crassulaceae, they are present as crystal sand in *Kalanchoe* (and the related *Adromischus* and *Cotyledon*, both of which are also included in Crassulaceae subfam. Kalanchooideae); crystal sand is considered an apomorphy for subfamily Kalanchooideae (Stevens, 2001 onwards).

According to Thiede & Eggli (2007), foliar hydathodes are present in many (perhaps all) Crassulaceae, with the Kalanchooideae typically having one (sub)apical hydathode. However, with the exception of *Kalanchoe crenata*, we did not find any mention of hydathodes being present in the indigenous southern African kalanchoes. In some of these species, we did notice what appears to be a rudimentary hydathode (scar?) apically on some (but not all) leaves, but this still requires anatomical confirmation. The presence of hydathodes has, however, been confirmed in plants obtained from the Botanical Garden of Rio de Janeiro and attributed to the African *K. crenata* and Madagascan *K. pinnata* (Moreira et al., 2012). In both species, several hydathodes were observed along the leaf blade margins and confirmed anatomically. These hydathodes have a well-developed epithem and are functional as the secretion of droplets of water from these pores has been recorded when plants were kept in a humid enclosure. The possible presence of hydathodes in more, if not all, of the indigenous southern African kalanchoes requires further study.

Relatively, little is known about variation in the structure of hairs found on the external surfaces of various organs (e.g. leaves and peduncles) of kalanchoes. We have noted multicellular, simple, glandular hairs in some indigenous species (e.g. *Kalanchoe lanceolata*, *K. hirta*) and multicellular nonglandular 3-armed stellate hairs (illustrated in Schulte et al., 2009) in some of the Madagascan species (e.g. *K. beharensis*, *K. tomentosa*). Further investigation of the indumentum in kalanchoes, including the functional significance of the glandular hairs (see also Chapter 7), could prove illuminating.

Among species with photosynthetic succulent leaves, two main types of anatomical arrangement were recognised by Ihlenfeldt (1985). In the first type, water is stored in enlarged parenchyma cells that also contain chloroplasts. The photosynthetic tissue (chlorenchyma) therefore also serves as water-storage tissue. Referred to as *Allzellsukkulente* ("all-cell succulents"), this is the type of leaf anatomy found in *Kalanchoe*. In the second type, called *Speichersukkulente* ("storage succulents"), there is a division of labour between photosynthetic tissues (chlorenchyma) and specialised water-storage tissues (hydrenchyma). This type of arrangement is, for example, found in members of the genus *Aloe* (Smith & Van Wyk, 2008a).

There was a general tendency for past anatomical studies on Crassulaceae in general and *Kalanchoe* in particular, to be mainly descriptive. Although there is still a dearth of basic descriptive information on the anatomy of

Kalanchoe, especially for those species indigenous to southern Africa, future studies should also attempt to more deliberately interpret observed structure in terms of potential functional and adaptive significance. As highlighted by Males (2017), considerable potential exists for the exploration of mechanistic relationships between anatomical structure and physiological function to improve our understanding of the constraints that have shaped the evolution of succulence. An example of such a study is provided by Males & Griffiths (2017), who compared the anatomy of the stomatal complex between several CAM and C3 lineages.

Biocultural Significance and Toxicity

FIG. 9.1 Flowers of *Kalanchoe paniculata*, a poisonous plant that can cause paralysis in domestic livestock. Photograph: Abraham E. van Wyk.

INTRODUCTION

As primary producers, green plants are not only capable of photosynthesis but also structurally the products of this process by which they use sunlight to synthesise nutrients from carbon dioxide and water. Essentially comprising a potential source of food, their largely sessile way of life makes plants an easy target for numerous consumers in the environment. If it was not for various defence mechanisms evolved by plants, they would not have been able to survive. One defence strategy is to produce so-called secondary metabolites, chemicals that render them distasteful or poisonous to potential consumers (Swain, 1977). Secondary metabolites also protect against abiotic stresses such as radiation (see Chapter 8 under section "Anthocyanins"), assist in interplant competition, and attract pollinators and seed-dispersing animals (Wink, 1999).

The metabolic processes associated with growth, development, and reproduction are very similar in all plants; hence, they share similar chemical compounds essential for these basic life functions. Called primary metabolites, these molecules are vital for basic metabolism and include proteins, carbohydrates, and lipids. Secondary metabolites, on the other hand, are not indispensable for normal growth and development. They are biosynthetically derived from primary metabolites, primarily to serve survival functions for the plants producing them. Common classes of secondary metabolites include alkaloids, flavonoids, essential oils, glucosinolates, lignins, tannins, resins, steroids, and cyanogenic and cardiac glycosides. The metabolic pathways through which secondary metabolites are manufactured may be restricted to a particular species or a group of plants. They may also not be distributed evenly throughout the plant, but confined to the most vulnerable organs or be concentrated in epidermal tissue, as might be expected if these compounds function in defence.

Members of the genus *Kalanchoe* are particularly rich in a broad range of secondary metabolites (Pandurangan et al., 2015). Among these are compounds known as cardiac glycosides. There is no known or even hypothesised "primary" function for cardiac glycosides in terms of plant growth and development, and thus, they make

an excellent model of "secondary" metabolites to study (Agrawal et al., 2012). Not only are cardiac glycosides toxic to animals, but also they represent an important source of active pharmaceuticals (Verpoorte, 1998; Bourgaud et al., 2001; Kamboj et al., 2013).

Southern Africa not only has and exceptionally rich and diverse flora (Van Wyk & Smith, 2001) but also is rich in poisonous plants (Kellerman, 2009). It is estimated that the region contains well over 600 poisonous plant species (Vahrmeijer, 1981). Most important, from an economic point of view, are those species containing cardiac glycosides, including members of the genus *Kalanchoe* and related genera of the Crassulaceae (Fig. 9.1). Poisoning of domestic livestock with cardiac glycoside-containing plants, collectively, is the single most important plant poisoning in South Africa, annually accounting for about 33% of all such deaths in cattle (Kellerman, 2009; Kellerman et al., 1996). The two main classes of cardiac glycosides are the cardenolides and bufadienolides. All the important cardiac glycoside-containing plants in South Africa have bufadienolides as their active principles, as is the case with the genus *Kalanchoe* (Kellerman et al., 2005).

In the present chapter, we primarily focus on the structure of cardiac glycosides and the presence of these and other secondary metabolites in *Kalanchoe*. We will explain the mode of action responsible for the toxicity of these compounds, with emphasis on the kalanchoes and their involvement in animal poisoning in southern Africa. Bioculturally, kalanchoes have been widely used in traditional medicine, and we draw attention to some of these. Note, however, that the horticultural significance of the group is covered in Chapter 11. Finally, we highlight the potential of the genus *Kalanchoe* as a largely untapped resource of potentially novel compounds for applications in medicine.

CARDIAC GLYCOSIDES

The general structure of a cardiac glycoside consists of a steroid nucleus (the aglycone part) linked to a sugar (glycoside; the glycone part), with a variable lactone side chain attached to C17 of the aglycone (Fig. 9.2). This class

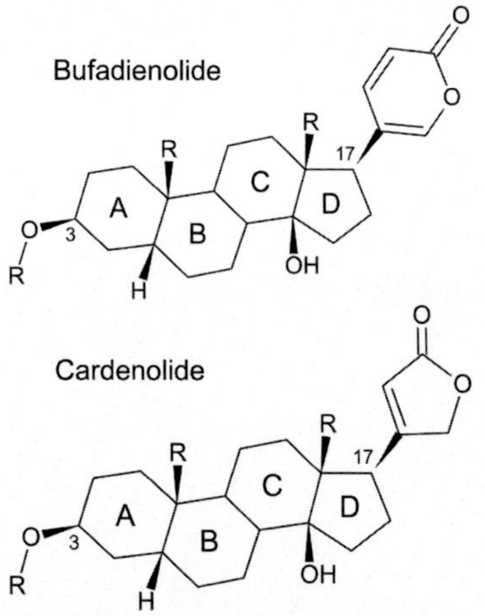

FIG. 9.2 Basic chemical structures of the aglycones of bufadienolides and cardenolides, both heart glycosides. In these compounds, the aglycone is composed of a core steroid (four fused rings: A, B, C, and D). Bufadienolides have a six-membered lactone ring at C17, whereas cardenolides have a five-membered lactone group at this position. A glycoside group (sugar) is typically attached to the aglycone at position C3 (R).

of plant organic compounds is known to act on the cellular sodium-potassium ATPase pump, resulting in an increase in the output force of the heart and a decrease in its rate of contractions, hence the use of "cardiac" as part of the name (Patel, 2016; Morsy, 2017).

Based on the side chain at the position C17, cardiac glycosides can be classified into two types, namely, cardenolides and bufadienolides; both, however, share a similarity in their biological activity on the sodium-potassium pump. Cardenolides have a five-membered lactone group in the 17β position, whereas the bufadienolides have a six-membered lactone ring at this position (Krenn & Kopp, 1998). Bufadienolides have first been derived from the venom of the cane toad *Bufo marinus*, now classified as *Rhinella marina*, hence the "bufa-" portion of the name.

Plants can produce both cardenolides and bufadienolides, as do various species of animals such as toads (Bufonidae) and insects (Lampyridae and Chrysomelidae) (Krenn & Kopp, 1998; Steyn & Van Heerden, 1998; Agrawal et al., 2012). In plants, cardiac glycosides seem to be confined to the flowering plants, with bufadienolides known from six families, namely, Crassulaceae, Hyacinthaceae, Iridaceae, Melianthaceae, Ranunculaceae, and Santalaceae. Cardiac glycosides have almost certainly evolved as a defence, against herbivores in plants and against predators in animals (Agrawal et al., 2012).

In Crassulaceae, bufadienolides are restricted to subfamily Kalanchooideae, a clade for which this class of compounds is a possible synapomorphy (Stevens, 2001 onwards). Hitherto, various bufadienolides have been reported from members of three of the four genera

comprising this subfamily, namely, *Cotyledon*, *Kalanchoe*, and *Tylecodon*. Bufadienolides can, however, also be expected in *Adromischus*, a group that is chemically still poorly known. Of these genera, *Adromischus* and *Tylecodon* are endemic to southern Africa, whereas *Cotyledon* has its centre of diversity in this part of the subcontinent (Tölken, 1985; Pilbeam et al., 1998; Van Jaarsveld & Koutnik, 2004).

TOXICITY

The cellular sodium-potassium pump is ubiquitous in animals and plays a crucial role in the functioning of, among others, the central nervous system and neuromuscular signalling. Because cardiac glycosides are strong inhibitors of this pump, the ingestion of these compounds by animals and humans can, depending on dosage, lead to paralysis of both striated and smooth muscles of the heart, lungs, and skeleton (Smith, 2004; Wink, 2010). Hence, cardiac glucosides are classified as neurotoxins.

Based on degree of toxicity of different poisons and toxins, The World Health Organisation (WHO) recognises four toxic classes, ranging from Class Ia that is extremely hazardous to Class III that is slightly hazardous. Cardiac glycosides are placed in Class Ia and have traditionally been described as "highly poisonous" (Wink, 2010; Wink & Van Wyk, 2008).

Since toxic bufadienolide cardiac glycosides are common in kalanchoes, we will in this section focus on the toxicity of this group of plants, especially as it manifests in southern Africa. Among the indigenous species hitherto tested for toxicity, the presence of bufadienolides has been confirmed in *Kalanchoe crenata* (Nguelefack et al., 2006), *K. lanceolata* (Anderson et al., 1983, 1984), *K. paniculata* (Steyn & Van der Walt, 1941), *K. rotundifolia* (Van der Walt & Steyn, 1941), and *K. thyrsiflora* (Steyn & Van der Walt, 1941). The presence of bufadienolides has also been confirmed in the following Madagascan kalanchoes introduced to southern Africa: *Kalanchoe diagremontiana*, *K. pinnata*, *K. tomentosa*, and *K. tubiflora* (Krenn & Kopp, 1998, and references therein). Although bufadienolides are claimed to be absent in the commonly cultivated *Kalanchoe blossfeldiana* (Wink & Van Wyk, 2008), it has been assumed (by inference to the presence of these toxins in other members of the group) to be poisonous to pets (Smith, 2004) (Fig. 9.3). However, experimental administering of an extract of this plant to dogs, followed by a relatively short period of clinical monitoring (12 hours), did not produce any detectable ill effects (Teixeira et al., 2010). On the other hand, this species has been implicated in the death of cattle in Brazil, with clinical signs and lesions consistent with poisoning by plants that contain cardiac glycosides (Mendonça et al., 2018).

FIG. 9.3 The widely cultivated *Kalanchoe blossfeldiana* has been implicated in the death of cattle with symptoms consistent with acute bufadienolide poisoning; despite reports to the contrary, it should be considered toxic to both humans and animals. Photograph: Abraham E. van Wyk.

FIG. 9.4 Typical posture of a sheep with krimpsiekte, a chronic poisoning with bufadienolides. The animal is standing with the back arched, feet together, and head hanging down. Photograph: Christo Botha, Department of Paraclinical Sciences, University of Pretoria.

Studies to isolate bufadienolides from kalanchoes have focused mostly on the aerial parts (mainly leaves and stems) of the plants, but in at least some species, these compounds are also present in the roots (Kolodziejczyk-Czepas & Stochmal, 2017, and references therein) and, in the case of *Kalanchoe tubiflora*, also the flowers (McKenzie et al., 1987). It is noteworthy that in *K. tubiflora*, the flowers and roots turned out to be the more poisonous parts of the plant. According to McKenzie et al. (1987), the median lethal doses for cattle in this species of flowers, roots, and leaves plus stems were 0.7, 2.3, and 5.0 g dry matter/kg live weight, respectively (7, 7, and 40 g wet weight/kg).

In a historically important veterinary article, Soga (1891) provided the first *definite experimental evidence* to prove that a plant is toxic to livestock in South Africa. In former years, there were merely anecdotal reports of plants being suspected of causing mortality (Curson, 1928; Botha, 2013). The poisoning studied by Soga was that caused in goats by *Tylecodon ventricosus* (Burm.f.) Toelken (=*Cotyledon ventricosa* Burm.f.), a member of Crassulaceae subfam. Kalanchooideae. Historically, this condition was known in the vernacular as "nenta", a name derived from Khoekhoe, and first documented in

1775 by the Swedish botanist and explorer, Carl Peter Thunberg (Smith, 1966; Forbes, 1986). This condition is also known as "*cotyledonosis*", but currently is usually referred to as "krimpsiekte" (from the Afrikaans; literally meaning "shrink disease"). The latter name is derived from the arching of the back and bending of the neck to one side (torticollis) displayed by poisoned stock (Fig. 9.4), as if brought about by the shrinkage or contraction of the muscles (Smith, 1966). Subsequently, it was shown that a similar type of poisoning is caused by species of *Cotyledon* and *Kalanchoe* (Vahrmeijer, 1981; Kellerman et al., 2005; Van Wyk et al., 2002).

Poisoning with bufadienolides may be either acute or chronic, depending on whether the bufadienolides have a cumulative effect or not. Acute poisoning affects the respiratory, cardiovascular, gastrointestinal, and nervous systems, usually leading to either death or recovery within days (Kellerman et al., 2005). Krimpsiekte is a *chronic* form of poisoning caused by the cumulative neurotoxic effect of bufadienolides brought about by repeated ingestion of small quantities of the plants over an extended period of time (Naudé, 1977; Vahrmeijer, 1981). It is a neurological syndrome and involves mainly the nervous system, sometimes resulting in sudden death or, more commonly, prolonged muscular weakness or paralysis (Kellerman et al., 2005). Cases of bufadienolide poisoning reported in Australia, Brazil, and Zimbabwe, for example, are acute cardiac glycoside intoxication and not krimpsiekte (McKenzie et al., 1987; Masvingwe & Mavenyengwa, 1997; Mendonça et al., 2018). Species of *Kalanchoe* implicated in poisoning of livestock in southern Africa include *K. brachyloba*, *K. crenata*, *K. paniculata*, *K. rotundifolia*, and *K. thyrsiflora*; general symptoms include paralyses and haemorrhagic diarrhoea (Steyn, 1949; Kellerman et al., 2005; Mannheimer et al., 2012).

Krimpsiekte is regarded as one of the most important poisonings of sheep and goats in South Africa. Horses, dogs, and fowls have also been diagnosed with this type of poisoning (Kellerman et al., 2005; Botha & Penrith, 2009) but cattle only rarely so. Fatally poisoned sheep become paralysed and lie on their sides until they either die or are put down (for a comprehensive account of the condition, see Kellerman et al., 2005). Krimpsiekte is perhaps the most important plant poisoning of small stock in the Klein Karoo and southern fringes of the Groot Karoo (Kellerman et al., 1996). It also occurs in Succulent Karoo along the Atlantic coastline of the Northern Cape province (see maps in Kellerman et al., 2005: 142; Kellerman, 2009). All of these are areas which fall well outside the range of native members of *Kalanchoe*. Hence, in South Africa, most cases of krimpsiekte poisoning are ascribed to species of *Tylecodon* and *Cotyledon*. Furthermore, krimpsiekte is the only plant poisoning in southern Africa reputed to cause secondary poisoning; humans and animals that eat the meat of krimpsiekte carcases (even when cooked) may themselves become affected (Henning, 1926; Botha, 2016).

BIOCULTURAL SIGNIFICANCE

Because of their toxicity, plants containing cardiac glycosides have been known to humans since ancient times. Drugs prepared from cardiac glycoside-containing plants are widely used in traditional medicine and has also been the subject of many chemical and clinical studies in mainstream medicine (Patel, 2016). This class of secondary metabolites is neurotoxins and affects not only humans but also other animals with a nervous system (see section "Toxicity" as discussed earlier). Hence, nature has also seen the coevolution of specialist feeders adapted to these toxins. Consequently, much of the chemical ecology research to date has concentrated on plant–animal interactions involving this class of compounds (Harborne, 1994; Agrawal et al., 2012).

By far, the greatest research interest in cardiac glycosides has been in the medical field. Most of the clinical attention was directed to the cardenolides (see section "Cardiac glycosides" as discussed earlier) owing to their therapeutic use (Agrawal et al., 2012). Digitoxin and digoxin are two classical plant-derived prescription cardenolides, the former originally isolated from foxglove, *Digitalis purpurea* L. (Plantaginaceae/Scrophulariaceae). Of the two, digoxin is the one still most frequently used in cardiac medication at present; it is commercially extracted from *Digitalis lanata* Ehrh. (Mastenbroek,

1985). Digoxin is approved by the US Food and Drug Administration to control ventricular response in patients with chronic atrial fibrillation and is said to be the only oral agent that increases the strength of heart muscular contraction used in the management of mild to moderate heart failure (Ehle et al., 2011). Digoxin is also on the WHO core list of essential medicines; these are the minimum medicines needed for a basic healthcare system (World Health Organization, 2017).

Compared with the cardenolides, research on the medical significance of the bufadienolides has been somewhat neglected, despite them sharing rather similar biological activities. However, drugs prepared from bufadienolide-containing plants and toads are also widely used in traditional medicine but mainly outside southern Africa. The efficacy of these substances was already known by the ancient Egyptians who used squill, *Drimia maritima* (L.) Stearn (Hyacinthaceae), a bufadienolide-containing plant, in the treatment of heart diseases (Krenn & Kopp, 1998).

In Chapter 12, we have mentioned under each *Kalanchoe* species' taxonomic treatment some of the ethnobotanical uses (if available) reported for it in the literature (under heading "Additional notes and discussion", subheading "Uses"), with emphasis on applications in southern Africa. Most uses reported are for medicinal and magical purposes. Although we have not tried to account for all uses hitherto recorded in the literature, the group is apparently not as extensively used as one would expect considering the potent bioactive compounds known from some of the species. No kalanchoe is, for example, mentioned in a scientific review of all the most important plant-derived products of special cultural and/or commercial significance in southern Africa (Van Wyk & Gericke, 2000). This apparent limited cultural use is also reflected by the fact that kalanchoes are not widely traded commercially in local traditional medicinal plant markets. Material of *K. crenata* was, however, recorded in some of these markets in Gauteng, South Africa (Williams et al., 2001). In southern Africa, most of the interest in the indigenous kalanchoes has so far centred on the problems some are causing in agriculture when stock is poisoned (see section "Toxicity" as discussed earlier).

An informative review of the medical utility of bufadienolides in general is provided by Kamboj et al. (2013). Plants containing these compounds are used in traditional remedies for the treatment of ailments such as infections, rheumatism, inflammation, disorders associated with the central nervous system, tumours, and cancer. A limitation of some of these compounds is their toxicity, but it has been shown that structural changes in functionality could significantly alter their cytotoxic activities. According to Kamboj et al. (2013), this group of compounds holds particular promise for the development of new antitumor drugs. Kolodziejczyk-Czepas &

Stochmal (2017) provide an overview of 31 bufadienolides known from species of *Kalanchoe*, focusing on their chemical structure, biological activity, and prospects for therapeutic use.

In addition to bufadienolides, members of *Kalanchoe* evidently contain many more other secondary constituents of potential use to humans. In a comprehensive overview of the known chemistry of the well-studied Madagascan *Kalanchoe pinnata*, Pandurangan et al. (2015) showed that in addition to bufadienolides, the plant contains alkaloids, phenols, flavonoids, tannins, anthocyanins, glycosides, saponins, coumarins, sitosterols, quinines, carotenoids, tocopherol, and lectins. Many pharmacological effects have been reported for some of these compounds, including anticancer, antioxidant, immunomodulating, antibacterial, anthelmintic, antiprotozoal, antiinflammatory, analgesic, diuretic, antiurolithiatic, hepatoprotective, antipeptic ulcer, antidiabetic, and wound-healing activity.

A review of the ethnomedicinal, botanical, chemical, and pharmacological properties of *Kalanchoe* as a genus indicated that the diversity in chemistry and pharmacological properties reported in *K. pinnata* also extends to other members of the genus. Considering the size of the group, only a handful of species has hitherto been studied for their biological activities (Milad et al., 2014).

Of the southern African indigenous kalanchoes, crude extracts of the widespread *Kalanchoe laciniata* were shown to exhibit antioxidant activity, in vitro spasmogenic and spasmolytic activities on the gut in the rabbit, and antimicrobial activity against especially Gram-positive bacteria (Iqbal et al., 2016; Manan et al., 2016). Leaf extracts of *K. crenata* tested on mice and rats have shown significant peripheral and central analgesic activities as well as an anticonvulsant effect (Nguelefack et al., 2006). Other activities demonstrated for *K. crenata* include, to mention but a few, antimicrobial (Akinsulire et al., 2007), antiinflammatory and antiarthritic (Dimo et al., 2006), antihyperglycaemic (Kamgang et al., 2008), antidyslipidemic, and antioxidant (Fondjo et al., 2012).

With the exception of *Kalanchoe laciniata* and *K. crenata*, the chemistry and biological activity of hardly any of the other indigenous southern African kalanchoes have been studied for anything else than their toxicity to livestock. Therefore, the southern African species of *Kalanchoe*, as is the case with many members of the group elsewhere, still offer a vast untapped potential resource for chemical and pharmaceutical biobeneficiation.

10

Invasiveness

FIG. 10.1   A naturalised stand of *Kalanchoe tubiflora* growing with the indigenous *Aloe spicata* L.f. on a granite outcrop in Swaziland in eastern southern Africa, about 12 km north of Piggs Peak. Photograph: Gideon F. Smith.

111

INTRODUCTION

In general, a comparatively mild climate is prevalent over much of the southern African subcontinent. Only above the fairly narrow coastal belt does the region in places experience somewhat harsher climatic conditions that include regular winter frosts and long-lasting snow on the high mountains. The central and eastern parts of the subcontinent receive most of their rain in the warm to hot summer months, while the southwest and west coasts and a narrow strip to the interior have a Mediterranean-type climate, with maximum rainfall recorded in winter.

The vast majority of the kalanchoes indigenous to southern Africa are concentrated in the eastern and north-central parts of the region, with *Kalanchoe rotundifolia* having the most westerly distribution range along the southeastern Cape coast of South Africa, and somewhat further northwest, in parts of the Groot Karoo and into Namibia. Only very few species, for example, *K. paniculata* and *K. thyrsiflora*, occur on the Drakensberg, which is one of the coldest regions of the subcontinent. These two species also occur in parts of eastern and northeastern southern Africa in places with a much milder climate.

Several species of *Kalanchoe* are commonly cultivated in mild-climate parts of the world, with most of these originating from Madagascar (Figs. 10.1–10.4). Some of these species have become naturalised and ultimately weedy in countries where they are grown (Silva et al., 2015; Smith et al., 2015b). For example, a range of *Kalanchoe* species has been recorded as cultivated in parts of tropical and south-tropical Africa (Wickens 1987: 27; Newton & Mbugua, 1993: 46), where, in Zimbabwe, some are noted as naturalised escapees. Elsewhere in the world, kalanchoes have been recorded as introduced in the Azores (Borges et al., 2005) and Madeira (Hansen & Sunding, 1993) in Macaronesia, south-coastal Texas (USDA https://plants.usda.gov) and Florida (Ward, 2008) in USA, Mexico (Breedlove, 1986; Villaseñor & Espinosa-Garcia, 2004), Australia (http://weeds.brisbane.qld.gov.au), New Zealand (Given, 1984: 191 [*K. pinnata*, as *Bryophyllum pinnatum*]), Hawaii (Flora of the Hawaiian Islands. http://botany.si.edu/pacificislandbiodiversity/hawaiianflora/query2.cfm), and China (Wang et al., 2016).

Over the past few years, a concerted effort to document the exotic succulent species that are becoming naturalised, weedy, and invasive in South Africa has confirmed the remarkable success with which these species are spreading across arid, semiarid, and mesic landscapes of the country (Walters et al., 2011). While none of the historically and recently recorded alien succulent plant species has yet become as problematic as, for example, *Opuntia ficus-indica* (L.) Mill. (Cactaceae), the prickly pear, during the early-20th century when it infested large

FIG. 10.2   *Kalanchoe tubiflora* is a popular garden subject, until it starts taking over a rockery, that is. Material then discarded irresponsibly will easily spread into natural vegetation. Photograph: Gideon F. Smith.

parts of South Africa's Groot Karoo, the mounting presence of several alien species in South Africa remains a cause for environmental concern.

In 1985, a single species of *Kalanchoe* from Madagascar, *K. tubiflora* (Figs. 10.1–10.3), which was at the time still treated as *Bryophyllum delagoense*, was reported as a garden escape from 'frost-free areas of the eastern Transvaal [Mpumalanga]' (Tölken, 1985: 73). In the past 30 years, the number of exotic kalanchoes variously established in southern Africa has increased eightfold (Table 10.1). Given the preponderance of indigenous *Kalanchoe* species in parts of southern Africa where a subtropical climate prevails, it is not surprising that the Madagascan kalanchoes that have become naturalised in the subcontinent are also concentrated in the eastern parts of the subcontinent, specifically in South Africa (Table 10.1). Most recently, *K. fedtschenkoi* was for the first time recorded with confirmed localities as naturalised in the Western Cape province in South Africa's Klein Karoo, which is a region of high indigenous succulent plant diversity and a recognised Centre of Endemism (Van Wyk & Smith, 2001). Given the presence of the exotic kalanchoes

FIG. 10.3 A multitude of plantlets are formed at the tips of the cylindrical leaves of *Kalanchoe tubiflora*. These easily become detached and root where they fall. Photograph: Gideon F. Smith.

TABLE 10.1 Cultivated and naturalised, exotic Crassulaceae in southern Africa (see Walters et al., 2011: 232–259; Smith & Figueiredo, 2017c)

No.	Species	Region in southern Africa
A.	Recorded as naturalised	
1.	*Aeonium arboretum* (L.) Webb & Berthel.	Southwestern Cape
2.	*Kalanchoe tubiflora* (Harv.) Raym.-Hamet (Figs. 10.1–10.3)	Widely in southern Africa
3.	*Kalanchoe fedtschenkoi* Raym.-Hamet & H. Perrier (Figs. 10.7–10.10)	Western Cape, South Africa
4.	*Kalanchoe pinnata* (Lam.) Pers. (Fig. 10.13)	KwaZulu-Natal, Swaziland
5.	*Kalanchoe prolifera* (Bowie ex Hook.) Raym.-Hamet	KwaZulu-Natal
B.	Recorded as widely cultivated, escaping, and to be monitored	
1.	*Kalanchoe beharensis* Drake (Figs. 10.5 and 10.6)	Kruger National Park
2.	*Kalanchoe daigremontiana* Raym.-Hamet & H.Perrier	Cultivated, to be monitored
3.	*Kalanchoe gastonis-bonnieri* Raym.-Hamet & H.Perrier (Figs. 10.11 and 10.12)	Cultivated, to be monitored

FIG. 10.5 Often planted as a small- to medium-sized, decorative tree, *Kalanchoe beharensis* has large, felty leaves. Photograph: Gideon F. Smith.

FIG. 10.4 Small plants of *Kalanchoe daigremontiana* are becoming established around the larger specimen in the centre, which produces a large number of bulbils along its leaf margins. Photograph: Gideon F. Smith.

predominantly in the climatically mild eastern parts of South Africa, especially in KwaZulu-Natal, here, confirming the occurrence of *K. fedtschenkoi* in the Klein Karoo is especially significant (Figs. 10.7–10.10).

With the exception of *Aeonium arboreum* Webb & Berthel. from the Canary Islands, Spain, all the exotic Crassulaceae naturalised in southern Africa belong to

FIG. 10.6   Plantlets developing on the cut edge of a petiole of *Kalanchoe beharensis*. Photograph: Gideon F. Smith.

FIG. 10.7   *Kalanchoe fedtschenkoi* growing in a garden bed in Tshipise, Limpopo province, South Africa. Note how small plantlets that became detached from the leaves of the large plants are becoming established. Photograph: Gideon F. Smith.

FIG. 10.8   Discarded material of *Kalanchoe fedtschenkoi* easily strikes root where it is deposited. Photograph: Gideon F. Smith.

the genus *Kalanchoe*, and all of them originate from Madagascar (Table 10.1).

THE ESTABLISHMENT SUCCESS OF KALANCHOE SPECIES IN PLACES REMOTE FROM THEIR NATURAL HABITATS

The species of *Kalanchoe* that Walters et al. (2011: 232–259) recorded as naturalised in southern Africa all have multiple mechanisms through which they reproduce and spread. Most noticeably, they are able to regenerate from discarded stems and leaves and produce bulbils (plantlets) on inflorescences and detached leaves, as well as basal shoots, and seed copiously. The seed of most species of *Kalanchoe* is fine and dustlike and easily spreads through wind dispersal. Many Madagascan kalanchoes are quick growing and rapidly reach flowering and seeding maturity. These multiple propagation mechanisms and rapid growth rates make them especially suited to spread into and become established in various suitable niches and habitats in and near places where they are cultivated.

Kalanchoes, including exotic ones, are drought tolerant and often grown as waterwise alternatives to regular garden plants with higher irrigation requirements. At times of drought, some species are prone to shedding the lower leaves on their stems before the leaves are entirely desiccated; these detached leaves habitually and rapidly form small plantlets on their margins. The plantlets will easily become rooted and established where they fall, so often creating dense, unwanted, and unsightly stands in gardens. Once the plants have overstayed their welcome in gardens, surplus material is often not destroyed, but rather irresponsibly discarded and dumped, for example, in natural veld, and will quickly form new populations.

Of the kalanchoes naturalised in southern Africa, *Kalanchoe tubiflora* is by far the most widely grown and naturalised. Fortunately, the highly invasive *Kalanchoe ×houghtonii* (Figs. 10.15 and 10.16), a hybrid between *K. daigremontiana* and *K. tubiflora*, which is commonly known as Houghton's Hybrid, is unknown in South Africa. This hybrid has been recorded from several places as widely separated as Mediterranean Europe, for example Portugal and Spain, and Florida in USA (Guillot Ortiz et al., 2009, 2014). In the horticultural trade, this hybrid is also variously known as *Bryophyllum tubimontanum* Houghton, 'K. hybrida Jacobsen', and *Kalanchoe* 'Hybrida' Hort.

Two further species of Madagascan *Kalanchoe*, *K. daigremontiana* and *K. gastonis-bonnieri*, have been observed as widely cultivated in gardens in KwaZulu-Natal, and both are showing signs of escaping and moving into natural veld. The Madagascan *Kalanchoe synsepala* has similar tendencies (Fig. 10.14). It is therefore important that these

FIG. 10.9   Several alien succulent species have become established across the extensive landscape diversity, with its exceptional natural beauty, found in southern Africa. The foothills of the mountains north of Calitzdorp in the Klein Karoo are locations where *Kalanchoe fedtschenkoi* has been found to become established. Photograph: Gideon F. Smith.

FIG. 10.10   Like *Kalanchoe fedtschenkoi*, *K. serrata* Baker (shown here) produces small, perfectly formed plantlets along the margins of detached leaves. The plantlets root rapidly to form new plants. This species is hardly known in cultivation in southern Africa. Photograph: Gideon F. Smith.

FIG. 10.11   Plantlets that formed at the leaf tips of *Kalanchoe gastonis-bonnieri*. Photograph: Neil R. Crouch.

FIG. 10.12   A small population of *Kalanchoe gastonis-bonnieri* spreading from the mother plant that produced them. Photograph: Neil R. Crouch.

FIG. 10.14   Several species of Madagascan *Kalanchoe*, such as *K. synsepala*, reproduce vegetatively and rapidly so, forming dense stands. Photograph: Neil R. Crouch.

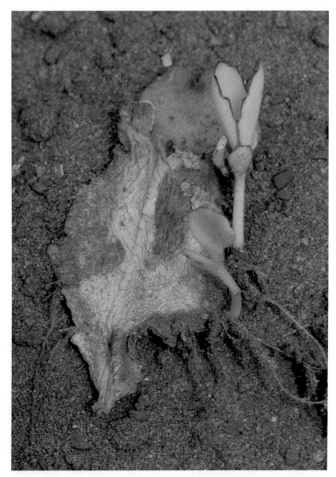

FIG. 10.13   Plantlets that formed on the margin of a leaf of *Kalanchoe pinnata* easily root where they touch the soil. Photograph: Neil R. Crouch.

FIG. 10.15   The invasive *Kalanchoe* ×*houghtonii* growing on a rock wall in the outskirts of Cascais, at Almoínhas Velhas, in central Portugal. This *Kalanchoe* hybrid has not been recorded in southern Africa. Photograph: Estrela Figueiredo.

FIG. 10.16   Dense stands rapidly develop where *Kalanchoe* ×*houghtonii* is grown in a garden. Photograph: Gideon F. Smith.

species be monitored and removed from gardens where possible.

The arborescent *Kalanchoe beharensis*, the donkey's ear, which is also native to Madagascar, is widely grown in gardens, even in rest camps in some of the National Parks of South Africa. Metabolically active, green leaves of this species that are severed from a tree will rapidly form plantlets at the proximal end of the petiole. This is the only species of locally cultivated, exotic *Kalanchoe* that uses this reproductive mechanism. However, the formation of plantlets on the margins of such leaves has not yet been observed.

11

Gardening

FIG. 11.1   Kalanchoes thrive in full sun and semishade, although the reds and yellows in the colour of the leaves become considerably more accentuated in direct sunlight, as here in the case of *Kalanchoe luciae*. Photograph: Gideon F. Smith.

'One day you may plant a little flower with your own hand; I planted a houseleek.'

Čapek (1931: 7).

INTRODUCTION

Given the variation found among species and cultivars of *Kalanchoe*, plants work well as a ground cover in open beds, as accent plants, or when planted in containers (Figs. 11.1–11.7 and 11.11–11.13). Plants are additionally waterwise and drought-resistant and grow quickly. Even when not in flower, many species provide interest as accent plants through their exquisitely shaped, colourful, and often textured foliage. Two of the southern African *Kalanchoe* species with enormous outdoor horticultural appeal, *K. luciae* (Figs. 11.1 and 11.5) and *K. thyrsiflora*, have thick and heavy, elongated-egg-shaped or round, soup plate-sized leaves with smooth margins. This growth form has even inspired the mass production of *Kalanchoe thyrsiflora* in plastic (Fig. 11.8); the resulting plastic plants are very decorative and come complete with red-tinged leaf margins and waxy leaf surfaces!

Several species of *Kalanchoe* are increasingly grown in gardens (Figs. 11.9 and 11.10) and specialist collections, as well as in general domestic and amenity landscaping in mild-climate zones. This also applies to southern African species, several of which are increasingly found in cultivation in and beyond the subregion.

FIG. 11.3   The horticultural uses and benefits of kalanchoes are manifold. Several species will easily serve one or more of multiple roles in a garden, for example, as a groundcover or as a specimen plant, as here with *Kalanchoe sexangularis*. It is cultivated together with the scrambling aloe, *Aloiampelos striatula* (Haw.) Klopper & Gideon F.Sm. Photograph: Gideon F. Smith.

The two other southern African kalanchoes that are popular in horticulture, *Kalanchoe sexangularis* (Figs. 11.9 and 11.10) and *K. longiflora* (Fig. 11.2), have variously coloured, especially red and turquoise orange, respectively, recurved and channelled, elongated, round, or flat leaves with irregularly scalloped margins. Other kalanchoes, especially the exquisitely beautiful *Kalanchoe blossfeldiana* cultivars, come into their own when in flower, explaining why Mendelson (2016: 183) stated, no doubt tongue in cheek, 'Does someone dye those kalanchoes to make them even uglier?'

Kalanchoes are generally hardy and thickset, with colourful, attractive, and broad dark or light green leaves that, in the case of many species, take on more intense shades of red or orange in winter so providing for splashes of colour throughout all the seasons (Figs. 11.14–11.16). In addition, in autumn and winter, clusters of small, yellow, red, or orange, tubular to cigar-shaped flowers with variously reflexed petal lobes are borne on erect, branched flowering stems (Figs. 11.17 and 11.20).

Always be aware that some exotic kalanchoes, especially those of Madagascan origin, do have invasive tendencies, as are sometimes found in a range of exotic species. Surplus material should therefore be responsibly disposed of and never be discarded where it could spread into natural vegetation.

CULTIVATION

In general, *Kalanchoe* species are horticulturally undemanding and easy to cultivate. They grow in virtually any soil type and can tolerate very low levels of rainfall and therefore irrigation. In cultivation, *Kalanchoe* species not only usually flourish in full sun but also will thrive in semishade given its propensity to grow in association

FIG. 11.2   Here, an unshakable devotion to waterwise gardening pays dividends as it cleverly combines an eclectic mix of highly identifiable leaf shapes and colours that effectively creates a visually flowing passage through the garden. The centre piece is the turquoise-leaved *Kalanchoe longiflora*, while ×*Graptoveria* 'Fantome' Aubé ex Gideon F. Sm. & Bischofb. grows in the right foreground. Photograph: Gideon F. Smith.

FIG. 11.4   With minimal effort and attention, a groundcover predominantly created by using succulents, such as *Kalanchoe sexangularis*, easily adds not only instantaneous but also lasting value for many seasons. These plants introduce fresh aesthetics that complement modern architectural styles. Photograph: Gideon F. Smith.

FIG. 11.5 Two of the horticulturally most popular kalanchoes *Kalanchoe luciae* and *K. sexangularis* var. *sexangularis* (right background) grown together in a garden bed. Photograph: Gideon F. Smith.

with bushveld trees and shrubs. In more shady situations, the leaves are less intensely coloured. A few species, such as *K. crundallii*, occur naturally and preferentially in the undergrowth of forests and should be grown in shady positions where they are ideally exposed to filtered sunlight only.

PROPAGATION

In general, *Kalanchoe* species are exceedingly easy to propagate from stem cuttings that can be planted directly in the spot where plants are intended to grow. Most species

grow well in virtually any soil type. Although kalanchoes produce seed copiously, seed propagation is hardly used; taking and planting cuttings are much easier and quicker ways to produce large volumes of rooted plants. However, where kalanchoes grow in a bed or container, self-seeding occurs invariably and results in the establishment of a multitude of seedlings in close proximity to the mother plant. The seedlings can be pricked out after a month or two and established elsewhere in a garden.

In the case of some species, leaf cuttings have been known to strike root with considerable ease. This applies to most Madagascan species, for example, *Kalanchoe beharensis* (Fig. 11.18). The southern African *K. rotundifolia* and south-tropical African *K. humilis* will also grow from leaves, as they readily strike root from the proximal end of a severed petiole. However, in the case of these African species, this is a much slower process than is the case with Madagascan kalanchoes, most of which grow much more readily from leaves.

If the inflorescence of a monocarpic species, such as *Kalanchoe luciae*, is removed early on, much smaller secondary inflorescences or small plantlets will sometimes develop around the base of the peduncle. The plantlets can be removed with a sharp knife, left to dry for a week to 10 days, and then planted where they are intended to grow.

Some species form plantlets in the axils of the leaves of diminishing size that grow on a developing inflorescence

FIG. 11.6   In a garden in Pretoria, Gauteng province, South Africa, the ovate to near-orbicular, wine-red leaves of *Kalanchoe sexangularis* from north-eastern southern Africa are juxtaposed with the fingerlike, light green ones of *Carpobrotus edulis* (L). N.E.Br. (Aizoaceae) from southwestern South Africa. *Kalanchoe sexangularis* is a knee-high, shrublike species, while *C. edulis* remains low growing with a densely creeping habit.   Photograph: Gideon F. Smith.

FIG. 11.7   Two potted *Kalanchoe* cultivars, *K.* 'Magic Bells' (prominently bell-shaped, light green calyx tubes) and *K.* 'Dorothy' (orange flowers), offered for sale in a garden shop in the United Kingdom. The plants are often sold with translucent sleeves around the foliage and the pots enclosed in decorative wrapping.   Photograph: Estrela Figueiredo.

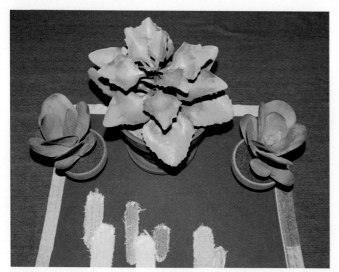

FIG. 11.8 Two true-to-life plastic specimens of *Kalanchoe thyrsiflora* flank a plastic plant of *Kalanchoe beharensis* 'Fang' in an indoor display. Photograph: Gideon F. Smith.

FIG. 11.10 Slightly larger potted specimens of *Kalanchoe sexangularis* var. *sexangularis* ready to be planted in a garden bed. Photograph: Gideon F. Smith.

FIG. 11.9 Kalanchoes are easy to propagate from stem cuttings. Here, such short, rooted cuttings of *Kalanchoe sexangularis* have been potted up in a seedling tray and are sold as a waterwise groundcover in a nursery in Pretoria, South Africa. Photograph: Gideon F. Smith.

FIG. 11.11 The regular, non-variegated-leaved form of *Kalanchoe fedtschenkoi* grown in an azure blue pot in Jardin Majorelle in Marrakech, Morocco. Photograph taken on 12 April 2017 by Estrela Figueiredo.

of some *Kalanchoe* species, such as *K. winteri*. These plantlets can be removed and planted in a seedling tray, for example, where they will easily strike root. In general, a severed inflorescence itself will not strike root; however, in the case of *K. leblanciae*, it was found that an inflorescence planted in a well-drained soil mixture that is kept moist throughout will develop into new plants.

LANDSCAPING

In not only especially mild-climate regions, such as southern Africa, but also other harsher climatic parts of the world, local authenticity in gardening and landscaping is increasingly achieved by thoughtful design that must nowadays embrace sound 'green gardening' principles (Fig. 11.19). In a world increasingly impacted by humans, waterwise succulents are very much suitable for establishing in a variety of garden settings. The modern tendency in garden and landscape development to appropriately use succulents in garden beds demonstrates that cultivated green environments of any size or shape can spark a feeling of warmth and enthusiasm. This is achieved inter alia through

FIG. 11.12   The non-variegated-leaved form of *Kalanchoe fedtschenkoi* growing in a discarded enamel pot. Photograph: Gideon F. Smith.

repeatedly emphasising the strong architectural outlines and colours of succulents and keeping repetitive designs simple and fitting for the region or location in which the garden is situated.

Design, establishment, development, and maintenance costs of waterwise gardens circumvent the need to annually renew plant material, making such a gardening approach more affordable and the material itself more durable.

Plant material serves many purposes in domestic and public gardens. From a landscaping point of view, two of the most obvious are that plants must either function (1) as an often densely but evenly packed ground cover or as (2) accent plants that draw and focus the visual attention of the observer or visitor. Ground covers essentially serve as a 'canvas' against which other plants, such as architectural, accent specimens or even garden statues and other hardscaping embellishments, can be contrasted. Depending on the exact needs of a gardener or landscape architect, ground covers that work well generally have some, or all, of the following characteristics:

- Spread horizontally or near horizontally
- Grow rapidly
- Are of even height in vegetative and reproductive phases
- Have uniformly coloured leaves
- Can tolerate any or a combination of shade, dappled shade, or direct sun.

FIG. 11.13   A selection of potted *Kalanchoe* specimens (center, left) grown together with other plants in a commercial nursery in Marrakech, Morocco. Photograph: Estrela Figueiredo.

FIG. 11.14 In a large rustic bushveld garden, the look and feel of the developed and natural beds of indigenous grasses and forbs change depending on the time of day. Against this large rock aloes and narrow-leaved grasses are well complemented by the broad-leaved *Kalanchoe paniculata*. Photograph: Gideon F. Smith.

FIG. 11.15 The stark lines of two decommissioned cocopans used as plant containers provide a dramatic counterpoint to the soup plate-sized leaves of *Kalanchoe luciae* in the town of Cullinan in Gauteng, South Africa. Photograph: Gideon F. Smith.

FIG. 11.16   The cultivar *Kalanchoe* 'Fern Leaf', or *K. beharensis* 'Fern Leaf', is a putative hybrid of which the parentage has been suggested as having been *Kalanchoe beharensis* × *K. millotii*. Photograph: Gideon F. Smith.

FIG. 11.18   This form of *Kalanchoe beharensis*, here cultivated in the Shoenberg Temperate House in the Missouri Botanical Garden, St Louis, USA, has copper-coloured adaxial leaf surfaces. This colouration is most obvious on the surfaces of the young, emerging leaves. Photograph: Gideon F. Smith.

FIG. 11.17   Bright splashes of intense colour introduced by *Kalanchoe blossfeldiana* cultivars brighten up a shady spot in the Garden Shop of the Missouri Botanical Garden, St Louis, USA. The light-pink-flowered cultivar in the centre has been given the cultivar name *K. blossfeldiana* 'Selena Salmon Pink'. Photograph: Gideon F. Smith.

FIG. 11.19   *Kalanchoe thyrsiflora* grown in a raised bed in the Garden Block, which is maintained by the Denver Botanical Garden, in downtown Denver, Colorado, USA. Photograph: Gideon F. Smith.

HYBRIDS AND CULTIVARS

In the family Crassulaceae in general, a large number of artificially produced interspecific and even intergeneric hybrids have been recorded ever since these plants reached, especially, plant collectors and garden aficionados in the northern hemisphere (see Figs. 11.21–11.27). This also applies to the genus *Kalanchoe*, with Madagascan species being particularly popular as parents of hybrids from which literally hundreds of cultivars have been selected (Fig. 11.20). In most instances such dedicated breeding, programmes have resulted in the deliberate introduction into horticulture of visually pleasing hybrids and cultivars.

At least two species of southern African *Kalanchoe*, *K. sexangularis* and *K. longiflora*, make excellent ground covers. *Kalanchoe sexangularis* has bright red leaves when grown in exposed positions, while in similar positions, *K. longiflora* has orange-infused, turquoise-coloured leaves. Plants of both species are of low to medium height, especially when grown in direct sunlight, and their flowers are carried on long, erect peduncles. The flowers of these two species are yellow to yellowish green and borne from autumn to spring but peaking in winter.

FIG. 11.20 *Kalanchoe manginii* Raym.-Hamet & H.Perrier from Madagascar has been used as one parent in several *Kalanchoe* hybrids and cultivars, such as *K.* 'Tessa'. In some parts of the world, it is a favourite garden subject among growers of succulents in general and with gardeners who have little more than a passing interest in succulents. It is especially popular for growing in hanging baskets but will also thrive simply as a pot plant if given a well-drained, friable growing medium. Photograph: Gideon F. Smith.

A cultivar results from a deliberate selection, or experimentation, to breed plants with superior and desirable, especially horticultural, characteristics. The word is naturally derived from combining 'culti(vated)' and 'var (iety)'. For example, the cultivar *Kalanchoe* 'Fern Leaf', also referred to as *K. beharensis* 'Fern Leaf', is a putative hybrid of which the parentage has been suggested to be *K. beharensis* and *K. millotii* Raym.-Hamet & H.Perrier. This hybrid was raised by Mrs Hummel of Hummel's Exotic Gardens. A rather similar-looking plant, also raised by Mrs Hummel, is referred to as *Kalanchoe* 'Rose Leaf', or *K. beharensis* 'Rose Leaf', and has the same putative parentage (*K. beharensis* × *K. millotii*). Rather confusingly, the name 'Rose Leaf' has also been used for a hybrid of which *K. beharensis* and *K. pilosa* Baker are the parents. *Kalanchoe* 'Rose Leaf' and *Kalanchoe* 'Fern Leaf' can be easily confused as the differences in the size and shape of the leaves are rather subtle (see http://www.crassulaceae.ch).

Note though that given the plasticity of characters of the Madagascan *Kalanchoe beharensis* (Fig. 11.18), these putative hybrids could simply represent variation found in the species. Unless the suggested crossing (*K. beharensis* × *K. millotii*) is repeated, it will be difficult to determine whether these hybrids are simply variants of *K. beharensis* (M. Bischofberger personal communication).

Perhaps somewhat surprisingly, natural hybrids between southern African species of *Kalanchoe* are not as plentiful as would be predicted by their often overlapping natural geographical distribution ranges. A chance hybrid between the two southern African kalanchoes, *K. luciae* and *K. sexangularis*, that arose in cultivation and that has been named *K.* 'Vivien', is a striking landscape plant (Bischofberger, 2015).

OTHER HORTICULTURALLY IMPORTANT CRASSULACEAE

FIG. 11.21 Several superior horticultural forms, such as this one, known as *Hylotelephium* 'Herbstfreude', have been selected from hybrids between *Hylotelepium spectabile* from Korea and northern China and *H. telephium* (L.) H.Ohba subsp. *maximum* (L.) H.Ohba, which occurs widespread in continental Europe and western Asia. Photograph: Gideon F. Smith.

FIG. 11.22 *Aeonium arboreum* 'Zwartkop' cultivated in the Succulent Garden of the Huntington Botanical Gardens, California, USA. Photograph taken on 27 April 2011 by Gideon F. Smith.

FIG. 11.24 The nothospecies *Petrosedum ×estrelae* Gideon F.Sm. & R.Stephenson, a hybrid between *Petrosedum sediforme* and *Petrosedum forsterianum*, occurs naturally in central Portugal. Photograph: Gideon F. Smith.

FIG. 11.23 *Aeonium* 'Jack Catlin' was selected from hybrids that resulted from *Aeonium tabuliforme* × *A. arboreum* 'Zwartkop'. This hybrid was originally produced in California, USA. Photograph: Gideon F. Smith.

FIG. 11.25 This selection of *Petrosedum rupestre* (L.) P.V.Heath is widely referred to as 'Lemon Coral'. It has bright green, miniature pine needle-like leaves and makes a useful, low-growing groundcover. Photograph: Gideon F. Smith.

FIG. 11.26   This selection of *Phedimus floriferus* (Praeger) 't Hart from northern China is commonly seen labelled with the cultivar name 'Weihen-stephaner Gold'. The species was previously known as *Sedum floriferum* Praeger. Photograph: Gideon F. Smith.

FIG. 11.27   *Sedum ×rubrotinctum* Clausen with its bright red, jelly bean-shaped leaves is one of the most widely grown crassuloid hybrids glob-ally. It has been postulated (see, e.g. Stephenson, 1994: 231) that the parents of this hybrid are *S. pachyphyllum* Rose and *S. stahlii* Solms. Photograph: Gideon F. Smith.

12

Taxonomic Treatment

FIG. 12.1   *Kalanchoe luciae* is a southern African representative of *K.* subg. *Kalanchoe.* Photograph: Gideon F. Smith.

GENUS TREATMENT

***Kalanchoe* Adans.**, *Fam. Pl.* 2: 248 (1763) [as *Kalanchoè*]. De Candolle (1828: 394); Haworth (1829: 301); Harvey (1862: 378–380); Schönland (1891: 23); Hamet (1907: 869); Berger (1930: 402); Raymond-Hamet & Marnier-Lapostolle (1964: 1); Fernandes (1980: 325); Fernandes (1982a: 15); Fernandes (1983: 41); Tölken (1985: 61); Wickens (1987: 30); Boiteau & Allorge-Boiteau (1995: 17); Rauh (1995: 115); Descoings (2003: 143); Descoings (2006: 23); Smith & Figueiredo (2018a: 167). **Type:** *Kalanchoe laciniata* (L.) DC. [=*Cotyledon laciniata* L., *Sp. Pl.*: 430 (1753)].

Synonyms:

Vereia Andrews, *Bot. Repos.* 1: t. 21 (1798). **Type:** *Vereia crenata* Andrews.

Verea Willd., *Sp. Pl.*, ed. 4, 2(1): 471 (1799). Orthographic variant of *Vereia*.

Calanchoe Pers., *Syn. Pl.* 1: 445 (1805). Orthographic variant of *Kalanchoe*.

Bryophyllum Salisb., *Parad. Lond.* t. 3 (1805). **Type:** *Bryophyllum calycinum* Salisb.

Kalenchoe Haw., *Syn. Pl. Succ.*: 109 (1812). Orthographic variant of *Kalanchoe*.

Physocalycium Vest, *Flora* 3: 409 (1820), *nom. illeg.* **Type:** *Vereia crenata* Andrews.

Baumgartenia Trattinnick, *Auswahl Gartenpfl.* 1 109, t. 59 (1821). **Type:** *Bryophyllum calycinum* Salisb. [See Heath, 1997: 129.]

Meristostylus Klotzsch in Peters, *Naturwiss. Reise Mossambique, Bot.*: 267 (1861, '1862'). **Type:** not designated. [Amended to *Meristostylis* in the International Plant Name Index, and The Plant List.]

Geaya Costantin & Poiss. in *Compt. Rend. Hebd. Séances Acad. Sci. Paris* 147: 636 (1908). **Type:** *Geaya purpurea* Costantin & Poiss.

Nomenclatural note:

The designation 'Crassouvia Comm.' was not validly published. Following this 'name' being listed by Hamet (1907), it has been given in the synonymy of *Kalanchoe* by several authors. In the two works usually referenced as validating it (Lamarck, 1786: 141 and De Candolle, 1828: 395), it was listed as a synonym and therefore was not validly published.

Description:

Perennial, multiannual, biennial, rarely with individual plants annual or apparently monocarpic, smooth or tomentose, succulent shrubs, subshrubs, or shrublets, rarely treelike, often sprouting from the base. *Branches* erect to spreading to leaning, succulent, often weak, brittle, sometimes ± woody towards the base. *Leaves* opposite, free or basally slightly fused, fleshy, smooth or tomentose, persistent or deciduous; margins smooth or crenate with large, harmless teeth. *Inflorescence* a thyrse consisting of several dichasia, usually terminating in monochasia, few- to many-flowered, growing point often gradually transitioning into a peduncle with regular basal leaves and similar but much smaller bract-like leaves higher up. *Flowers* erect or spreading (*Kalanchoe* subg. *Kalanchoe*) or pendulous (*Kalanchoe* subg. *Bryophyllum*), 4-merous; *calyx* 4-partite, with sepals variously fused; *corolla* 4-partite, fused into a tube; *tube* usually much longer than the calyx and lobes (*Kalanchoe* subg. *Kalanchoe*) or with calyx very prominent and obscuring much of the tube (*Kalanchoe* subg. *Bryophyllum*); *corolla lobes* erect to variously reflexed; *stamens* 8, in one or ± two whorls; *filaments* glabrous, fused to corolla tube medially or in lower third; *anthers* included or exserted, with terminal ± spherical connective appendage (gland); *carpels* 4, free; *squamae* 4, variously shaped; *ovary* free, apically abruptly constricted; *style* filiform. *Seeds* ellipsoid, very small, variously ridged, grooved, and striated. *Chromosome number*: n = 17 (Raadts, 1985, 1989); n = 17, 18, and 20 (*fide* Tölken, 1985: 61).

Kalanchoe **Adans. subg.** *Kalanchoe*. **Type:** as for *Kalanchoe* Adans (see Figs. 12.1 to 12.6, 12.8, 12.10, and 12.11).

Synonym:

Kalanchoe subg. *Calophygia* Desc. in *J. Bot. Soc. Bot. France* 33: 24 (2006) pro parte. **Type:** *Kalanchoe arborescens* Humbert in *Bull. Mus. Natl. Hist. Nat.*, Ser. II. 5: 163 (1933).

Diagnosis:

Annual, multiannual, or perennial; herbaceous or woody arborescent; terrestrial. Leaf margins usually not bulbiliferous. *Flowers* erect or spreading; *floral tube* usually much longer than the calyx and lobes; *filaments* ± medially inserted in corolla tube; *anthers* included or very slightly exserted.

Kalanchoe **subg.** *Bryophyllum* **(Salisb.) Koorders** in *Bull. Jard. Bot. Buitenz.* 3. 1: 170 (1918). **Type:** as for *Bryophyllum* Salisb (see Figs. 12.7 and 12.11).

FIG. 12.2   In the two identification keys, deliberate use is made of characters that are easy to observe. For example, when studied in isolation, the shape of the flowers of the southern African species with large, soup plate-sized leaves look very similar, and such characters have been avoided in the keys. It is only once flowers of all these species are available for direct comparison that differences become apparent. From left to right: *Kalanchoe winteri*, *K. thyrsiflora*, *K. montana*, *K. luciae*, and an undescribed species. Photograph: Gideon F. Smith.

FIG. 12.3   Once flowers are dissected, further characters, such as those of the gynoecium, become apparent. This series clearly shows the differences in the shape and size of the carpels, stigmas, styles, and nectar scales (yellowish, toothlike structures at the base of the carpels). These characters have been avoided in the identification keys. From left to right: an undescribed species, *Kalanchoe luciae*, *K. montana*, *K. thyrsiflora*, and *K. winteri*. Photograph: Gideon F. Smith.

FIG. 12.4   *Kalanchoe bracteata* Baker is a Madagascan representative of *K.* subg. *Kalanchoe*. Photograph: Gideon F. Smith.

FIG. 12.5   Perhaps rather counter-intuitively, the leaf colour of some forms of *Kalanchoe orgyalis* tends to be reddish brown to coppery-red in dappled shade, while they turn a more greyish-green colour when exposed to direct sunlight. This specimen is offered for sale in the Garden Shop of the Missouri Botanical Garden, St. Louis, USA. Photograph: Gideon F. Smith.

FIG. 12.7   Like most representatives of *Kalanchoe* subg. *Bryophyllum*, *K. daigremontiana* has distinctly pendulous flowers. Photograph: Gideon F. Smith.

FIG. 12.6   *Kalanchoe tomentosa* is a further Madagascan representative of *K.* subg. *Kalanchoe*. This plant is reminiscent of the cultivar known as *K. tomentosa* 'Panda Bear'. The various cultivars of this species are widely cultivated in southern Africa and beyond. Photograph: Gideon F. Smith.

FIG. 12.8   Most representatives of *Kalanchoe* subg. *Kalanchoe*, such as *K. neglecta*, carry their flowers erect to horizontally spreading. Photograph: Gideon F. Smith.

± inserted in lower third of corolla tube; *anthers* ± exserted.

Synonyms:

Bryophyllum Salisb., *Parad. Lond.* t. 3 (1805). **Type:** *Bryophyllum calycinum* Salisb., *Parad. Lond.* t. 3 (1805).

Kalanchoe subg. *Calophygia* Desc. in *J. Bot. Soc. Bot. France* 33: 24 (2006) pro parte, excl. type.

Diagnosis:

Perennial or biennial; herbaceous; terrestrial, rarely epiphytic. Indentations between leaf marginal crenations often bulbiliferous. *Flowers* pendulous; *floral tube* with calyx often prominently obscuring much of tube; *filaments*

***Kalanchoe* subg. *Kitchingia* (Baker) Gideon F.Sm. & Figueiredo** in *Bradleya* 36: 169 (2018a). **Type:** as for *Kitchingia* Baker (see Fig. 12.9).

Synonyms:

Kitchingia Baker in *J. Linn. Soc., Bot.* 18: 268 (1881). **Lectotype:** *Kitchingia gracilipes* Baker in *J. Linn. Soc., Bot.* 18: 268 (1881), designated by Smith & Figueiredo (2018a: 169).

Kalanchoe subg. *Calophygia* Desc. in *J. Bot. Soc. Bot. France* 33: 24 (2006) pro parte, excl. type.

FIG. 12.10   A red-flowered, double-flower selection of *Kalanchoe blossfeldiana*. The various colour variations found in this so-called Rosebud Flaming Katy group are widely grown globally as an indoor plant. Photograph: Gideon F. Smith.

FIG. 12.11   Kalanchoes in outdoor cultivation in summer in the Denver Botanical Garden, Colorado, USA. *Kalanchoe luciae* (large bluish leaves) and *K. orgyalis* (copper-red leaves) were planted in the pot in the centre, while *K. fedtschenkoi* grows on the left of the pot on the right. Photograph: Gideon F. Smith.

FIG. 12.9   In its natural habitat in Madagascar, *Kalanchoe uniflora* Raym.-Hamet, a representative of *K.* subg. *Kitchingia*, is often found growing as an epiphyte on tree trunks. Here, it was photographed in the rainforest at Montagne D'Ambre, in northeastern Madagascar. In cultivation, plants do well when grown in hanging baskets. Photograph: Neil R. Crouch.

Diagnosis:

Perennial; herbaceous; terrestrial or epiphytic. Leaf margins usually not bulbiliferous. *Flowers* pendulous; *floral tube* with calyx not often prominently obscuring much of tube; *filaments* ± inserted in lower third of corolla tube; *anthers* ± exserted.

Of the three subgenera of *Kalanchoe*, only *Kalanchoe* subg. *Kalanchoe* occur in southern Africa, where 17 species have been recorded. Globally, about 150 species are recognised in *Kalanchoe*. The genus occurs naturally from Asia, through Madagascar and some of the Indian Ocean islands, to eastern and central Africa and southern Africa. In the subcontinent, kalanchoes predominantly occur in the summer-rainfall region.

Common names:

Afrikaans: kalanchoe and viltbosse.

English: felt bushes, kalanchoe, and kalenshoe (Adendorff, n.d.).

KEY TO KALANCHOES INDIGENOUS TO SOUTHERN AFRICA

Many southern African species of *Kalanchoe* are remarkably stable in the expression of their characters, making them comparatively easy to characterise and identify. Examples include the red-leaved *Kalanchoe sexangularis* N.E.Br. (Figueiredo et al., 2016), pyramid- to Christmas tree-shaped *K. brachyloba* Welw. ex

Britten (Smith & Figueiredo, 2017a), and turquoise-leaved *K. longiflora* Schltr. ex J.M. Wood (Smith & Figueiredo, 2017b).

However, in some species, vegetative and reproductive variation found across their natural geographical distribution ranges makes it challenging to construct a fool proof dichotomous key to facilitate their identification; *Kalanchoe rotundifolia* is an example of such a species (Smith & Figueiredo, 2017e).

In the key provided below, deliberate use is made of vegetative and reproductive characters that are easy to observe.

1a. Inflorescence dense, more or less cylindrical, and peduncle stout...2
1b. Inflorescence more or less open, rather diffuse; peduncle flimsy 6

2a. Flowers red, at least when exposed to the sun .... 4. *K. crundallii*
2b. Flowers white, pinkish-infused, greenish, or yellow 3

3a. Corolla lobes whitish, pale pinkish-infused, pale yellowish green, and rarely yellow....................................4
3b. Corolla lobes bright golden yellow..5

4a. Plants glabrous and calyx lobes 2.5–4.0 mm long...10. *K. luciae*
4b. Plants hairy, sometimes glabrous; calyx lobes 5–7 mm long ....11. *K. montana*

5a. Corolla lobes squarish, to 3 mm long; corolla tube cylindrical, hardly contracted below the mouth; four anthers exserted .... 16. *K. thyrsiflora*
5b. Corolla lobes triangular, >6 mm long; corolla tube ellipsoid (cigar-shaped), contracted below the mouth; eight anthers exserted... 17. *K. winteri*

6a. Leaves turquoise-orange ....9. *K. longiflora*
6b. Leaves green, red, or glaucous 7

7a. Corolla very densely tomentose with long, conspicuous hairs ....7. *K. lanceolata*
7b. Corolla glabrous or tomentose with short hairs 8

8a. Leaf blade deeply lobed, almost appearing once-compound .... 6. *K. laciniata*
8b. Leaf blade expanded, flat or recurved; margin entire; crenate; dentate; or, if lobed, never deeply so...................9

9a. Leaf blade basally cordate or peltate ....12. *K. neglecta*
9b. Leaf blade basally cuneate..10

10a. Leaf margins dentate, leaf blades usually strongly recurved, and peduncle to 1.6 m tall..................8. *K. leblanciae*
10b. Leaf margins entire or crenate/lobed, leaf blades usually ± horizontal to spreading, sometimes weakly recurved, and peduncle to 1.5 m tall..11

11a. Leaves dark green, infused with red or bright crimson red ....15. *K. sexangularis*
11b. Leaves light to dull green to glaucous, never red 12

12a. Leaf blades to 150 mm broad and leaf margins always entire...13. *K. paniculata*
12b. Leaf blades to 130 mm broad; leaf margins dentate, crenate, or entire...13

13a. Plants conspicuously tomentose with short hairs..5. *K. hirta*
13b. Plants completely glabrous or indistinctly hairy..14

14a. Flowers with corolla conspicuously twisted postanthesis.. 14. *K. rotundifolia*
14b. Flowers with corolla remaining straight or only weakly twisted postanthesis 15

15a. Leaf blade oblong-elliptical, to 280 mm long, usually strongly recurved; leaf surface glabrous, although with a somewhat velvety feel ....2. *K. brachyloba*
15b. Leaf blade oblanceolate to ovate to oblong, to 130 mm long, straight, and flat; leaf surface glabrous or sometimes somewhat short hairy...16

16a. Leaf margins entire, paler than blade or somewhat red-infused; leaves ± sessile; leaf surface dull and glabrous ....1. *K. alticola*
16b. Leaf margins dentate or crenate, same colour as blade or sometimes red-infused; leaves petiolate; leaf surface shiny, glabrous, or indistinctly hairy .... 3. *K. crenata*

KEY TO KALANCHOES VARIOUSLY NATURALISED IN SOUTHERN AFRICA

In general, all the species of *Kalanchoe* that are varyingly naturalised in southern Africa are easy to distinguish from the species of this genus that are indigenous to the subcontinent. All the naturalised species originate from Madagascar. With the exception of the arborescent *K. beharensis*, which belong to *K.* subg. *Kalanchoe*, all the species are shrubby and included in *K.* subg. *Bryophyllum*.

In the key provided below, deliberate use is made of vegetative and reproductive characters that are easy to observe.

1a. Plants arborescent and leaves densely pubescent.. 18. *K. beharensis*
1b. Plants low-growing to medium-sized shrubs and leaves glabrous 2

2a. Leaves abaxially distinctly mottled with purplish spots against a monochromatic, usually light green to yellowish green background 3
2b. Leaves abaxially monochromatic, lacking purple spots 4

3a. Calyx yellowish pruinose to greenish white pruinose and calyx tube 20–25 mm long ....21. *K. gastonis-bonnieri*
3b. Calyx shiny light green, strongly purple-infused; calyx tube 4–6 mm long ....19. *K. daigremontiana*

4a. Leaf blades narrowly cylindrical ....24. *K. tubiflora*
4b. Leaf blades expanded and flat..5

5a. Calyx tube box-shaped, distinctly 4-angled; sepals light green..23. *K. prolifera*
5b. Calyx tube cylindrical, not distinctly 4-angled tubular; sepals pale yellow to green or light greenish purple......6

6a. Leaves simple, glaucous to purplish to brownish green, and distinctly waxy; inflorescence erect, to 20 cm tall..20. *K. fedtschenkoi*
6b. Leaves pinnate, first leaves sometimes simple, midgreen, sometimes with purplish lines, and somewhat shiny; inflorescence erect to leaning, to 80 cm tall ....22. *K. pinnata*

SPECIES TREATMENTS

Indigenous species

1. *Kalanchoe alticola* Compton (Coin-leaved kalanchoe)
2. *Kalanchoe brachyloba* Welw. ex Britten (Velvet kalanchoe)
3. *Kalanchoe crenata* (Andrews) Haw. (Crowned kalanchoe)
4. *Kalanchoe crundallii* I.Verd. (Crundall's kalanchoe)
5. *Kalanchoe hirta* Harv. (Bush kalanchoe)
6. *Kalanchoe laciniata* (L.) DC. (Christmas tree kalanchoe)
7. *Kalanchoe lanceolata* (Forssk.) Pers. (Narrow-leaved kalanchoe)
8. *Kalanchoe leblanciae* Raym.-Hamet (Sand forest kalanchoe)
9. *Kalanchoe longiflora* Schltr. ex J.M.Wood (Turquoise kalanchoe)
10. *Kalanchoe luciae* Raym.-Hamet (Flipping flapjacks)
11. *Kalanchoe montana* Compton (Mountain kalanchoe)
12. *Kalanchoe neglecta* Toelken (Umbrella kalanchoe)
13. *Kalanchoe paniculata* Harv. (Rabbit's ear kalanchoe)
14. *Kalanchoe rotundifolia* (Haw.) Haw. (Common kalanchoe)
15. *Kalanchoe sexangularis* N.E.Br. (Red-leaved kalanchoe)
16. *Kalanchoe thyrsiflora* Harv. (White lady)
17. *Kalanchoe winteri* Gideon F.Sm., N.R.Crouch & Mich. Walters (Wolkberg kalanchoe)

Naturalised species and garden escapes

18. *Kalanchoe beharensis* Drake (Donkey's ear)
19. *Kalanchoe daigremontiana* Raym.-Hamet & H.Perrier (Mother of thousands)
20. *Kalanchoe fedtschenkoi* Raym.-Hamet & H.Perrier (Purple scallops)
21. *Kalanchoe gastonis-bonnieri* Raym.-Hamet & H.Perrier (Life plant)
22. *Kalanchoe pinnata* (Lam.) Pers. (Cathedral bells)
23. *Kalanchoe prolifera* (Bowie ex Hook.) Raym.-Hamet (Blooming boxes)
24. *Kalanchoe tubiflora* (Harv.) Raym.-Hamet (Chandelier plant)

Exotic *Kalanchoe* species of horticultural importance

25. *Kalanchoe blossfeldiana* Poelln. (Flaming Katy)
26. *Kalanchoe humilis* Britten (Zebra kalanchoe)
27. *Kalanchoe* 'Margrit's Magic' (Red chandelier plant)
28. *Kalanchoe porphyrocalyx* (Baker) Baill. (Red-cup kalanchoe)
29. *Kalanchoe pumila* Baker (Flower dust kalanchoe)
30. *Kalanchoe tomentosa* Baker (Panda plant)

1. KALANCHOE ALTICOLA COMPTON

Coin-leaved kalanchoe (Figs. 12.1.1 to 12.1.12)

FIG. 12.1.1 *Kalanchoe alticola* growing in rock rubble and plant debris at the foot of a hill in the Mlawula Nature Reserve in eastern Swaziland. Photograph: Gideon F. Smith.

Kalanchoe alticola **Compton** in *J. S. Afr. Bot.* 41: 47 (1975). Compton (1976: 219); Descoings (2003: 145); Smith et al. (2017b: 306). **Type**: Swaziland, Mukusini Hills, *c.* 1300 m above sea level, 2631 (Mbabane) (–AB), 29 May 1964, *Compton 32107* (NBG NBG0076374-0 holo-; K K000232855, PRE PRE0390372-0, PRE0523573-0 iso-).

Derivation of the scientific name:
From the Latin words 'altus' (high) and 'cola' (inhabiting), in reference to the occurrence of the species at higher altitudes in eastern southern Africa.

Description:
Perennial, few- to many-leaved, multi-branched, glabrous throughout, small, haphazardly branched succulent, and 0.1–0.25 m tall. *Stems* thin, usually leaning and tangled, light green to distinctively pinkish green, several arising from a shallow rootstock near ground level, and round in cross-section. *Leaves* opposite-decussate, leaf pairs widely spaced, hardly petiolate, light bluish green to glaucous, shiny but with a dull powdery bloom, succulent, and erect to slightly spreading; *petiole* usually absent, leaves clasping the stem, and almost linking with the leaf opposite; *blade* 2–4 × 0.9–1.5 cm, ovate-oblong to oblong, and flat; *base* cuneate; *apex* rounded, sometimes mucronate; *margins* somewhat bony and entire. *Inflorescence* erect, apically sparsely or closely packed, 5 cm in diameter, few-flowered, all parts with a distinctive waxy bloom, and rounded to flat-topped thyrse with several dichasia; *pedicels* 2–3 mm long and robust. *Flowers* erect, light yellow to sulphur yellow to light orange; *calyx* bluish to purplish green; *sepals* ± 5 mm long, elongated-lanceolate, ± separate, basally fused for ± 0.5 mm, acute, clasping the corolla tube lower down, succulent, and hardly

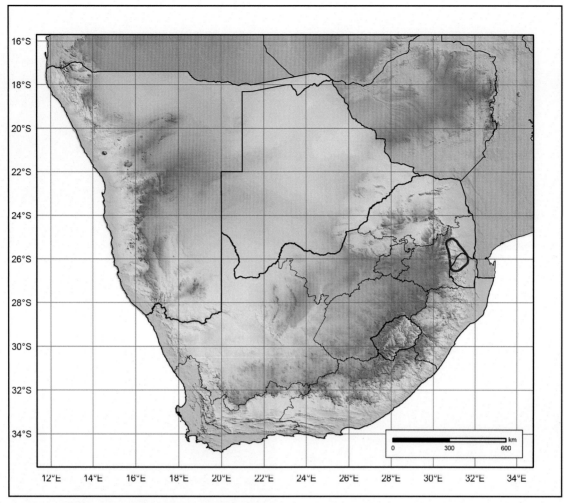

FIG. 12.1.2   Known geographical distribution range of *Kalanchoe alticola* in southern Africa.

FIG. 12.1.3   A form of *Kalanchoe alticola* with red leaf margins. Photograph: Gideon F. Smith.

FIG. 12.1.4   *Kalanchoe alticola* growing socially with *Urginavea epigea*, an important medicinal plant, in eastern southern Africa. Photograph: Gideon F. Smith.

FIG. 12.1.5   This form of *Kalanchoe alticola* has glaucous leaves. Note the dull powdery bloom on both leaf surfaces. Photograph: Gideon F. Smith.

FIG. 12.1.6   A form of *Kalanchoe alticola* with light green leaves. Photograph: Gideon F. Smith.

FIG. 12.1.7   A typical rocky habitat of *Kalanchoe alticola* in bushveld vegetation in eastern Swaziland. Photograph: Gideon F. Smith.

FIG. 12.1.8   The leaf pairs of *Kalanchoe alticola* are widely spaced and basally distinctly cuneate. Photograph: Gideon F. Smith.

FIG. 12.1.9   The margins of the hardly petiolate leaves of *Kalanchoe alticola* are conspicuous and have a somewhat bony appearance. Photograph: Gideon F. Smith.

FIG. 12.1.10   The inflorescences of *Kalanchoe alticola* are generally rather sparsely flowered. Photograph: Gideon F. Smith.

contrasting against yellowish green corolla tube; *corolla* 9–13 mm long, somewhat inflated and enlarged lower down, twisted apically after anthesis, light green below, and light yellow to sulphur yellow to light orange above; *tube* 8–12 mm long, very slightly 4-angled, rounded below and above, light green below, and turning light

FIG. 12.1.11   The corolla tube of *Kalanchoe alticola* is basally somewhat inflated and enlarged, but not to the same extent as those of *K. rotundifolia*. The tube is light green below, light yellow to sulphur yellow to light orange above, and twisted apically after anthesis. Photograph: Gideon F. Smith.

FIG. 12.1.12   Red-flowered form of *Kalanchoe alticola* from Swaziland. Photograph: Gideon F. Smith.

yellow to light orange above; *lobes* 5–6 × 2–4 mm, lanceolate, spreading to recurved, lanceolate, tapering to an acute at apex, light yellow to sulphur yellow to light orange, and sometimes diurnal. *Stamens* well included; *filaments* 3–4 mm long, flimsy, and inserted in two ranks at and above middle of corolla; *anthers* 0.6–0.8 mm long and visible at the mouth of the corolla tube. *Pistil* consisting of 4 carpels, 4–5 mm long; *scales* 3–4 mm long, linear. *Follicles* not seen. *Seeds* not seen. *Chromosome number*: unknown.

Flowering time:

May–September (southern hemisphere).

Illustrations:

Smith et al. (2017b: 306, bottom). Note that the illustration in Onderstall (1996: 81) likely is of *Kalanchoe sexangularis* and does not represent *K. alticola* as presently circumscribed.

Common names:

Afrikaans: Penniekalanchoe.

English: Coin kalanchoe.

Geographical distribution:

The species occurs not only in the eastern Highveld of southern Africa, in the mountainous grasslands of northwestern Swaziland, and the adjacent Mpumalanga province of South Africa, especially around Barberton, but also slightly further north, towards Mbombela (Fig. 12.1.2).

Distribution by country. South Africa (Mpumalanga) and (northwestern) Swaziland.

Habitat:

Kalanchoe alticola has been recorded from shallow peaty soils on granite rocks and in rock crevices (Compton, 1975: 48; Descoings, 2003: 145; Smith et al., 2017b: 306).

Conservation status:

Not threatened.

Additional notes and discussion:

Taxonomic history and nomenclature. Kalanchoe alticola is one of a set of species that Compton collected while he resided in Swaziland after his retirement from Kirstenbosch, one of the National Botanical Gardens (NBGs) maintained as a network of gardens by the present-day South African National Biodiversity Institute. In his retirement, Compton was funded to produce a Flora of Swaziland and within a period of about 20 years saw the publication of this Flora. Compton, along with some of his colleagues in Swaziland, was a prolific collector and additionally had more than a passing interest in *Kalanchoe*, as evidenced by his description of a number of new species from the small mountainous Kingdom. In addition to *K. alticola*, he also described *K. decumbens*, which is now treated as a synonym of *K. rotundifolia* (Smith and Figueiredo, 2017e), and *K. montana*, which was recently reinstated at species rank (Smith et al., 2016a), after it was treated as a subspecies of *K. luciae* for about 40 years following its description.

Identity and close allies. Kalanchoe alticola is a small-growing species, with elongated, somewhat coin-shaped leaves, hence its Afrikaans and English common names. The main characters that differentiate it from *Kalanchoe rotundifolia*, its closest relative, are the long, thick, prominent sepals and the long, narrow, lanceolate corolla lobes.

Cultivation. Kalanchoe alticola is easy to propagate from stem cuttings. Such cuttings can be taken at any time of year and will rapidly root in open beds or in containers. The species is easy in cultivation and will grow readily in any soil-based growing medium, even when it is rather clayey.

2. KALANCHOE BRACHYLOBA WELW. EX BRITTEN

Velvet kalanchoe (Figs. 12.2.1 to 12.2.7)

FIG. 12.2.1 A dense stand of *Kalanchoe brachyloba* growing in the shade of *Acacia nilotica* (L.) Willd. ex Delile subsp. *kraussiana* (Benth.) Brenan, a typical bushveld tree. Photograph: Gideon F. Smith.

Kalanchoe brachyloba **Welw. ex Britten** in Oliver (ed.), *Fl. Trop. Afr.* 2: 392 (1871). Hiern (1896: 326); Hamet (1907: 896); Friedrich (1968: 37); Jacobsen (1970: 283); Fernandes (1980: 329); Fernandes (1982a: 27); Fernandes (1983: 63); Tölken (1985: 67); Jacobsen (1986: 610); Retief & Herman (1997: 392); Descoings (2003: 149); Lebrun & Stork (2003: 212, 215); Court (2010: 91); Smith & Figueiredo (2017a: 2). **Type**: Lower Guinea [West Africa], [Angola], Benguela, Huilla [probably Huíla Province], *Welwitsch 2486* (LISU LISU218888, lecto-). *Welwitsch 2486* designated by Fernandes, 1980: 329 (as "holotype"), specimen LISU218888 designated by Smith & Figueiredo (2017a: 8) in a second-step lectotypification.

Synonyms:

Kalanchoe multiflora Schinz in *Verh. Bot. Vereins Prov. Brandenburg* 30: 172 (1888). Brown (1909: 110); Dinter (1922: 434). **Type**: Botswana, southwest of Lake Ngami, *Schinz 177* (Z Z-000004323 & Z-000004324, K K000232900).

Kalanchoe paniculata sensu Baker f. in *J. Bot. (Lond.)* 37: 434 (1899). Engler (1906: 878); Burtt Davy (1926: 144) pro parte; Raymond-Hamet (1963: t. 94–95), non Harvey (1862: 380).

Kalanchoe baumii Engl. & Gilg in Warburg, *Kunene-Sambesi-Exped.*: 242 (1903). Hamet (1907: 895) pro parte; Hamet (1908b: 255); Hamet (1910a: 21); Jacobsen (1970: 282); Jacobsen (1986: 606). **Type**: [Angola], Goudkopje [English: Gold Hillock], 22 May 1900, 1300 m, *Baum 938* (B†, holo-).

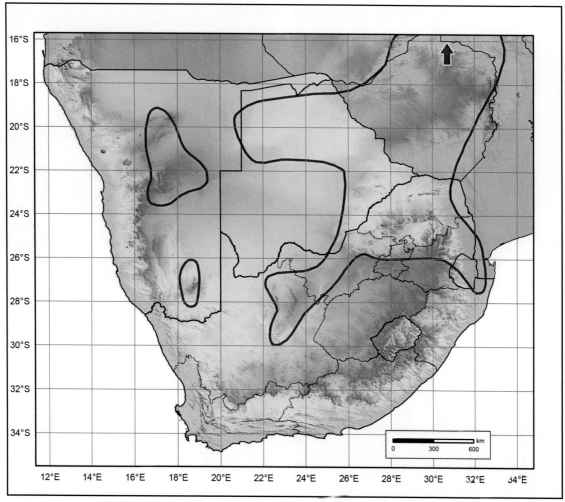

FIG. 12.2.2   Known geographical distribution range of *Kalanchoe brachyloba* in southern and south-tropical Africa.

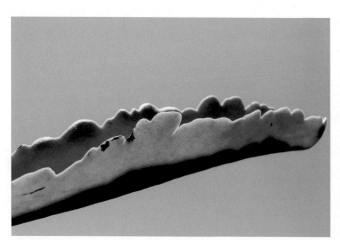

FIG. 12.2.3   The velvety, oblong-lanceolate leaves of *Kalanchoe brachyloba* are longitudinally folded upward along the midrib. The leaf margins are light greenish white-edged and have conspicuous crenations. Photograph: Gideon F. Smith.

FIG. 12.2.4   *Kalanchoe brachyloba* has uniformly light bluish green stems and leaves. Its leaves are sessile and often drooping when plants approach maturity. Photograph: Gideon F. Smith.

FIG. 12.2.7 *Kalanchoe brachyloba* (background) and *K. rotundifolia* (foreground) sometimes grow sympatrically in the shade of bushveld trees in gritty, sandy soils, as here in South Africa's North-West province. Photograph: Gideon F. Smith.

FIG. 12.2.5 The inflorescence branches of *Kalanchoe brachyloba* are often very densely covered in a thick, white, waxy substance. Photograph: Gideon F. Smith.

FIG. 12.2.6 This inflorescence of *Kalanchoe brachyloba* carries flowers at all stages of development, from buds to wilted flowers; in most of them, the corolla lobes remain connivent. Photograph: Gideon F. Smith.

Kalanchoe pyramidalis Schönland in *Rec. Albany Mus.* 2: 154 (1907). **Type:** Material originally collected in September 1903 at Serowe, Botswana, in the "N.E. Kalahari" fide Schonland (1907), cultivated and flowered in Grahamstown in May 1904 and in 1905, *Schonland s.n.* (GRA GRA0001094-0 holo-; K K000232901, iso-).

Kalanchoe pruinosa Dinter in *Repert. Spec. Nov. Regni Veg.* 18: 434 (1922). Dinter (1923a: 147). **Type:** [Namibia/South West Africa], Hereroland, Auasberge (Auas Mountains), "in Lichtenstein blühend [flowered on the farm 'Lichtenstein' (south of Windhoek, the capital of Namibia)], 9 February 1923, auch im nördlichen Namaland häufig [also common in northern Namaland]", [M.] *K. Dinter 4432* (B†).

'Kalanchoe praesidentis-vervoerdii Raym.-Hamet' (Raymond-Hamet, 1963: 81–83), *nomen nudum*. [Epithet spelled "*praesidentis-verwoerdii*" in Lebrun & Stork, *Trop. Afr. Flow. Pl. Ecol. Distrib.* 1: 212 (2003)].

Derivation of the scientific name:
A combination of the Greek words *brachys* (= short) + *-lobos* (= lobe), in reference to the length of the corolla lobes. The lobes are not particularly short though.

Description:
Perennial or biennial; few-leaved; solitary or, rarely, sparsely branched; glabrous but with a velvety feel to the surface; robust; subrosulate; succulent; and 0.6–2.0 m tall. *Stems* light bluish green, usually one arising from a rosette, unbranched or rarely once-branched, erect to slightly leaning, terete, and not angled. *Leaves* opposite-decussate, subrosulate, sessile, the lowest the largest, the uppermost becoming smaller and bract-like, mostly uniformly greyish green to dull light green, rarely midgreen, succulent throughout, straight when young, becoming gracefully arched to recurved, lower, oldest ones becoming variously floppy, velvety smooth, drying papery chartaceous, and sometimes caducous; *petiole* absent; *blade* 10–20 × 4–7 cm, narrowly oblong-lanceolate, usually distinctly folded lengthwise along the midrib, and cuneate; *base* subpetiolar, tapering towards midblade, mostly clasping the stem, and sometimes slightly

auriculate basally; *apex* rounded-obtuse, often curved downwards; *margins* light greenish white, coarsely and irregularly serrate-lobed to crenate or undulate-crenate into irregular, rounded, harmless crenations, rarely almost smooth, and sometimes lightly reddish-tinted. *Inflorescence* 30–150 cm tall; erect to slightly leaning; apically subdense; few- to many-flowered; flat-topped thyrse with several subdense dichasia; round to ellipsoid in outline when viewed from above; branches opposite; projecting erectly away from main axis at 45 degrees; subtended by small leaflike bracts; leafy branchlets in axils mostly lacking; axis light green or covered in a dense, white, waxy substance that obscures the colour; *pedicels* 5–15 mm long; and slender. *Flowers* erect, greenish yellow (tube) and light yellow (lobes); *calyx* succulent, light bluish green; *sepals* 2–5 × 1.5–3.0 mm, lanceolate to triangular-lanceolate, basally fused, acute, and distinctly contrasting against corolla tube; *corolla* 13–17 mm long, somewhat enlarged lower down, often twisted apically at and after anthesis, and yellowish green (tube) and light yellow (lobes); *tube* 10–15 mm long, quadrangular-cylindrical, distinctly 4-angled, obtusely rounded at bottom, box-shaped-square when viewed from below, longitudinally fluted above, yellowish green, and becoming creamy white to reddish brown and scarious when dry; *lobes* 2–5 × 2–4 mm, ovate to subcircular, acute, apiculate, usually connivent, slightly twisted with upper part of tube, becoming slightly spreading depending on time of day, light yellow, and faintly white-margined. *Stamens* appearing almost sessile, inserted in 2 obscure ranks ± above middle of corolla tube, and often included because of lobe connivence; *filaments* short, 1.0–1.5 mm long; *anthers* 0.5–1.0 mm long. *Pistil* consisting of 4 carpels; *carpels* 8–12 mm long; *styles* 0.5–1.5 mm long; *scales* up to 4 mm long, linear-lanceolate. *Follicles* 6–7 mm long, brittle, grass spikelet-like, enveloped in dry greyish black remains of corolla, and dark brownish black. *Seeds* 0.75–1.00 mm long, rectangular to slightly banana-shape curved, tapering at both ends, and dark brown to black. *Chromosome number*: unknown.

Flowering time:
February–May, peaking in March and April.

Illustrations:
Tölken (1985: 68, figure 8.2). Note that in Pooley (1998: 254, 255), an image of *Kalanchoe leblanciae* is mistakenly labelled as *K. brachyloba* (see Crouch et al., 2016a: 73 and Smith et al., 2017a, b).

Common names:
Afrikaans: krimpsiektebos, plakkiesblom, and plakkie (leaves), all recorded under the name *Kalanchoe pyramidalis*, a synonym of *K. brachyloba* (see Smith, 1966: 565 and Barkhuizen, 1978: 80).
English: velvet kalanchoe.
Tshivenḓa: tshinyanyu (Mabogo, 1990).

Notes. In Angola, the species is known as kombaluva by the two tribes Quilengues Musos and Quilengues Humbes (Figueiredo & Smith, 2012, 2017a).

This is one of the only species of *Kalanchoe* that has a distribution range that approaches the Karoo proper in the arid central South Africa; however, Powrie (2004) did not record any common names for the species in that region of South Africa.

In Zimbabwe, *Kalanchoe* species in general, including *Kalanchoe brachyloba*, are known as chifumuro, chikohwa, and chikowa in Shona and as Intelezi in Ndebele. In that country, no common names have been recorded for *K. brachyloba* in Tonga (Kwembeya & Takawira, n.d.; Kwembeya & Takawira, 2002). Wild et al. (1972) also did not record common names for *K. brachyloba* in Zimbabwe.

Geographical distribution:
Kalanchoe brachyloba is a very common species in the western, central, northern, and eastern parts of southern Africa, as well as further north towards central Africa. It has a very broad geographical distribution range south of the equator that stretches across the African continent in a broad, west-east direction, in southern, south-tropical, and parts of tropical Africa. The southern limit of the species is a southward arching line that connects south-central Namibia in the west to southern Mozambique in the east. Summer rainfall predominates throughout the distribution range of *K. brachyloba*. The species is noticeably absent from the western, Mediterranean parts of the southern African subcontinent. Remarkably, in central South Africa, it has been recorded from as far afield as the dry bushveld just north of the arid, karroid Northern Cape Province and from northeastern KwaZulu-Natal in the east of the country, just southeast of Swaziland and Mozambique, which has a subtropical climate. Few succulents display such a wide ecological amplitude.

Unlike several other southern African *Kalanchoe* species that have distribution ranges restricted to Centres of Endemism formally recognised in the region (Van Wyk & Smith, 2001), *K. brachyloba* is therefore a generalist with a very broad range.

North of southern Africa, *Kalanchoe brachyloba* occurs in Angola and throughout all the *Flora zambesiaca* countries [Botswana, Zambia, Zimbabwe, Malawi, and Mozambique] (Fernandes, 1983) and slightly further northwards into central Africa. Huíla in southern Angola, whence the type material of *Kalanchoe brachyloba* originates, is a well-known collecting locality that also yielded the type material of several other succulents.

Distribution by country. Angola, Botswana, Democratic Republic of Congo [formerly Belgian Congo; Zaire], Malawi, Mozambique (southern and central), Namibia, South Africa (Northern Cape, North-West, Gauteng,

Limpopo, Mpumalanga, and northeastern KwaZulu-Natal), Swaziland likely in southern Tanzania (*fide* Wickens, 1987: 38), Zambia, and Zimbabwe.

Habitat:

Given the extensive geographical distribution range across which *Kalanchoe brachyloba* has been recorded, it occurs in a large variety of habitats, including marginally karroid; subtropical; grassland; and, most commonly, savannah habitats. It often grows within the drip line of the canopies of bushveld trees with low-bending branches. At times of drought and towards the end of the rainy (summer) season, when plants come into flower, specimens could be very flimsy and slender with apparently wilted leaves but will still produce inflorescences.

Conservation status:

In South Africa, the typical variety is regarded as 'Least concern' and not threatened with extinction.

Additional notes and discussion:

Taxonomic history and nomenclature. Kalanchoe brachyloba was first described in the *Flora of Tropical Africa* by Britten (1871: 392) from material collected at "Benguela and Huilla (Huíla)" in Angola. The *Flora of Tropical Africa* was a serial publication that appeared over a period of 34 years and covered the countries south of the equator, up to, and including, Angola in the west and Mozambique in the east. For Angola, this broad floristic work has been surpassed by the country level but incomplete, *Conspectus florae angolensis*.

The original specimens on which the name *K. brachyloba* is based were collected by Friedrich Welwitsch (1806–72), with Britten using the manuscript name that had been suggested by Welwitsch, indicated as "*Welw. mss*" after "5. **K. brachyloba**" (Britten, 1871: 392). Welwitsch was the most prolific collector of botanical artefacts in Angola (Figueiredo & Smith, 2008: 2) and is credited with having discovered about 1000 species new to science. Benguela and Huíla, cited by Britten (1871), are today provinces of Angola. Even though both were cited in the protologue of *K. brachyloba* (Britten 1871), Hiern (1896: 372) in his *Catalogue of the African plants collected by Dr Friedrich Welwitsch* refers only to Huíla, which is indeed the locality that appears on the specimen labels of the type material of *K. brachyloba*.

Almost 20 years after *Kalanchoe brachyloba* was described, Hans Schinz (1888) redescribed the species as *K. multiflora* Schinz, so creating the first synonym of the name *K. brachyloba*. The type material of Schinz's name was collected at Lake Ngami in north-central Botswana, a considerable distance southeast of Huíla, Angola, the origin of the type specimen of the name *K. brachyloba*. It is perhaps not surprising that Schinz published this synonym, as, in the late 19th century, there

seems to have been little regard for the possibility of a single species having a distribution range that can, in some cases, stretch over several thousand kilometres. Schinz (1888) distinguished his species, *K. multiflora*, on the longer, dull-coloured calyx segments and linear scales (also referred to as squamae or glands) that are three times longer than those of *K. brachyloba* from Huíla.

A further 10 years later, confusion first arose between *Kalanchoe brachyloba* and *K. paniculata*. Baker (1899) treated what was clearly *K. brachyloba* under the name *K. paniculata*, and his interpretation was perpetuated in several works on the botany of South Africa, including in Burtt Davy (1926; but also see the earlier Burtt Davy & Pott-Leendertz 1912, and Pott 1920), which appeared over the ensuing more or less 30 years.

In the meantime, a further synonym of *Kalanchoe brachyloba*, *K. baumii*, was described by Engler & Gilg in Warburg (1903). The material was collected by Hugo Baum (1867–1950) during an expedition to Angola, led by a Dutchman, Pieter van der Kellen, that took place from 1899 to 1900 (Figueiredo et al., 2009). The expedition left Moçâmedes (Namibe) on 11 August 1899 in an easterly direction, through the provinces of Cunene and Cuando Cubango, ultimately reaching the Cuando River in March 1900. After turning around, the expedition finally arrived back at Moçâmedes on 26 June 1900, more than 10 months later. In terms of type material, the Baum collection of over 1000 botanical specimens is one of the most important among those that originated from Angola and is only surpassed by that of Welwitsch (Albuquerque et al., 2009) and John Gossweiler (Figueiredo & Smith, 2012). "Goudkopje" [English: Gold Hillock], the type locality of *Kalanchoe baumii*, is a toponym no longer in extant use for a locality near Cassinga, which is *c.* 400 km inland from the southwestern Angolan coastal town of Namibe.

In 1907, four years after *Kalanchoe baumii* was described by Engler & Gilg, Selmar Schonland (15 August 1860–22 April 1940) described *K. pyramidalis* (Schonland, 1907). Schonland had the plant on which he based the name *K. pyramidalis* in cultivation in Grahamstown in the Eastern Cape Province in South Africa, where he was at the time the first professor of botany at the Rhodes University College [now Rhodes University], and Director of the Albany Museum (Gunn & Codd, 1981: 318). The material was originally collected at Serowe in east-central Botswana and apparently later preserved as the type specimen, although the protologue does not mention any specimens. Schonland (1907) noted that the material that flowered in Grahamstown did not survive, clearly indicating that it was monocarpic and, likely, biennial, as is the case with a number of *Kalanchoe* species. The large leaves (up to 35–45 cm long according to

Schonland, 1907) and pyramidal appearance of the plant when coming into flower (largest leaves low-down, becoming smaller upwards) prompted him to regard it as a distinct taxon. However, Fernandes (1980) synonymised the name with *K. brachyloba*, a view with which we agree.

The final synonym of *Kalanchoe brachyloba* that was validly published, *K. pruinosa* Dinter, appeared in print in the early 1920s (Dinter, 1922), 15 years after Schonland established his *K. pyramidalis*. A year later, in a more comprehensive treatment of the species, Dinter (1923a) expanded on his rather brief earlier establishment of the name. Dinter was a remarkable collector of botanical and other specimens and contributed extensively to the present-day knowledge of the flora of Namibia (Gunn & Codd, 1981: 131–135; Figueiredo et al., 2013). In his 1923 publication, Dinter admitted that his species, material of which apparently became known in 1917, was in all respects very similar to *K. multiflora*, a name also regarded as belonging in the synonymy of *K. brachyloba* (see above), except in the stem-clasping nature of the leaves that are shaped like arrowheads. It is interesting that material on which he based the name *K. pruinosa* became known in 1917 (Dinter, 1923a, b), the penultimate year of WWI, as Dinter departed from Namibia for Germany in March 1914, four months before WWI hostilities started, and only returned to Namibia in 1922. Although the material on which *K. pruinosa* was based was collected in the Auas Mountains, Omaheke region, of east-central Namibia, material flowered, probably in cultivation, on the farm 'Lichtenstein', just south of Windhoek. Dinter also noted that his *K. pruinosa* was to be found in Hereroland that is further northeast from the Auas Mountains along the Namibia-Botswana border and in Namaland, more or less halfway between Windhoek and Namibia's southern border with South Africa.

'Kalanchoe praesidentis-vervoerdii', a *nomen nudum*, is the designation given by Raymond-Hamet (1963) to a plant illustrated but not described in Fascicle 5 of his *Crassulacearum icones selectae*. The illustrations (photographs and drawings) of this entity refer to a collection from Zimbabwe (Matabele County, Shasha River, *Holub s.n.* [K K000232902]). Fernandes (1980) noted that the inflorescences of the specimen belong to *K. brachyloba* but that the leaves (which are detached) do not appear to belong to any species of *Kalanchoe*.

Interestingly, the type specimens of all the synonyms *of Kalanchoe brachyloba* were based on material collected in the rather arid western or central savannah, or grassland, regions of southern Africa. Even though botanical collecting activities in the eastern parts of the subcontinent and further north in south-tropical Africa had already

commenced in earnest at that time, material of *K. brachyloba* from the more tropical climes was not at any stage regarded as warranting description as new.

Identity and close allies. *Kalanchoe brachyloba* has in the past been confused (see Baker, 1899) with *K. paniculata* from which it differs in several respects. Although having affinities with *K. paniculata*, *K. brachyloba* is a distinctive species characterised by its sessile, arched, oblong-lanceolate leaves that are lengthwise-folded upwards along the midrib and have a velvety feel to the surface (hence the common name 'velvet kalanchoe'). The leaf margins are coarsely and irregularly serrate-lobed to crenate, very rarely entire as in *K. paniculata*, which has flat to variously floppily folded, subcircular to broadly ovate leaves. The stems of *K. brachyloba* are uniformly light bluish green, while those of *K. paniculata* are usually strongly red-infused. In the case of *K. paniculata*, the mature leaves are stalked, straight, broadly ovate to subcircular, and flat, almost consistently with glabrous leaf surfaces and margins (Raymond-Hamet & Marnier-Lapostolle, 1964; Germishuizen & Fabian, 1982: 116, plate 53b, 1997: 150, plate 67b). The inflorescence branches of *K. brachyloba* are often covered in a very dense, almost pure white, waxy substance that obscures its underlying light bluish green colour, while the inflorescences of *K. paniculata* are light green infused with red. In addition, *K. brachyloba* is one of the few southern African species of which the flower apices are often twisted below the free portions (lobes) of the petals. *K. rotundifolia* (Haw.) Haw. is the best known southern African representative of *Kalanchoe* that has its flowers very distinctly twisted towards the petal lobes; this character is especially evident very soon after anthesis. The flower apices of *K. paniculata* do not show this twisting character.

In the case of *Kalanchoe brachyloba*, the flowers initially dry to a creamy-white colour, eventually fading to reddish brown, so contrasting strongly against the dark brown, dry follicles. In most species of *Kalanchoe*, the flowers rapidly dry to reddish brown.

Uses. Mabogo (1990) records that the Vhavenḓa use the leaves to treat mental illness.

Cultivation. *Kalanchoe brachyloba* is exceedingly easy to propagate from seed that can be broadcast evenly on top of the soil in a seedling tray. Once the seedlings have two sets of leaves, they can be transplanted to the spot where they are intended to grow; it grows well in virtually any soil type. Leaf cuttings have not been known to strike root. In cultivation, it usually not only flourishes in full sun but also will thrive in semishade given that it often occurs in the shade of nurse plants such as grasses, forbs, and acacioid bushveld trees and shrubs.

3. *KALANCHOE CRENATA* (ANDREWS) HAW.

Crowned kalanchoe (Figs. 12.3.1 to 12.3.10)

FIG. 12.3.1   The leaves of *Kalanchoe crenata* are often red-infused along the margins. Photograph: Gideon F. Smith.

Kalanchoe crenata (**Andrews**) **Haw.**, *Syn. pl. succ.*: 109 (1812); Haworth (1829: 303); Harvey (1862: 378) pro parte; Batten & Bokelmann (1966: 73, and plate 62,3); Jacobsen (1977: 286); Raadts (1977: 126); Fernandes (1980: 336); Fernandes (1982a: 20); Fernandes (1983: 47); Tölken (1985: 65) pro parte; Jacobsen (1986: 612); Wickens (1987: 42); Descoings (2003: 152) pro parte; Lebrun & Stork (2003: 214); Smith & Crouch (2009: 84); Court (2010: 91); Smith et al. (2017b: 308). **Type:** "Hort. Vere 1798 (e Sierra Leone)", *s.c. s.n.* (BM BM000649706, holo-).

Synonyms:
Vereia crenata Andrews in *Bot. Repos.*: t. 21 (1798). **Type:** as above.

Cotyledon crenata (Andrews) Vent., *Jard. Malmaison*: t. 49 (1804). **Type:** as above.
Cotyledon verea Jacq., *Pl. rar. hort. Schoenbr.* 4: 17, t. 435 (1804), nom. illegit.
Kalanchoe verea (Jacq.) Pers., *Syn. pl.* 1: 446 (1805), nom. illegit.
Kalanchoe afzeliana Britten in Oliver (ed.), *Fl. Trop. Afr.* 2: 393 (1871), nom. illegit. **Type:** as above.
Kalanchoe coccinea Welw. ex Britten in Oliver (ed.), *Fl. Trop. Afr.* 2: 395 (1871). **Syntypes:** Angola, Golungo Alto, *Welwitsch 2487* (LISU LISU209304 & LISU209305), *Welwitsch 2487* (BM BM000649705), *Welwitsch 2487* (K K000232918), *Welwitsch 2487* (P P00374104), *Welwitsch 2487* (B B 10 0153749, fragment).

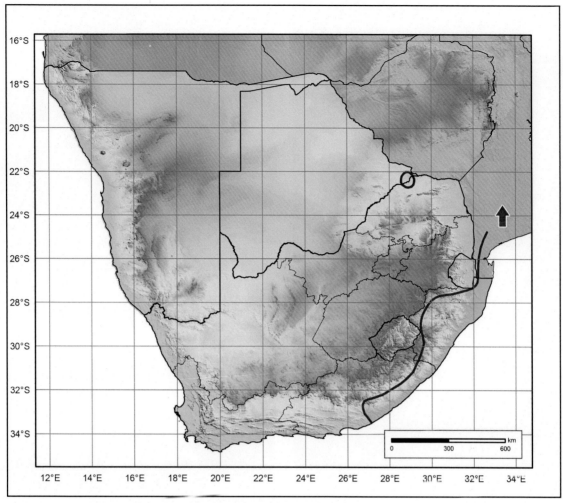

FIG. 12.3.2   Known geographical distribution range of *Kalanchoe crenata* in southern Africa.

FIG. 12.3.3   The rootstock of *Kalanchoe crenata* can become thickened and somewhat woody. Photograph: Gideon F. Smith.

FIG. 12.3.4   Kalanchoe *crenata* often grows in deep shade. Note the general lack of pubescence. Photograph: Neil R. Crouch.

FIG. 12.3.5   A plant of *Kalanchoe crenata* growing as a near-lithophyte on a large rock. Photograph: Neil R. Crouch.

FIG. 12.3.6   The leaf margins of *Kalanchoe crenata* are crenate to dentate. Photograph: Gideon F. Smith.

Kalanchoe brittenii Raym.-Hamet in *J. Bot.* 54, Suppl. 1: 3 (1916). **Type**: Kenya, Changamve, 14.3.1902, *Kassner 258a* (BM BM000649704, holo-).

Kalanchoe connata Sprague, *Bull. Misc. Inform. Kew* 1923 (5): 183 (1923). **Type**: Angola, *Dawe 263-21* (K K000232919 holo-; K K000232920, iso-).

FIG. 12.3.7   As is the case with most species of *Kalanchoe*, the leaves along an elongating inflorescence of *K. crenata* are very similar to the basal, rosulate ones but diminish in size higher up along the peduncle. Photograph: Gideon F. Smith.

FIG. 12.3.8   Flower colour in *Kalanchoe crenata* varies from bright orange, as here, to bright yellow.

Kalanchoe crenata var. *coccinea* (Welw. ex Britten) Cufod. in *Bull. Jard. Bot. État Bruxelles* 27: 717, figure 70 (1957). **Type** as for *Kalanchoe coccinea.*

Kalanchoe crenata var. *verea* (Jacq.) Cufod. in *Bull. Jard. Bot. État Bruxelles* 27: 714, figure 69 (1957). nom. illeg.

Kalanchoe integra var. *verea* (Jacq.) Cufod. in *Österr. Bot. Zeit.* 116: 317 (1969), nom. illeg.

Kalanchoe integra var. *crenato-rubra* Cufod. in *Österr. Bot. Zeit.* 116: 320 (1969). **Type**: Tanzania, Iringa, Mporotos, Mbeya District, edge of upland forest, 10. 9. 1954, *Smith 1280* (K K000232744, holo-; EA EA000002389, iso-).

Kalanchoe integra var. *crenata* (Andrews) Cufod. in *Österr. Bot. Zeit.* 116: 320 (1969). **Type** as for *Kalanchoe crenata.*

FIG. 12.3.9   In the case of very large-growing specimens, the inflorescences of *Kalanchoe crenata* can be densely branched and flowered. Photograph: Neil R. Crouch.

Derivation of the scientific name:

From the Latin *crenatus*, referring to the indentations on the leaf margin, a character though shared with many other species in the genus.

Description:

Perennial or multiannual, solitary or sparsely branched, often clumped, glabrous or velvety, flimsy or robust succulent, and 0.3–2.0 m tall when flowering. *Stems* light green to reddish green, erect, usually simple, sometimes branched, round, fleshy, terete, glabrous or velvety towards base, glabrous or glandular-pubescent above, and hairs short. *Leaves* opposite-decussate, subrosulate, succulent throughout, horizontal to variously deflexed, petiolate, the lowest the largest, the uppermost becoming smaller and bract-like, not very crowded below, mostly uniformly light green to dull greyish-light green, and rarely midgreen; upper bract-like leaves shield-shaped-attenuate, drying papery chartaceous, and sometimes caducous; *petiole* up to 4 cm long, flattened, grooved above, slightly broadened at base, often clasping stem, and sometimes connate with opposite one; *blade* 45–300 × 15–200 mm and ovate or obovate-oblong to spathulate; *base* cuneate and glabrous throughout or sometimes upper ones sparsely glandular-pubescent; *margins* irregularly crenate to doubly crenate to sublobed, and sometimes edged with red. *Inflorescence* terminal, a flat-topped to rounded thyrse, usually with many dichasia ending in monochasia, up to 30 cm tall, and glabrous or glandular-pubescent; branches at ± 45 degrees with axis; *pedicel* to 10 mm long and glabrous or glandular-pubescent. *Flowers* erect, basally greenish becoming light orange upwards (tube) and light orange longitudinally infused with yellow (lobes); *calyx* succulent, light green infused with red, *sepals* 4–5 × 0.7–1.5 mm, lanceolate to linear-lanceolate, free to scarcely connate at base, acute, and hardly contrasting against corolla tube; *corolla* 8–16(–22) mm long; distinctly enlarged lower down; often twisted apically at and after anthesis; glabrous or glandular-pubescent; orange (in southern Africa), bright salmon to red, deep red, or brick-coloured elsewhere; lower part pale or greenish orange or whitish (in southern Africa); sulphur-yellow to bright or deep yellow elsewhere (tube) and light orange longitudinally infused with yellow (lobes); *tube* 7–21 mm long; quadrangular-cylindrical; obscurely 4-angled; obtusely rounded at bottom; indistinctly box-shaped-square when viewed from below; orange; becoming scarious when dry; papery and somewhat rigid in

FIG. 12.3.10   The fruit of *Kalanchoe crenata* is enveloped in the flimsy, white remains of the desiccating corolla.   Photograph: Gideon F. Smith.

fruit; *lobes* 4–8 × 2.5–5.0 mm; oblong-lanceolate to elliptic; acute or subacute; with apicular mucro 0.5–1.0 mm long; light orange longitudinally infused with yellow; and becoming fully spreading depending on time of day. *Stamens* inserted in 2 distinct ranks ± at and just above middle of corolla tube, both whorls included; *filaments* glabrous and short, 1.0–2.5 mm long; *anthers* 0.5–1.0 mm long and broadly oblong. *Pistil* consisting of 4 carpels; *carpels* 5–10 mm long; *styles* 1–5 mm long; *scales* up to 4 mm long and linear. *Fruit* 5–7 mm long, brittle, grass spikelet-like, enveloped in whitish remains of the corolla, and ultimately turning dull brown. *Seeds* not seen. *Chromosome number*: $n = 51$ (http://ccdb.tau.ac.il/search/), $2n = 102$ Baldwin (1938, 576), and $2n = c$. 102 (Raadts, 1985: 162). See Raadts (1989) for a discussion.

Flowering time:
May–July, peaking in June (southern hemisphere).

Illustrations:
Andrews (1798: t. 21), Ventenat (1804: t. 49), Sims (1812: Plate 1436), Wickens (1987: 44, figure 6); Crouch & Smith (2009: Plate 2249); Smith & Crouch (2009: 84), Smith et al. (2017b: 308).

Common names:
Afrikaans: kroonplakkie, plakkie (Smith et al., 2017b: 308), woudplakkie, and woudkalanchoe.

English: crowned kalanchoe, orange forest kalanchoe (Smith et al., 2017b: 308), and scalloped kalanchoe.

Kikuyu (Kenya): mahuithia (Sapieha, n.d.: 52).

isiZulu: mathongwe, mahogwe (Durban, KwaZulu-Natal); umahogwe (Johannesburg, Gauteng) (Williams et al., 2001).

isiXhosa: uquwe (Dold & Cocks, 1999).

Geographical distribution:
The species has one of the widest distribution ranges of all the kalanchoes, not only the African ones, occurring from Chad in the north to the east coast of South Africa in the south. It covers a broad west-east-west range that straddles the African equator, on the western side of the continent stretching down to Angola. For example, a subspecies, *Kalanchoe crenata* subsp. *bieensis*, was described as an endemic from Bié Province in the centre of this south-tropical African country. However, as far as is known, *K. crenata* does not cross into the much more arid Namibia, while for Botswana, there is a single record known from

the northeastern border of the country. *Kalanchoe crenata* rather occurs in a southeasterly sweep across the continent through several eastern African countries to South Africa's south coast (KwaZulu-Natal and Eastern Cape Provinces) (Fig. 12.3.2).

Distribution by country. Angola, Benin, Burkina Faso, Burundi, Cameroon, Chad, Central African Republic, Democratic Republic of Congo [formerly Belgian Congo; Zaire], Equatorial Guinea, Gabon, Ghana, Ivory Coast, Kenya, Malawi, Mozambique, Niger, Nigeria, Rwanda, Sierra Leone, South Africa, Tanzania, Uganda, Zambia, and Zimbabwe. The species has also been recorded as indigenous on the Arabian Peninsula and has become naturalised in Central and South America, India, and Malaysia.

Habitat:

As one of the common names, Orange forest kalanchoe, of *Kalanchoe crenata* indicates, this is a species that often occurs in the undergrowth of forests and moist savannas. Most kalanchoes prefer open, fully exposed, or dappled shady positions, but *K. crenata* also often thrives in deep shade, for example, along forest margins. Where *K. crenata* occurs, it is a prolific seeder and can locally appear in dense stands. However, in southern Africa its distribution range is often rather broken with large tracts lacking any plants between sites of occurrence. Its preference for moist forest margins and moist sheltered scrub may account for its relative obscurity. Plants are sometimes found growing well on rotting logs in the understorey, or in the in often considerable leaf-fall that overly the soil substrate (Crouch & Smith, 2009).

Conservation status:

In South Africa, the species is regarded as 'Least concern' and not threatened with extinction.

Additional notes and discussion:

Taxonomic history and nomenclature. *Kalanchoe crenata* was described in 1798 by Andrews, as *Vereia crenata*, from a plant cultivated in the garden of James Vere, a wealthy merchant at Kensington Gore, London, England (Desmond, 1994). The gardener in charge was William Anderson (1766–1846). Vere was commemorated in the genus name *Vereia* Andrews, a synonym of *Kalanchoe*. The plant that was cultivated in his garden originated from Sierra Leone, where it had been collected in 1793 by Adam Afzelius (1750–1837), a Swedish botanist. Nevertheless, as noted by Fernandes (1980: 338), it is possible that the plant was known in Europe long before that date. A plate published by Plukenet (1692: Plate 228, figure 3) that he described as 'Telephium maximum *Africanum* flore aurantiaco, ex Cod. *Renting. Phytogr. Tal. 228. fig. 3* Item ex Hollandia habuimus' (Plukenet, 1696: 362) and that Haworth (1819: 303) considered a synonym of *K. crenata* could represent that species and thus indicate that it was already known in Europe (the Netherlands) in the 17th century.

It was only some 70 years later, in 1871, that the distribution range of the species in West Africa was extended beyond its original collecting locality (Sierra Leone) by Britten (1871: 394), who additionally recorded it as occurring also in the mountains of Cameroon and in Angola. However, earlier, it had been recorded from South Africa by Harvey (1862: 379) who cited collections from the Northern and Eastern Cape, made by Ecklon & Zeyher ("Mountain sides near Phillipstown, Caffr."), by Bowker ("Kreilis Country"), and by Drege ("between the Kei and the Gekau"). Philipstown is about 80 km (50 miles) north west of Colesberg in the Northern Cape province. The place name "Kreilis", which was recorded by Bowker, could be an early corruption of "Kreli's Kraal", a village near Cofimvaba in the Eastern Cape (Raper et al., 2014: 252), and a known locality of *K. crenata*. "Gekau" refers to Gaikau Mission Station (alternatively "Aftrek", or "Middeldoortrek"), which is near Butterworth in the Eastern Cape (Gunn & Codd, 1981: 138).

Several synonyms have been listed for *Kalanchoe crenata*. Some of these names are illegitimate while others were eventually omitted from synonymy because they represent insufficiently known taxa (see Fernandes, 1980, for a discussion of these names). Some listed synonyms were described on account of flower colour variation, a character that was later found to be very variable in the species and of no taxonomic consequence. For example, *Kalanchoe coccinea* was described by Britten based on material collected by Welwitsch in Angola that had scarlet or bright orange flowers. As happens with many names based on Welwitsch material, its typification is not straightforward. In the publication of the name, Britten (1871) cited a single locality: "Angola, Golungo Alto, Dr Welwitsch". The Welwitsch material of *K. coccinea* consists of collection number *2487*. This collection was made in Golungo Alto (Cuanza Norte); however, it includes specimens collected in separate localities (between Muria and Calolo, between Xixe and Muria. from Calolo to Sange, and near the Congo River) and on distinct dates (1854–57). For this reason, even though the number is the same, the specimens are not duplicates and must be considered as syntypes. Welwitsch grouped specimens from different localities under the same number when he thought they belonged to the same taxon. Hence, Welwitsch's collection numbers often do not correspond to single gatherings, and duplicates are treated as syntypes. Note also that while *Kalanchoe coccinea* is considered a synonym of *K. crenata*, *K. coccinea* var. *subsessilis* Britten is a synonym of another species, *Kalanchoe lateritia* Engl.

Interestingly, Hamet, one of the principal workers on *Kalanchoe* and who monographed the genus in the early 1900s (Hamet, 1907, 1908a, b), treated *Kalanchoe crenata* as a synonym of *Kalanchoe laciniata*, a similarly variable species with a very broad distribution range (Hamet, 1907: 897). We regard these two species as distinct.

Kalanchoe connata, a species that was described by Sprague, 1923, is only known from the type collection, a plant collected by Morley Thomas Dawe (1880–1943) in Angola at an unknown locality; material was sent to Kew and was grown there. Fernandes (1982a) considered it an insufficiently known species, noting that it differed from *K. crenata* on account of its glabrousness and apparently sessile leaves (only two remain attached to the specimen) but that it was very close to that species (Fernandes, 1980: 336). The name remains questioned in synonymy.

Fernandes (1978, 1980, 1982a, 1983) recognised three subspecies in *Kalanchoe crenata*: the widespread subsp. *crenata* and subsp. *nyassensis* [*Kalanchoe crenata* subsp. *nyassensis* R.Fern. in Bol. Soc. Brot. sér. 2, 52: 199 (1978). **Type**: Malawi, Mafinga Hills, 28.8.1962, *Tyrer 585* (BM BM000649703 holo-; BR, SRGH SRGH0106779-0 iso-)], distributed in Malawi and Tanzania, and subsp. *bieensis* [*Kalanchoe crenata* subsp. *bieensis* R.Fern. in Bol. Soc. Brot., sér. 2, 53: 361 (1980). **Type**: Angola, Bié, Ceilunga, 6.7.1963, *Murta 207* (COI COI00005659 holo-; LISC LISC002338 & LISC002337, LUA iso-)] that is restricted to Angola. It must be noted that the latter is often incorrectly listed in databases as "subsp. biennis". The correct epithet (as published) is 'bieensis', deriving from Bié (a province of Angola). Subsequent authors in general did not accept these three subspecies that Fernandes (1982a, 1983) had distinguished mostly on account of flower and fruit sizes and whether anthers are included or exserted. Tölken (1985) went a step further, as not only did he not accept the infraspecific taxa but also he synonymised *K. hirta* with *K. crenata*. This view was subsequently widely followed (see, e.g. Descoings, 2003: 152), but we prefer to treat *K. hirta* as a species separate from *K. crenata* (Smith et al., 2018a). Fernandes, (1983: 46) and Descoings (2003: 152) cite *K. hirta* as occurring in South Africa and in Zimbabwe. However, it seems that the only specimen known from Zimbabwe consists of material grown in Pretoria ("collected in Zimbabwe and grown at Division of Plant Industry, Pretoria, s.d., *P. Koch 8666* (K, PRE)" fide Fernandes, 1983: 46).

Identity and close allies. Kalanchoe crenata is highly variable as to flower colour, ranging from yellow, to orange, to red, with these primary colours further varying in intensity, from light to dark. In southern Africa, *K. crenata* is quite easy to distinguish from all other kalanchoes that occur in the region. Especially, its habitat preference for fairly moist, shady habitats, along forest and bushveld margins, combined with its orange flowers make it easy to distinguish. At least in the region, an affinity may be sought with *K. lanceolata*, which also has orange flowers, and the distinction that plants are glabrous when young but become pubescent higher up, as plants enter the reproductive phase. This trend in increased hairiness

on the inflorescences and flowers of *K. lanceolata* has also been observed in eastern African material of *K. crenata* (Raadts, 1977: 129–131). In eastern Africa, Raadts (1977) and Wickens (1987: 45) suggest that *K. crenata* could be confused, at least morphologically, with *K. densiflora* Rolfe.

Regardless of variation found in *K. crenata* with regard to flower colour, dimensions of the flowers, leaf size and proportions, and hairiness, the species is rather constant throughout its wide distribution range in terms of the linear-lanceolate shape of the sepals and the position of the two-ranked anthers, all eight of which are included within the corolla tube.

Uses. It has been suggested that ethnomedicinal uses recorded for *K. crenata* by Hutchings et al. (1996) apply rather to *K. hirta*, given the broad species concepts adopted by these authors (see, e.g. Hulme, 1954 and Crouch & Smith, 2009). Furthermore, reports of *K. crenata* being used in Madagascar (Hutchings et al., 1996) are questionable given that the species has not been reported as naturalised on that Island. Possible confusion in this respect might relate to the presence there of the indigenous *Bryophyllum crenatum* Baker (Rauh, 1995), which is at present widely treated as a synonym of *K. laxiflora* Baker (Raymond-Hamet & Marnier-Lapostolle, 1964; Descoings, 2003).

Cultivation. Kalanchoe crenata is not common in general horticulture, at least in southern Africa. However, it will make an interesting addition to forest and even bushveld (savanna) gardens. *Kalanchoe crenata* flourishes under low-light conditions and during the short days of winter provides splashes of orange flowers, often brightening up the undergrowth of savannas and forests. One of the reasons why the species may be less often cultivated is that its flowering stems die back to a contracted, persistent base after the fruiting phase; in a garden, this implies that plants require active management, with the unsightly, dead flowering stems having to be cut off to make for a more pleasant horticultural subject. Fortunately, come summer, plants resprout from the persistent base.

Plants set seed freely, and a multitude of seedlings will become established where a specimen flowered. During the warm summer months, plants can also be propagated by stem or leaf cuttings.

Springate (1995) included *Kalanchoe crenata* amongst nearly 50 species recommended for growing under glass in Europe, a necessity for, as pointed out by Brickell (1990), most species in the genus are frost-tender, requiring a minimum of 7–15°C to survive during winter. Some southern African kalanchoes, for example, *K. luciae* and *K. paniculata*, can indeed tolerate lower temperatures, but this is not the case for the largely tropical African *K. crenata*.

4. *KALANCHOE CRUNDALLII* I.VERD.

Crundall's kalanchoe (Figs. 12.4.1 to 12.4.9)

FIG. 12.4.1   A large clump of *Kalanchoe crundallii* growing wedged between two large boulders in deep shade. Photograph: Gideon F. Smith.

Kalanchoe crundallii **I.Verd.** in *The Flowering Plants of Africa* 25: plate 967 (1946). Tölken, (1985: 69); Retief & Herman (1997: 392); Descoings (2003: 152); Court (2010: 92). **Type**: [South Africa], Transvaal [Limpopo Province], Zoutspansberg [Soutspansberg] district, Mount Lejuma, no date, *A.H. Crundall s.n. sub* PRE 27157 (PRE PRE0457400-0, holo-).

Derivation of scientific name:

The species was named for Albert Henry Crundall (24 September 1889–22 May 1975). Crundall, an indefatigable botanical collector, also collected the type of *Aloe soutpansbergensis* I.Verd. (Gunn & Codd, 1981: 125; Hardy & Fabian, 1992: xx).

Description:

Perennial or rarely multiannual, few- to many-leaved, rarely branched, glabrous throughout, sparse succulent, and to 1m tall when in flower but usually much shorter. *Stems* usually solitary, erect to leaning, rooting where they touch the ground, supported by thin wiry roots. *Leaves* spreading to patent-erect to down-curved, often arranged in four distinct vertical rows, succulent, petiolate, ± flattened above to concave, slightly convex below, green to yellowish green, and faintly flaky waxy with age; *petiole* 5–20mm long, not stem-clasping; *blade* 3–7 × 2–5cm and suborbicular to broadly oblong, not folded lengthwise; *base* cuneate; *apex* rounded-obtuse; *margins* crenate-dentate, rarely entire, and infused with a reddish tint. *Inflorescence* a slender, erect, densely flowered, shortly cylindrical to club-shaped, flat-topped thyrse consisting of several dichasia, to 80cm tall; *pedicels* 10–12mm long. *Flowers* erect at bud stage to spreading or pendent at anthesis, dry flowers erect, buds yellowish to yellowish green, red at anthesis, and drying dark reddish brown; *calyx* light green, remaining so postanthesis, and contrasting against corolla tube; *sepals* 2.0–2.5 × 1mm, deltoid-lanceolate, almost free to the base, acute, and

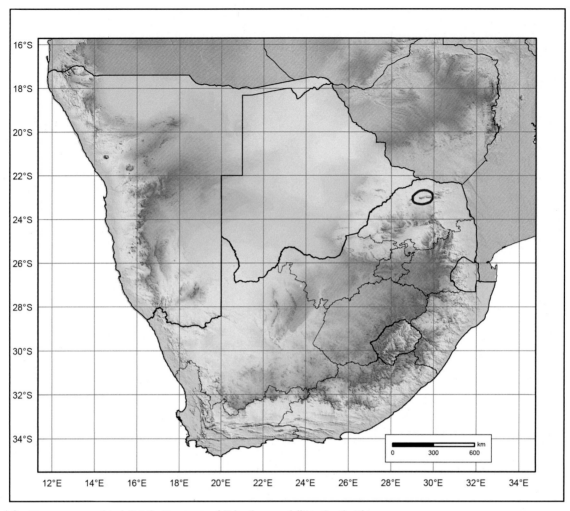

FIG. 12.4.2   Known geographical distribution range of *Kalanchoe crundallii* in South Africa.

FIG. 12.4.3   The leaves of *Kalanchoe crundallii* are often arranged in four, distinct vertical rows. Photograph: Gideon F. Smith.

FIG. 12.4.4   Small plants of *Kalanchoe crundallii* growing as lithophytes on a vertical cliff face. Photograph: Gideon F. Smith.

FIG. 12.4.5   The natural habitat of *Kalanchoe crundallii* on the slopes of the Soutpansberg, Limpopo Province, South Africa, is typically densely wooded, with the species thriving in the understorey. Photograph: Gideon F. Smith.

FIG. 12.4.8   The flowers of *Kalanchoe crundallii* tend to be a dullish, light green colour basally when not exposed to sunlight. Photograph: Gideon F. Smith.

FIG. 12.4.6   Close-up of an inflorescence of *Kalanchoe crundallii*. Photograph: Gideon F. Smith.

FIG. 12.4.7   The leaves of *Kalanchoe crundallii* are reminiscent of those of *K. rotundifolia*. Photograph: Gideon F. Smith

FIG. 12.4.9   The desiccating corolla tube that surrounds the fruit of *Kalanchoe crundallii* rapidly dries to a dark brown colour. Photograph: Gideon F. Smith.

red-tipped and slightly so along margins; *corolla* 11–15 mm long, more or less cigar-shaped-cylindrical yellowish to yellowish green especially lower down, and uniformly red at anthesis but parts not exposed to direct sunlight remaining yellowish green; *tube* 10–14 mm long, more or less cylindrical, somewhat cigar-shaped-thickened in the middle, and slightly 4-angled especially postanthesis; *lobes* 3–4 × 3 mm, ovate to slightly square, straight to very slightly curved outwards, tapering to a point, minutely apiculate, yellow within, and reddish without. *Stamens* included, arranged in 2 obscure ranks just above middle of corolla tube; *filaments* 3 mm long, thin, and light yellow; *anthers* 0.5–1.5 mm long and oblong to arrowhead-shaped. *Pistil* consisting of 4 carpels; *carpels* 6–7 mm long and light green; *styles* 3 mm long and light yellow; *stigmas* slightly capitate; *scales* ± 2.5–3.0 mm long, linear oblong, and light yellow. *Follicles* 5–6 mm long, brittle, grass spikelet-like, enveloped in shiny brown remains of the corolla, remaining shiny brown for a long time, and ultimately turning dull brownish black. *Seeds* 0.4–0.5 mm long and light to dark brownish black. *Chromosome number*: unknown.

Flowering time:
December–March (southern hemisphere). This is one of very few southern African kalanchoes that flowers in summer. Some forms of the predominantly winter-flowering *K. rotundifolia* also have been observed to rarely flower in summer.

Illustrations:
Verdoorn (1946: Plate 967); Hardy & Fabian (1992: Plate 21b); Van Wyk & Smith (2001: figure 321); Hahn (2002: figures 151–154).

Kalanchoe crundallii was the second species of *Kalanchoe* to be illustrated in the serial publication initially entitled *The Flowering Plants of South Africa*. In 1945, therefore as from volume 25 onwards, this journal was renamed as *The Flowering Plants of Africa* (Killick & Du Plessis, n.d.). *Kalanchoe crundallii* was described in the first volume named *The Flowering Plants of Africa*.

Common names:
Afrikaans: Crundall-se-plakkie.
English: Crundall's kalanchoe.

Geographical distribution:
Distribution by country. South Africa (Limpopo).
Kalanchoe crundallii is restricted to the Soutpansberg Centre of Endemism (Van Wyk & Smith, 2001) (Fig. 12.4.2).

Habitat:
Retief & Herman, (1997: 392) recorded *Kalanchoe crundallii* from "forests". Hahn (2002: 136, 2017: 328) more specifically mentions the species as being "…found in the mist-belt regions of the western [Soutpansberg] mountain where it grows in woodlands in humus-rich soils". Tölken (1985: 69) notes that the species is found among boulders in forested areas.

Conservation status:
Kalanchoe crundallii is treated as 'Rare'.

Additional notes and discussion:
Taxonomic history and nomenclature. Living material that was eventually described as *Kalanchoe crundallii* was collected by Albert Henry Crundall from Mount Lejuma on the western extremity of the Zoutpansberg [Soutpansberg] in July 1938. He took material with him to Pretoria, where he cultivated the plants in his garden. When a specimen flowered there in March 1943, he took it to South Africa's National Herbarium, which is also based in Pretoria. Dr Inez Verdoorn, who became the head of the Herbarium one year later, in 1944, recognised the species as new and drew up a description and had a botanical painting of it prepared by Mrs E.K. Burges. Three years after the species flowered in Pretoria, the Burges plate and accompanying text appeared in *The Flowering Plants of Africa*.

Identity and close allies. Kalanchoe crundallii has flowers that can vary in colour depending on the exposure of the flowers to light: they are light yellowish green when not exposed to direct sunlight, while those exposed to sun are red. Furthermore, the yellowish buds are carried erectly, while the reddish open flowers are spreading to pendent. Once the flowers dry postanthesis, their disposition reverts to erect, and they are a dark reddish-brown colour.

In its natural habitat, *Kalanchoe crundallii* sometimes can be confused for the variable *K. sexangularis* var. *sexangularis*. However, *K. crundallii* flowers earlier (summer to very early autumn) than *K. sexangularis* var. *sexangularis* (autumn to winter and even in spring) and has red, not yellow, flowers.

Although it has been suggested that in terms of general vegetative and inflorescence reproductive morphology, the closest relative of *Kalanchoe crundallii* may be *K. thyrsiflora*, this needs further investigation. While both species have dense, club-shaped, somewhat densely flowered inflorescences, *K. thyrsiflora* has much larger greyish-green leaves. In addition, the flowers of *K. thyrsiflora* are very waxy and have yellow petal lobes, unlike those of *K. crundallii* that are red. When in flower, these two species are therefore easy to distinguish.

Cultivation. The species is poorly known in cultivation, and since its description in the mid-1940s, just after WWII, has not been introduced into general horticulture. It is though a worthy subject to grow, especially because of its reddish flowers, a deviation from the shades of yellow and greenish yellow presented by most large-growing southern African kalanchoes.

In habitat, plants often grow in very thin layers of humus-rich soil on rocks and boulders or even as an epiphyte on tree branches. It is a shade-loving species that should be protected from direct sunlight.

Uses. None recorded.

5. KALANCHOE HIRTA HARV.

Bush kalanchoe (Figs. 12.5.1 to 12.5.14)

FIG. 12.5.1 During the summer rainy season, plants of *Kalanchoe hirta* take on a light green colour in their natural habitat. The leaves are light green, rather than strongly yellowish-infused, and flattened-expanded. Photograph: Gideon F. Smith.

Kalanchoe hirta Harv. in Harvey & Sonder (eds) *Fl. cap*. 2: 379 (1862). Wood (1907: 46); Hamet (1908b: 36); Wood (1909: 155); Bews (1921: 98); Jacobsen (1970: 285); Ross (1972: 179); Fernandes (1983: 46); Jacobsen (1986: 615); Smith & Crouch (2009: 84); Smith et al. (2017b: 308); Smith et al. (2018a: 45). **Type**: [South Africa. Eastern Cape Province], Olifantshoek, and Uitenhage [C.P.; Cape Province], *K.L.P Zeyher s.n.* (S S-G-3472, holo-).

Synonyms:
None recorded.

Derivation of the scientific name:
From the Latin *hirtus*, referring to the hairs present on all plant parts. Various grades of pubescence are character though shared with many other species in the genus.

Description:
Perennial or more rarely multiannual, solitary or sparsely branched, sometimes clumped, conspicuously pubescent throughout, medium-sized to robust succulent, and to 1.5 m tall when flowering. *Roots* tuberous. *Stems* light green, erect from a curved base, usually simple, sometimes branched, round, somewhat fleshy, terete, pubescent, hairs prominent, and to 1 mm long. *Leaves* opposite-decussate, subrosulate, succulent throughout, more or less horizontal, distinctly petiolate to subpetiolate, sessile higher up, the lowest the largest, the uppermost becoming smaller, crowded below when young, and mostly uniformly light green to yellowish light green; drying papery chartaceous, sometimes caducous; *petiole* up to 4 cm long, flattened, grooved above, slightly broadened at base, often clasping stem, sometimes connate with

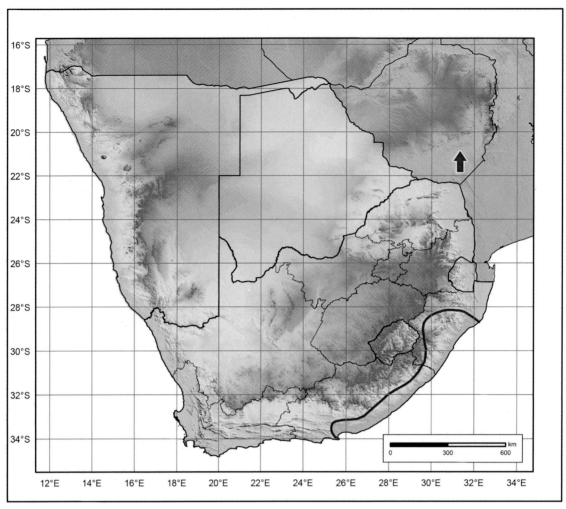

FIG. 12.5.2   Known geographical distribution range of *Kalanchoe hirta* in South Africa. A single record of the species is known from Zimbabwe but without an exact locality (see text for explanation).

FIG. 12.5.3   The large, soft leaves of *Kalanchoe hirta* are densely hairy, conspicuously so along the margins. Photograph: Gideon F. Smith.

FIG. 12.5.4   The tuberous rootstock of *Kalanchoe hirta* is often succulent, can grow quite large, and tends to become woody eventually, adapted as this species is to essentially drought-stressed ecological niches in bushveld (savanna) vegetation. Photograph: Gideon F. Smith.

FIG. 12.5.5 The stems of *Kalanchoe hirta* are densely tomentose. Photograph: Gideon F. Smith.

FIG. 12.5.7 The inflorescences of *Kalanchoe hirta* grow quite tall, easily reaching a length of well over 40 cm. Often only very few flowers are open simultaneously. Photograph: Gideon F. Smith.

FIG. 12.5.6 As is the case with several species of Crassulaceae (e.g. *Petrosedum forsterianum*), the developing inflorescences of *Kalanchoe hirta* often remain nodding, only straightening once the first flowers open. Photograph: Gideon F. Smith.

opposite one; *blade* 8–11 × 6–9 cm and ovate or obovate-oblong, flat, succulent but not very thick; *base* cuneate and densely glandular-hairy on both surfaces; *margins* irregularly dentate. *Inflorescence* terminal, a flat-topped to rounded thyrse, usually with many dichasia ending in monochasia, up to 40 cm tall, and glandular-hairy; branches at ± 45 degrees with axis, gracefully curved upwards; *pedicels* to 6 mm long, prominent, and glandular-hairy. *Flowers* erect, basally greenish becoming light yellow upwards (tube) and yellow (lobes); *calyx* light green; *sepals* 3–6 mm long, lanceolate-triangular, free to connate for up to 1.5 mm at base, acute, and hardly contrasting against corolla tube; *corolla* 11–17 mm long, distinctly enlarged lower down, finely glandular-pubescent,

FIG. 12.5.8   An inflorescence of *Kalanchoe hirta* showing the branching pattern. Photograph: Gideon F. Smith.

FIG. 12.5.9   All the flower parts of *Kalanchoe hirta* are finely and densely pubescent. Often, only a few flowers are open at a time. Photograph: Gideon F. Smith.

FIG. 12.5.10   The most common form of *Kalanchoe hirta* has light yellow flowers. Photograph: Neil R. Crouch.

FIG. 12.5.11   Some forms of *Kalanchoe hirta* have orange flowers, here densely arranged in a much-branched inflorescence. Photograph: Neil R. Crouch.

light yellow (tube) and yellow (lobes); *tube* 9–15mm long, quadrangular-cylindrical, lengthwise fluted to obscurely 4-angled, obtusely rounded at bottom, very indistinctly angled, yellowish, becoming scarious when dry, papery and somewhat rigid in fruit; *lobes* 4–8 × 2.5–3.0mm, oblong-lanceolate to elliptic, acute, apiculate, erect, becoming fully spreading depending on time of day, and yellow. *Stamens* included, with upper rank inserted just above middle of corolla tube and lower rank inserted just below middle; *filaments* short, ± 1mm long; *anthers* ± 0.5mm long. *Pistil* consisting of 4 carpels; *carpels* 4–6mm long; *styles* 1–2mm long; *scales* up to 2–3mm long, linear. *Follicles* brittle, grass spikelet-like, enveloped in dry light whitish-grey remains of corolla, remaining dull light green for a

long time, ultimately turning dull brownish black, and 5–6mm long. *Seeds* 0.4–0.6mm long and dark brownish black. *Chromosome number*: unknown.

Flowering time:
April–July, peaking in June (southern hemisphere).
Illustrations:
Smith & Crouch (2009: 84), Smith et al. (2017b: 309). The image illustrating *Kalanchoe crenata* in Pooley (1998: 255, top row, second from left) likely represents *K. hirta*.
Common names:
Afrikaans: plakkie (Smith et al., 2017b: 309).
English: bush kalanchoe and yellow hairy kalanchoe (Smith et al., 2017b: 309).
isiZulu: mbohlolololeshate (Crouch & Smith, 2009: 66).
Geographical distribution:
Kalanchoe hirta has a fairly restricted geographical distribution range within southern Africa, especially South

FIG. 12.5.12   The fruit of *Kalanchoe hirta*, in this case of the yellow-flowered form, is initially a deep green colour, before fading to white. Photograph: Gideon F. Smith.

FIG. 12.5.13   The ripe fruit of *Kalanchoe hirta* is enveloped in the flimsy, white remains of the corolla tubes. Photograph: Gideon F. Smith.

FIG. 12.5.14   *Kalanchoe hirta* in full flower. Photograph: Neil R. Crouch.

Africa, but enters Zimbabwe in the northeast. The species is endemic to southern and south-tropical Africa, but is not restricted to a single region or centre of endemism (Van Wyk & Smith, 2001). Fernandes (1983: 46) and Descoings (2003: 152) both note the species from Zimbabwe in south-tropical Africa, based on living material

"Collected at Zimbabwe", presumably Great Zimbabwe near Masvingo, and grown on in Pretoria, of which preserved specimens are held at PRE and K. The original herbarium specimen label states: "Grown at Div. [Division] of Plant Industry, Pretoria". The specimen is undated but was collected by *Pieter Koch s.n.* and was deposited under National Herbarium, Pretoria, No. 8666. Determination slips attached to the specimen show that it was identified as *K. hirta* by both Raymond-Hamet, as part of his work towards a treatment for "Das Pflanzenreich" and by Fernandes (1983: 46) as part of her work on the Crassulaceae for the *Flora zambesiaca* project.

The original collecting label further states "figured for The Flowering Plants of South Africa (F.P.S.A.) by C. Letty No. 152". However, Plate 152 is a painting of *Crocosmia* ×*crocosmiiflora* (G.Nicholson) N.E.Br. [=*Montbretia crocosmiiflora* Hort. ex Morren], a member of the Iridaceae. As far as we know, the plate prepared by Letty, if it was indeed prepared and still exists, has not been published in *The Flowering Plants of South Africa* nor in [*The*] *Flowering Plants of Africa*, as this series was later known (Killick & Du Plessis, n.d.). In years gone by, it was common practice to grow material collected further afield, including of *Kalanchoe* species, in the garden of the National Herbarium at its various historical locations in Pretoria where it was brought into flower, before having it illustrated for [The] *Flowering Plants of Africa* (see, e.g. Bruce, 1948: t. 1049 [on *Kalanchoe marmorata* Baker], 1949: t. 1052 [on *Kalanchoe brachycalyx* A.Rich. (=*Kalanchoe lanceolata* (Forssk.) Pers.], 1950: t. 1089 [on *Kalanchoe densiflora* Rolfe], and Dyer, 1979: t. 1783 [on *Kalanchoe robusta* Balf.f.]).

We concur that the *Pieter Koch s.n.* specimen held at PRE represents *Kalanchoe hirta*. Based on holdings in PRE, Koch's herbarium collections, mostly without collector's number, are very few, and as a consequence, he is not mentioned in Gunn & Codd (1981). By the time that Smith et al. (2018a) was accepted for publication, those authors had been unable to trace information about him nor of his collecting activities and localities. In the meantime, it had been determined that Pieter Koch was an agronomist in the service of the Division of Plant Industry, Department of Agriculture and Forestry of the then Union of South Africa, from 1913 to December 1947. Among others, he contributed significantly towards the establishment of the cotton and tobacco industries in southern Africa (Anonymous, 1948) and also conducted research on potential fibre-producing members of the family Malvaceae. Anonymous (1948) further notes that "On several occasions, he was sent to neighbouring states on government missions to investigate the fibre industry". It was likely on one of these trips that he collected the material of *Kalanchoe hirta* held at PRE as "National Herbarium, Pretoria, No. 8666".

Kalanchoe hirta and *K. crenata* can occur sympatrically in parts of eastern South Africa, especially in KwaZulu-Natal. In such instances of cooccurrence, intermediates have yet to be observed.

Distribution by country. South Africa and Zimbabwe.

Habitat:

Kalanchoe hirta is an essentially savanna (bushveld) species that has a distinct preference for drier habitats in this vegetation type, being often found in grassy patches on exposed rocky hillsides. It has also been recorded from the slightly more arid thickets of the Eastern Cape, even in close proximity to the sea.

Conservation status:

In South Africa, the species should be regarded as 'Least concern' and is not threatened with extinction.

Additional notes and discussion:

Taxonomic history and nomenclature. Of all the names available for species of *Kalanchoe* in southern Africa, the application of the name *K. hirta* has probably caused the most confusion. The species was originally described by Harvey (1862: 379), based on a single Carl [Karl P.L.] Zeyher specimen from Olifantshoek in the Uitenhage district in South Africa's Eastern Cape Province, which, as far as we could ascertain, is the most southwestern locality where the species occur. In this regard, it is

interesting that the type locality of *K. rotundifolia*, a much smaller-growing species, is also near Uitenhage, but the material in which this name is based was collected by James Bowie (Figueiredo & Smith, 2017a, b, c; Smith & Figueiredo, 2017a, b, c, d, e). *Kalanchoe rotundifolia* though has a much more extensive geographical distribution range that stretches northwards from South Africa through southwestern south-tropical Africa and northwards through southeastern and east-tropical Africa to Socotra.

Interestingly, *Kalanchoe hirta* was widely accepted as a good species for more than 120 years, but following Tölken (1985), who applied a broad species concept to *K. crenata* in his account of the *Flora of Southern Africa*, *K. hirta* was largely included in the synonymy of *K. crenata* (see, e.g. Descoings, 2003: 152). Tölken (1985) did, however, acknowledge the need for further studies to more fully resolve the relationship between *K. hirta* and *K. crenata*.

Identity and close allies. Kalanchoe hirta was described by Harvey (1862), who distinguished it from *K. crenata* on the basis of its densely hairy appearance; even the calyces of the yellow-flowered *K. hirta* are externally covered with short, rigid hairs. A further distinguishing character is the shape of the calyx lobes that are ovate rather than linear-lanceolate in the case of *K. hirta*. Furthermore, *K. hirta* is generally a more robust plant of dry, open bushveld. Sometimes, the leaves but especially the flowers are much more yellow-infused and in South Africa are usually taller (up to 1.5 m) than the generally glabrous, orange-flowered *K. crenata*. In South Africa, *K. crenata* hardly ever exceeds 0.6 m in height. However, also shorter forms of *K. hirta* have been documented (Court, 2010: 91), as have sparsely hairy forms of *K. crenata* (Fernandes, 1983: 46). The existence of such intermediate forms and variation in plant size and hairiness has added to confusion over species boundaries between *K. hirta* and *K. crenata*.

Uses. The ethnomedicinal uses attributed to *K. crenata* by Hutchings et al. (1996) may well rather be applicable to *K. hirta*, given that these authors followed the broad species concepts adopted by Tölken (1985). However, whether the uses of *K. hirta* (e.g. see Hulme, 1954) and *K. crenata* are indeed identical, or at least similar, remain open to speculation. In KwaZulu-Natal, decoctions of the crushed leaves of *K. hirta* are used as an anti-inflammatory to treat sprains and swellings (Hulme, 1954). In addition, plants of *K. hirta* are crushed and added to water for sprinkling to ward off evil spirits known as *mkovu* (Watt & Breyer-Brandwijk, 1962).

Cultivation. Kalanchoe hirta does not present any challenges in cultivation, but some forms tend to be rather slow-growing, much more so than many other southern African kalanchoes. In the case of other clones, plants grow so rapidly that the stems tend to topple over under the weight of their foliage.

6. *KALANCHOE LACINIATA* (L.) DC.

Christmas tree kalanchoe (Figs. 12.6.1 to 12.6.7)

FIG. 12.6.1 The inflorescences of *Kalanchoe laciniata* are apically dense, flat-topped, and few- to many-flowered. They are borne erectly or may lean to one side under the weight of the terminal flower clusters. All parts are finely and densely glandular-hairy. Inflorescences can reach a height of over 1 m. Photograph: Abraham E. van Wyk.

Kalanchoe laciniata **(L.) DC.**, *Pl. hist. succ.*: t. 100 and two preceding text pages (1802) [June–July 1802]. De Candolle (1828: 395); Haworth, (1829: 302); Raymond-Hamet & Marnier-Lapostolle (1964: 76, Plate XXVI, figures 84–85); Friedrich (1968: 38); Fernandes (1980: 368–378); Fernandes (1982a: 23); Fernandes (1983: 51); Wickens (1982: 672); Collenette (1985: 190); Tölken (1985: 65); Wickens (1987: 52); Thulin (1993; updated 2008); Descoings (2003: 160); Lebrun & Stork (2003: 214, 216). **Type**: Cultivated plant from Africa, "Cotyledon afra, folio crasso, lato laciniato, and flosculo aureo laciniata", *Hortus Cliffortianus 175* (BM BM000628567, lecto-). Designated by Fernandes (1980: 376).

Synonyms:
Cotyledon laciniata L., *Sp. pl.* 1: 430 (1753) [1 May 1753]. **Type**: as above.

Verea laciniata (L.) Willd., *Enum. pl.*: 433 (1809). **Type**: as above.

Kalanchoe schweinfurthii Penzig in *Atti Congr. Bot. Genova* 1892: 341 (1893). **Type**: Ethiopia, Eritrea, Gheleb, 31. 3. 1891, *Penzig s.n.* (GE, holo-; K K000232764, fragment iso-).

Kalanchoe rohlfsii Engl. in *Annuario del Reale Istituto Botanico di Roma* 9: 252 (1902). **Type**: Abyssinia, Katz, 2450 m, 30 December 1880, *Rohlfs & Stecker s.n.* (B B 10 0153761, lecto-). Designated by Raadts (1977: 144).

Kalanchoe gloveri Cufod. in *Webbia* 19: 738 (1965). **Type**: British Somaliland [Somalia], Molidere, 29 January 1945, *Glover & Gilliland 787* (K K000232763, holo-).

Kalanchoe lentiginosa Cufod. in *Senckenberg. biol.* 39: 120, figure 8 (1958). **Type**: Culta in Horto Botanico Vindobonensi, 2 June 1896, florens, *Wettstein s.n.* (WU, holo- not seen).

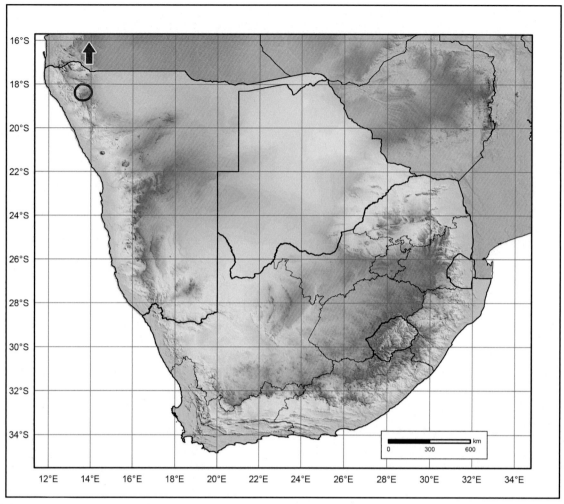

FIG. 12.6.2   Known geographical distribution range of *Kalanchoe laciniata* in southern Africa.

Derivation of scientific name:

From the Latin *laciniatus*, meaning divided into narrow lobes, usually referring to leaves.

Description:

Perennial or biennial, many-leaved, unbranched to sometimes once-branched, hairy to rarely glabrous, flimsy to robust succulent, and 0.3–1.2 m tall. *Stems* one to few, erect, mid- to light green to reddish-infused, somewhat glabrous lower down, and minutely and densely hairy higher up. *Leaves* petiolate, green, succulent, basal ones sometimes smooth, and upper leaves minutely hairy; *petiole* 1–10 mm long, flattened, canaliculate, and ± amplexicaul; *blade* variously shaped, dissected, trifid-lobed, rarely lacking incisions, 20–120 × 20–80 mm, leaflets ovate, and lanceolate or linear; *base* attenuate towards the base; *apex* obtuse-attenuate; *margins* crenate or dentate and rarely entire. *Inflorescence* 40–80 cm tall; erect to slightly leaning; apically dense; flat-topped; few- to many-flowered; arranged in cymes that form corymbs; rather elliptical in outline when viewed from above; branches opposite; growing upwards at a 45 degrees angle; subtended by small, undivided,

leaflike bracts; with very small leafy shoots in axils, axis light green, and all parts finely and densely glandular-hairy; *pedicels* 2–10 mm long and slender. *Flowers* erect to slanted sideways, greenish yellow (tube) and light yellow to creamy yellow (lobes), and sparsely glandular-hairy; *calyx* light green, glandular-hairy, *sepals* ± 4 × 2 mm, triangular-lanceolate, ± separate, basally adnate, acute, tips recurved, and hardly contrasting against corolla tube; *corolla* 9–12 mm long, somewhat enlarged lower down and in the middle, very slightly twisted apically after anthesis, and greenish yellow (tube) and light yellow to creamy yellow (tube); *tube* 8–11 mm long, indistinctly 4-angled, rounded when viewed from below, narrowly cylindrical above, greenish yellow, *lobes* 3.0–3.5 × 2–3 mm, ovate to oblong, acuminate towards, apiculate, light yellow to creamy yellow, and faintly green-tipped. *Stamens* inserted in two ranks well below the middle of the corolla tube, well included; *filaments* 2–4 mm long, thin, and whitish green; *anthers* 0.5–1 mm long and greenish yellow. *Pistil* consisting of 4 carpels; *carpels* 4–5 mm long and light yellowish green; *styles* 2–3 mm long; *stigmas* very slightly capitate and whitish green; *scales* ±

FIG. 12.6.3   The leaves of most forms of *Kalanchoe laciniata* are deeply incised. Photograph: Abraham E. van Wyk.

FIG. 12.6.4   The light green, sometimes red-infused, flowering stems of *Kalanchoe laciniata* are minutely hairy, more densely so higher up. The inflorescence branches are subtended by small, usually undivided, leaflike bracts, with very small leafy shoots in the axils. Photograph: Gideon F. Smith.

4 mm long, linear, whitish green, sometimes apically bifid. *Follicles* not seen. *Seeds* not seen. *Chromosome number*: $2n = 34$ and 68 (Sharma & Ghosh, 1967; Van Voorst & Arends, 1982; Raadts, 1989).

Flowering time:

April in southern Africa, elsewhere from August to September.

Illustrations:

De Candolle (1802: t. 100); Raymond-Hamet & Marnier-Lapostolle (1964: Plate XXVI, figures 84–85); Collenette (1985: 190, bottom right).

Common names:

English. Christmas tree kalanchoe and Christmas tree plant.

Geographical distribution:

Kalanchoe laciniata has a vast geographical distribution range that stretches across several thousand kilometres, from northwestern southern Africa to the Arabian Peninsula and beyond. In southern Africa, it barely crosses into the northern parts of Namibia but is

seemingly more common across the border in southern Angola. This species is therefore uncommon in southern Africa (Fig. 12.6.2).

Distribution by country. Angola, Ethiopia, Kenya, Malawi, Morocco [according to Descoings (2003: 160) a possible ancient introduction], Mozambique, Namibia, Somalia, Tanzania, Uganda, and Zimbabwe. Also in Arabia [included in Collenette (1985), but not in Lipscombe Vincett (1984)], India, and Thailand.

Habitat:

Kalanchoe laciniata occurs in grassland and bushveld (savanna), in stony and shaded places and in sandy or humus-rich soil. It was collected in the Kaokoveld Reserve on a weathered dolomite koppie [*De Winter & Leistner 5579* (PRE)]. Collenette (1985: 190) records it from a well-wooded ravine in Saudi Arabia.

Conservation status:

Not threatened.

Additional notes and discussion:

Taxonomic history and nomenclature. The identity of *Kalanchoe laciniata* was unclear for many decades as a result of its treatment as an exceedingly polymorphic entity by Raymond-Hamet in the first part of his monograph of *Kalanchoe* that was published in the early 1900s; in this work, he included over 50 names in the synonymy of *K. laciniata* (Hamet, 1907: 897–900). For the ensuing 50 years, this broad circumscription of *K. laciniata* caused considerable uncertainty on how the species should be interpreted, both morphologically and geographically. Hamet (1907) at least partly influenced how the species was treated by Jacobsen (1970: 286 pro parte, likely excluding glabrous material possibly referable to *K. crenata*), Cufodontis (1957: 710), and

FIG. 12.6.5   The yellow flowers of *Kalanchoe laciniata* are carried erectly to erectly spreading in dense or sparse inflorescences. Photograph: Gideon F. Smith.

FIG. 12.6.6   All parts of the flowers of *Kalanchoe laciniata*, including calyx and corolla, are finely and densely hairy. Photograph: Gideon F. Smith.

Jacobsen (1986: 617 pro parte, likely excluding glabrous material possibly referable to *K. crenata*, which Jacobsen recognised in this instance). Interestingly, some 50 years later, the views of Raymond-Hamet & Marnier-Lapostolle (1964: 76, Plate XXVI, figures 84–85) came to more closely reflect the current circumscription of *K. laciniata*.

Cufodontis (1965: 732), and later Raadts (1977), while rightly contesting Hamet's views of 1907 on *K. laciniata*, restricted this species to the glabrous material that occurs

FIG. 12.6.7 Close-up of the flowers and young, developing fruit of *Kalanchoe laciniata*. Photograph: Abraham E. van Wyk.

in the Far East and suggested that the name *K. schweinfurthii* should be applied to the African material that today is regarded as *K. laciniata*. Cufodontis (1965: 732), whose concept of *K. laciniata* therefore did not include *K. schweinfurthii*, based this assumption on the specimen 594-5 in the Linnaeus Herbarium (LINN) that had been identified as *K. laciniata* by Linnaeus and that Cufodontis considered to be the type of the name. However, that specimen is glabrous and very much unlike the African material. Fernandes (1980) discussed the early nomenclature of *K. laciniata* at length and, in consultation with G.E. Wickens, noted that the type of the name *K. laciniata* should be the pubescent specimen in the Hortus Cliffortianus herbarium. She argued that the protologue of *K. laciniata* (Linnaeus, 1753) repeats with very small changes the description Linnaeus (1738) had earlier published in his *Hortus cliffortianus*; therefore, the original material should be material in *Hortus Siccus Cliffortianus* not in LINN. According to Wickens (as cited by Fernandes 1980: 373), the specimen 594-5 likely corresponds to a species close to *K. gracilis* Hance (=*K. ceratophylla* Haw.), from India and China. Wickens (1982) held that this specimen had misled Linnaeus and subsequent authors, as Linnaeus had not seen the *Hortus cliffortianus* material again after completing that work.

As a consequence of this typification, Fernandes (1980: 379) synonymised *K. schweinfurthii* with *K. laciniata*, a view that has been followed since. Fernandes did not see the holotype of *K. schweinfurthii* [*Penzig s.n.* (GE)] and at present, the specimen is also not available for examination, as the holdings of the GE Herbarium (Genoa, Italy) are not accessible online. Even though

Penzig's original herbarium was partly destroyed at GE (Stafleu & Cowan, 1983), this holotype is listed as having been examined by Raadts (1977).

Kalanchoe laciniata is currently considered to be a species that extends from India through East Africa to Angola and marginally enters Namibia in the north. Although it was recorded as occurring in Nigeria and Cameroon, in West Africa, by Keay (1954: 117) and later by Burkill (1985), according to Fernandes (1980: 380), those records correspond to *K. crenata*. The confusion may have resulted from this West African material being glabrous, which is how Keay (1954) interpreted *K. laciniata* (Fernandes, 1980: 380).

Identity and close allies. *Kalanchoe laciniata* may have affinities with the partially hirsute, orange-flowered *K. lanceolata*, which in southern Africa is a predominantly eastern and north-central element. The geographical distributions of *K. laciniata* and *K. lanceolata* are very similar, with both species having extraordinarily wide ranges that stretch from southwestern south-tropical Africa to the Orient, through East Africa and Arabia. The possibility of a link between *K. laciniata* and *K. crenata* cannot be ruled out but seems somewhat more remote. The latter two species also occur sympatrically over parts of their distribution ranges (see, e.g. Sapieha, n.d.: 52).

Horticulture and cultivation. The species is largely unknown in cultivation in southern Africa. However, it presents few challenges to grow successfully and after flowering will self-seed with ease in a greenhouse, as at least some clones of *K. laciniata* are self-fertile.

Uses. No uses for the species have been recorded in southern Africa.

7. *KALANCHOE LANCEOLATA* (FORSSK.) PERS.

Narrow-leaved kalanchoe (Figs. 12.7.1 to 12.7.10)

FIG. 12.7.1   In southern Africa, the most common form of *Kalanchoe lanceolata* carries bright orange flowers.   Photograph: Gideon F. Smith.

Kalanchoe lanceolata (**Forssk.**) **Pers.**, *Syn. pl.* 1: 446 (1805), as *"Calanchoe lanceolata"*. De Candolle (1828: 395); Haworth (1829: 304); Hamet (1908b: 32); Hutchinson (1946: 459, 465, 471, 484, 488) [see Spalding (1953: 28)]; Raymond-Hamet & Marnier-Lapostolle (1964: 77, Plate XXVI, figure 86, and Plate XXVII, figures 87–88); Compton (1966: 44); Friedrich (1968: 38); Jacobsen (1970: 286); Compton (1976: 219); Raadts (1977: 139); Fernandes (1980: 381); Fernandes (1982a: 17); Germishuizen & Fabian (1982: 120, plate 55c); Tölken (1982: Plate 1848); Fernandes (1983: 42); Kemp (1983: 31); Collenette (1985: 191); Tölken (1985: 66); Jacobsen (1986: 617); Wickens (1987: 49); Boiteau & Allorge-Boiteau (1995: 193); Bolnick (1995: 12, Plate 6, figure 37); Germishuizen & Fabian (1997: 154, plate 69c); Retief & Herman (1997: 392); Pooley (1998: 54); Descoings (2003: 160); Lebrun & Stork (2003: 214, 216); Braun et al. (2004: 37); Court (2010: 91); Gill & Engelbrecht (2012: 88); Kirby (2013: 230); Smith et al. (2017b: 309). **Type**: Arabia [Yemen], Kurma, *Herb. Forsskål 689* (C C10002080, holo-).

Synonyms:

Cotyledon lanceolata Forssk., *Fl. Ægypt.-Arab.*: CXI & 89 (1775). **Type**: as above.

Verea lanceolata (Forssk.) Spreng., *Syst. Veg.* ed. 16, 2: 260 (1825). **Type**: as above.

Kalanchoe glandulosa Hochst. ex A.Rich., *Tent. Fl. Abyss.* 1: 312 (1848). **Type**: Ethiopia, Dscheladscheranne, *Schimper 904* (P P00374094, B B 10 0153764, BR BR0000008251183 & BR0000008885746, K K000232767, LE LE00013295 & LE00013296 & LE00013294, LG LG0000090028144, M M0108096, MPU MPU015134,

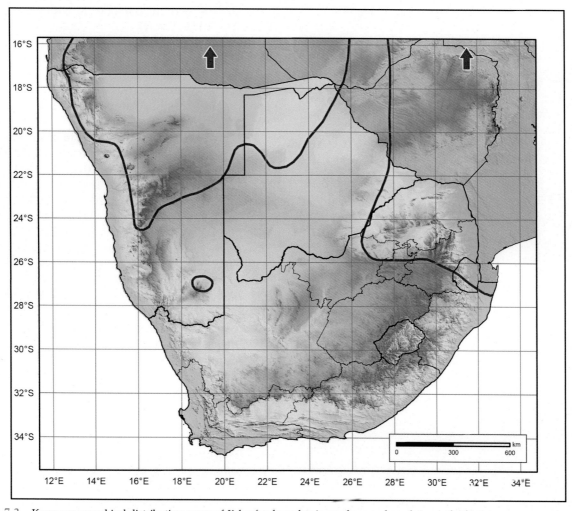

FIG. 12.7.2   Known geographical distribution range of *Kalanchoe lanceolata* in southern and south-tropical Africa.

FIG. 12.7.3   A dense stand of *Kalanchoe lanceolata*. In the natural habitat of the species, the leaves become floppy during the heat of midday. Photograph: Gideon F. Smith.

FIG. 12.7.4   When young, plants of *Kalanchoe lanceolata* are completely glabrous. Photograph: Gideon F. Smith.

FIG. 12.7.5 The margins of the lanceolate leaves of *Kalanchoe lanceolata* are usually variously crenate to dentate. Photograph: Gideon F. Smith.

FIG. 12.7.6 Small plantlets form in the axils of the leaves that are carried on the tomentose peduncle. Photograph: Gideon F. Smith.

P P00374093, S S-G-10754, TUB TUB000851 & TUB000852).

Kalanchoe ritchieana Dalzell in *Hooker's J. Bot. Kew Gard. Misc.* 4: 346 (1852). **Type**: India, Belgaum, December,

Ritchie 329 (K without barcode, K without barcode, E E00393033 & E00393034).

Meristostylus macrocalyx Klotzsch in Peters, *Reise Mossamb., Bot.* 1: 269 (1862). **Type**: Mozambique, Boror, *Peters s.n.* (B B_10_0241957, holo-).

Kalanchoe modesta Kotschy & Peyr., *Pl. Tinn.*: 18 (1867). **Type**: [Sudan] Dembo, Bahr-Djur, 11.1863, *Herb. Caes. Palat. Vindob. Exped. Tinn. 42* (W?).

Kalanchoe platysepala Welw. ex Britten in Oliver (ed.), *Fl. Trop. Afr.* 2: 393 (1871). Jacobsen (1970: 288); Jacobsen (1986: 624), and Court (2010: 91). **Type**: Angola, Distr. Huilla, Lopollo, fl. et fr. April et May 1860, *Welwitsch 2484* (LISU LISU209306, LISU209307, K K000232732, COI COI00005660, BM BM000649711). Designated by Fernandes (1982a: 20) as 'holotype'.

Kalanchoe glandulosa var. *benguelensis* Engl., *Hochgebirgsfl. Trop. Afr.*: 233 (1892). **Type**: Angola, Distr. Huilla [Huíla], Morro de Lopollo, fl. April et May 1860, *Welwitsch 2485* (BM BM000649710, BR, COI, K K000232731, LISU, P).

Kalanchoe crenata var. *collina* Engl., *Pflanzenw. Ost-Afr.* C: 189 (1895) pro parte. **Type**: Usambara, Kwa Mshusa [Kwa Mshuza], 5.8.1893, *Holst 8907* (B B 10 0153763, lecto-; G G00023457, HBG HBG505732, K K000232737, LE, M M0108100, US US00008079). Designated by Raadts (1977).

Kalanchoe pilosa Baker in *Bull. Misc. Inform. Kew* 1895: 289 (1895). **Type**: Zambia, Mwero Plateau, west of Lake Tanganyika, 1894, *Carson 3* (K K000232736, holo-).

Kalanchoe pentheri Schltr. in *J. Bot.* 35: 341 (1897). Burtt-Davy & Pott-Leendertz (1912: 144). **Type**: [South Africa], and Transvaal, [Limpopo Province], Buiskop, 28 May 1895, *Penther s. n.* (B B 10 0153766, holo-).

Kalanchoe glandulosa var. *rhodesica* Baker f. in *J. Bot.* 37: 434 (1899). **Type**: Rhodesia [Zimbabwe], Salisbury, and Mashonaland, 7.1898, *Rand 465* (BM BM000649707, holo-).

Kalanchoe glandulosa var. *tomentosa* Keissl. in Zahlbruckner in *Ann. Naturh. Mus. Wien* 15: 36 (1900). Pott (1920: 126). **Type**: Transvaal, Buiskop, 29 May 1895, *Krook sub Penther 2338* (W W18970009001, holo-; BOL).

Kalanchoe goetzei Engl. in *Bot. Jahrb. Syst.* 30: 312 (1901). **Type**: Nyassa-See und Kinga-Gebirgs-Exp., Ussangu, am Msimasi-Sumpf, 10.6.1899, *Goetze 1025* (B B 10 0153765, lecto-; BM without barcode, BR BR0000008886378 & BR0000008886361, E00217842, isolecto-).

Kalanchoe diversa N.E. Br. in *Gard. Chron.* 32: 210 (1902). **Type**: Brought from Somaliland by Mrs Lort Phillips, grown at Kew, 26 April 1902, *Brown s.n.* (K K000232765, holo-).

Kalanchoe junodii Schinz in *Vierteljahrsschr. Naturf. Ges. Zürich* 57: 556 (1912). Pott (1920: 126). **Type**: [South Africa] Transvaal, presumably from Shiluvane [also spelled 'Shilovane'], east-central Limpopo

FIG. 12.7.8 The inflorescences and all the plant parts of *Kalanchoe lanceolata* are densely tomentose. Photograph: Gideon F. Smith.

FIG. 12.7.7 This form of *Kalanchoe lanceolata* has bright orange flowers. Photograph: Gideon F. Smith.

Province, grown from seed sent by Junod, *Schinz 5537* (Z Z-000004320, holo-).

Kalanchoe ellacombei N.E.Br. in *Bull. Misc. Inform. Kew* 1912: 329 (1912). **Syntypes**: Rhodesia [Zambia], on the northern bank of River Zambezi, *Ellacombe s.n.* (K K000232734) & Livingstone, *Ellacombe s.n.* (K K000232733).

Kalanchoe homblei De Wild. in *Repert. Spec. Nov. Regni Veg.* 12: 298 (1913). **Type**: [Democratic Republic of Congo], Ober-Katanga, Elisabethville, 05.1912, *Homblé*

FIG. 12.7.9　In southern Africa, this yellow-flowered form of *Kalanchoe lanceolata* is much less commonly encountered than the orange-flowered form. Photograph: Gideon F. Smith.

FIG. 12.7.10　Plants of *Kalanchoe lanceolata* that have reached flowering maturity, such as this one, usually die after the flowers are spent. Photograph: Gideon F. Smith.

656 pro parte (BR BR0000008885791, BR0000008885807 pro parte, BR0000008885753 pro parte).

Kalanchoe homblei forma *reducta* De Wild. in *Ann. Soc. Sci. Brux.* 40: 89 (1921). **Types (syntypes)**: [Democratic Republic of Congo], Elisabethville, 1912, *Homblé 656* pro parte (BR0000008885807 pro parte & BR0000008885753 pro parte) & *Bequaert 452* (BR).

Kalanchoe laciniata var. *brachycalyx* (A.Rich.) Chiov., *Risult. Sci. Miss. Stef.-Paoli*: 75 (1916), as "Calanchoe".

Kalanchoe gregaria Dinter in *Repert. spec. nov. regni veg.* 18: 433 (1922). **Type**: Deutsch Südwestafrika [Namibia], Windhoek, Osona, *Dinter s.n.* (B†?).

Kalanchoe lanceolata var. *lanceolata* Cufod. in *Webbia* 19 (2): 729 (1965).

Kalanchoe lanceolata var. *glandulosa* (Hochst. ex A.Rich.) Cufod. in *Webbia* 19(2): 730 (1965). Jacobsen (1970: 286). **Type**: As for *Kalanchoe glandulosa* Hochst. ex A.Rich.

?Kalanchoe brachycalyx A.Rich., *Tent. Fl. Abyss.* 1: 312 (1848) [26 February 1848]. **Type**: Abyssinia [Ethiopia], Maye-Gouagoua, *Quartin Dillon s.n.* (P P00374103, holo-).

Derivation of scientific name:

From the Latin *lanceolatus*, meaning spear-shaped, that is, widest below the middle.

Description:

Annual or biennial, many-leaved, unbranched, glabrous (vegetative phase) to hairy (reproductive organs), flimsy to robust succulent, and to 1.5 m tall. *Stems* one, erect, green, glabrous lower down, glandular-hairy higher up, and often quadrangular-angled lower down. *Leaves* sessile, green to yellowish green, succulent, longitudinally folded, often floppy, caducous when inflorescence develops, basal ones glabrous, upper leaves glandular-hairy, and drying papery; *petiole* absent; *blade* 5–20 × 2–10 cm, variously shaped, narrowly oblong, lanceolate, elliptic, and ovate-spathulate; *base* attenuate to cuneate and stem-clasping; *apex* obtuse, often indented, and more rarely acute; *margins* variably crenate or dentate and rarely entire. *Inflorescence* 30–50 cm tall, erect, apically dense, flat-topped, many-flowered, arranged in an elongated thyrse consisting of a number of dichasia ending in elongated monochasia that can be recurved or coiled that is especially evident in bud and fruit, branches opposite, growing upwards at a 30 degrees angle, subtended by small leaflike bracts, sometimes with very small leafy shoots in axils, axis light to midgreen, and all parts finely and densely glandular-hairy; *pedicels* 1–8 mm long and slender. *Flowers* erect to slanted sideways, rarely pendent, yellowish orange (tube) and bright to yellowy orange (lobes), basal swollen part mid- to light green, and densely glandular-hairy; *calyx* midgreen and glandular-hairy; *sepals* basally fused in 1–8 mm long, separate above into lobes ± 3–12 × 2–5 mm, triangular-lanceolate or ovate-lanceolate, acute, curved away from corolla, obscuring and hardly contrasting against basal, and swollen part of corolla; *corolla* 6–13 mm long, distinctly enlarged lower down, very slightly twisted apically after anthesis, yellowy orange (tube) and bright to yellowy orange (lobes), and glandular-hairy; *tube* 5–12 mm long, distinctly to indistinctly 4-angled fluted, rounded when viewed from below, narrowly cylindrical above, and yellowish orange; *lobes* 2–6 × 1–4 mm, obovate or lanceolate, acuminate towards apex, apiculate-mucronate, and bright to yellowish orange. *Stamens* inserted in ± two ranks at about the middle of the corolla tube, well included; *filaments* 2–4 mm long, thin, and yellowish green; *anthers* 0.5–1 mm long, yellowish grey, and oblong. *Pistil* consisting of 4 carpels; *carpels* 4–5 mm long and light yellowish green; *styles* 1–3 mm long; *stigmas* very slightly capitate and whitish green; *scales* 2–4 mm long, linear, and yellowish green. *Follicles* 5–8 mm long; brittle; grass spikelet-like; and enveloped in dry, straw-coloured remains of corolla. *Seeds* 0.25–0.5 mm long, cigar-shaped to rectangular, and light to dark brown. *Chromosome number*: $2n = 34$ (Raadts, 1989: 171).

Flowering time:

April–June, peaking in May (southern hemisphere).

Illustrations:

Raymond-Hamet & Marnier-Lapostolle (1964: Plate XXVI, figure 86, and Plate XXVII, figures 87–88); Fernandes (1983: 44, Tab. 6); Wickens (1987: 50, figure 7 [reproduced from Fernandes, 1983: 44]); Germishuizen & Fabian (1982: 120, plate 55c); Tölken (1982: Plate 1848); Bolnick (1995: Plate 6, figure 37); Germishuizen & Fabian (1997: 154, plate 69c); Smith (1997: x [reproduced from Tölken, 1982: Plate 1848]); Sajeva & Costanzo (2000: 173, middle right); Gill & Engelbrecht (2012: 89, top, centre); Kirby (2013: 230); Smith et al. (2017b: 309, bottom).

Common names:

Afrikaans: wolplakkie and wolkalanchoe.

English: kalanchoe (Bolnick, 1995: 12), narrow-leaved kalanchoe (Pooley, 1998: 54), pig's ears (Long, 2005), and woolly kalanchoe.

Kunda (Zambia): jelijeli (Astle et al., 1997b: 194).

Setswana: moethimodiso, nyêthi, and semonye (Kirby, 2013: 230).

Geographical distribution:

In South Africa, *Kalanchoe lanceolata* is widely distributed in the northern and eastern provinces. It also occurs widely in Botswana and Namibia, extending further north through Zambia (Astle et al., 1997a: 144, b: 194; Smith, 1997: 232) and Zimbabwe to eastern Africa. North of the equator, it also occurs in an east-west band, as far west as Mali. Beyond [eastern] Africa, the species has been reported from Yemen, Saudi Arabia [recorded by Collenette, 1985, but not by Lipscombe Vincett, 1984], and India. Boiteau & Allorge-Boiteau (1995: 193–194, Map 22) also recorded the species from southwestern Madagascar (Fig. 12.7.2).

Distribution by country. Angola, Botswana, Cameroon, Central African Republic, Democratic Republic of Congo, Ethiopia, Ghana, Kenya, Madagascar, Malawi, Mali, Mozambique, Namibia, Nigeria, Somalia, South Africa, Sudan, Swaziland, Tanzania, Uganda, Zambia, and Zimbabwe. Also in India, Saudi Arabia, and Yemen.

Habitat:

In southern Africa, *Kalanchoe lanceolata* is a typical savanna (bushveld) species that often occurs in the shade of trees with sparse to dense canopies. In these habitats, plants of *Kalanchoe lanceolata* often grow in dense swathes with comparatively few other grass and forb species providing competition. In the dry season, plants appear decidedly desiccated, and the leaves turn variously floppy, often exposing their abaxial surfaces; recovery is quick after rain. *Kalanchoe lanceolata* will also be found in fairly dense stands of a variety of tall-growing grass species.

Conservation status:

The species is regarded as being of 'Least concern' and not threatened with extinction.

Additional notes and discussion:

Taxonomic history and nomenclature. *Kalanchoe lanceolata* was first described as *Cotyledon lanceolata*, by Forsskal in 1775. It was based on a plant from Kurma, Arabia (the west coast of Yemen). As the extensive synonymy of *K. lanceolata* indicates, since then, this species has been described several times, under different names, when plants were collected in other countries. A number of varieties and at least 12 species were described from material from Angola, Democratic Republic of Congo, Ethiopia, India, Kenya, Mozambique, Namibia, Somalia, South Africa, and Zambia, all of which are now considered synonyms of the name *K. lanceolata*, so reflecting the variability that is found in this widespread species.

Harvey (1862) in his treatment of *Kalanchoe* for the *Flora capensis* project did not mention *Kalanchoe lanceolata*; at the time of his floristic work in South Africa, the species had not yet been recorded for the country. As far as we could determine, material today referable to *Kalanchoe lanceolata* was for the first time mentioned from South

Africa (Transvaal, [Limpopo Province], Buiskop), under one of its synonyms, *K. pentheri* (Schlechter, 1897), an occurrence repeated by Burtt-Davy & Pott-Leendertz (1912: 144). Further records of what is today regarded as *K. lanceolata* were noted under the names *K. glandulosa* var. *tomentosa* and *K. junodii* by Pott (1920: 126) from the southern part of the Waterberg near Nylstroom (nowadays known as Modimolle) and Shiluvane [as 'Shilovane'] in the Limpopo Province, respectively. For the next 70 years, *K. lanceolata* was most often referred to in southern Africa as one of its synonyms, *K. pentheri*.

In the subtropical eastern parts of southern Africa, especially in KwaZulu-Natal, South Africa, *Kalanchoe lanceolata* remained undetected for a long time. In this region, the genus *Kalanchoe* has diversified extensively, and *K. lanceolata* is today known to occur there. Ross (1972: 179) omitted the species from his Flora of Natal [KwaZulu-Natal], and Pooley (1998: 54) only recorded it as occurring as far south as Swaziland. However, there is a possibility that the orange-flowered '*Kalanchoe* sp.' illustrated by Jeppe (1975: 50, Plate 28c) in fact represents *K. lanceolata*, but it could as well be a form of *K. rotundifolia*.

Kalanchoe platysepala was described from material collected in Angola and Malawi. Of the two syntypes ('Benguela, Huilla, Dr Welwitsch' and 'Shire Valley, Zambesi land, Dr Kirk'), Fernandes designated the Angolan collection [*Welwitsch 2484* (LISU)] as 'holotype' (Fernandes 1982a). However, there are two specimens of that collection at LISU. Court (2010: 91) upheld the species, but it is nowadays generally included in the synonymy of *K. lanceolata*.

'*Kalanchoe pubescens* R.Br. ex Oliv. in the *Fl. Trop. Afr.* [Oliver et al.] 2: 396. 1871' that is listed in the International Plant Names Index (IPNI) is not a name as it was not validated by Oliver. Although it was referred to by Britten (1871), it was then mentioned as a synonym of *K. brachycalyx* and not validated. *Kalanchoe brachycalyx* was there questioned as a possible synonym of *K. lanceolata*. Fernandes (1983) thought it could be a synonym of *K. laciniata* instead. Fernandes (1983) also questioned a further synonym, *Kalanchoe diversa*, which, in her view, should ideally remain as an unknown taxon. However, Thulin (1993) confirmed this name as a synonym.

IPNI also lists '*Cotyledon hirsuta* Herb.Heyne ex C.B. Clarke in Hooker, *Fl. Brit. India* 2(5): 414 (1878) [Jul 1878]'. This is not a published name. In this reference, it was listed as a synonym under *Kalanchoe glandulosa*, so it was not accepted by Clarke (1878: 414).

In certain cases, the existence of a mixture of plant parts from multiple species in type material has created some nomenclatural problems. For example, in *Kalanchoe crenata* var. *collina*, according to Raadts (1977), the type material seen by Engler partly belongs to *K. lanceolata*, and partly to *K. crenata*. Likewise, De Wildeman (1921) noted that

for *K. homblei* that he had described in 1914, the type, *Homble 656*, consisted of two different plants. These are mixed in two of the sheets. *Kalanchoe homblei* is the larger plant, with the lower part of the plants in sheet BR0000008885791 and the top (inflorescences) in sheet BR0000008885753 appearing together with some smaller plants that are referable to *K. homblei* forma *reducta*. Sheet BR0000008885807 includes parts of both formas of the species.

There is no holotype for *Kalanchoe ritchieana*. Even though one of the specimens at K has a label stating it is the holotype, there is no indication that a holotype was originally cited.

The type of *Kalanchoe gregaria* was probably destroyed at Berlin during WWII. Although there is a collection by Dinter identified as *K. gregaria* at BM and Z, it is from a later date (*Dinter 7400* from 1934). An earlier collection, *Dinter 3491*, kept at BM, was collected on 24 April 1922 before the publication of the name, which dates from December 1922. However, *Dinter 3491* is from Okapanka and not Windhoek or Osona as cited in the protologue, so it cannot be the type.

The type of *Kalanchoe pentheri* has been given wrongly in the literature (e.g. Tölken, 1985: 66) as being a *Krook sub Penther s.n.* specimen held at W. It should be *Penther s. n.* (B B 10 0153766, holo-).

Identity and close allies. In *Kalanchoe lanceolata*, the stem, which becomes roughly angled with age, remains erect and can reach a length of well over 1 m when in flower. The leaves are stem-clasping, sometimes even basally auriculate, and lanceolate to rather narrowly obovate to elliptical. In the vegetative phase, plants are virtually hairless, but when in flower, all the upper plant parts are densely covered in fine but conspicuous hairs, including the calyx and corolla. The species likely has some affinities with the often even more hirsute *K. laciniata*, which in southern Africa is restricted to northern Namibia, however.

Horticulture and cultivation. *Kalanchoe lanceolata* is not favoured by gardeners, and if grown at all, it serves as little more than a curiosity plant, given its smooth basal, somewhat rosulately arranged leaves, and tomentose peduncles and flowers.

The species tends to be a biennial and dies quite rapidly after having flowered. The light green leaves desiccate to a light-brown colour and papery texture. Regeneration is predominantly from the fine, almost dustlike, seed. In some cases, especially if moisture is maintained in the rooting zone, plants will form suckers.

Uses. Kirby (2013: 230) records the species as being used in traditional medicine for the treatment of stiff joints and rheumatism, as well as for snuff.

Poisonous properties. Plants are known to be poisonous (Kirby, 2013: 231), apparently causing a form of heart glycoside poisoning. The flowers may be more poisonous than the leaves (Long, 2005).

8. *KALANCHOE LEBLANCIAE* RAYM.-HAMET

Sand forest kalanchoe (Figs. 12.8.1 to 12.8.11)

FIG. 12.8.1 A dense clump of *Kalanchoe leblanciae* with distinctly recurved leaves arranged along the length of the erect to leaning stems. Photograph: Neil R. Crouch.

Kalanchoe leblanciae **Raym.-Hamet** in *Repert. Spec. Nov. Regni Veg.* 11: 294 (1912), as '*leblancae*'. Fernandes (1980: 329); Fernandes (1983: 64); Descoings (2003: 162); Bandeira et al. (2007: 83, 138) as '*leblancae*'; Crouch et al. (2016a: 70); Smith et al. (2017a: 57). **Type**: Delagoa Bay [Mozambique], 1893, *H. Junod 443* (G G00418140, holo- in 2 sheets; G G00418141, P not seen, image not available online, BR BR0000008886750, Z Z-000101954).

Synonyms:
None recorded.

Derivation of the scientific name:
The species is named after Alice Leblanc, an intimate friend of Raymond-Hamet. She is commemorated in several species of Crassulaceae, including in *Kalanchoe aliciae* Raym.-Hamet (see Raymond-Hamet & Marnier-Lapostolle, 1964: 29, Plate VIII [figures 18–19]). Somewhat playfully, Alice Leblanc is a co-author with Raymond-Hamet of *Kalanchoe mitejea* Leblanc & Raym.-Hamet ['mitejea' being an anagram of *je t'aime*] (see Raymond-Hamet & Marnier-Lapostolle, 1964: 81, Plate XXIX [figures 96–97]; Figueiredo & Smith 2010). The original spelling of the epithet *leblancae* has been corrected to *leblanciae* under the provisions of Article 60, Recommendation 60C.1 of McNeill et al. (2012).

Description:
Perennial, glabrous, succulent herb, 0.3–1.6 m tall, and with slight whitish bloom on preserved specimens. *Stems* simple or once-branched, erect to leaning, woody at the base, ± distinctly 4-angled, 5–9 mm in diameter, and tapering upwards. *Leaves* up to 10 × 1–3 cm, ± sessile, narrowing basally to a subpetiolar base, not amplexicaul,

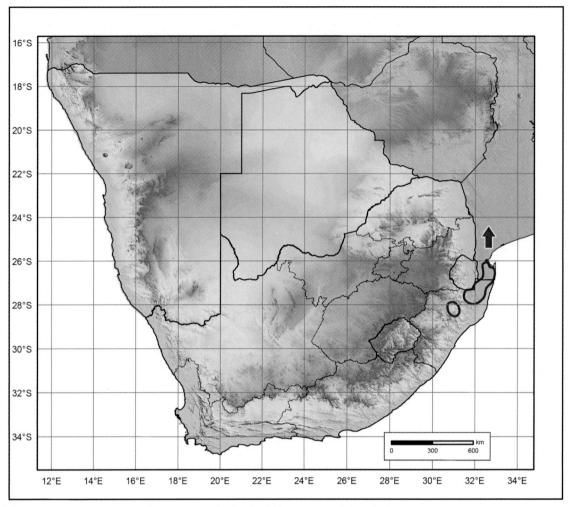

FIG. 12.8.2   Known geographical distribution range of *Kalanchoe leblanciae* in southern Africa.

FIG. 12.8.3   The leaves of *Kalanchoe leblanciae* are not petiolate, unlike those of *K. sexangularis*, a close relative, and generally do not take on the reddish colour characteristic of the latter species. Leaves remain light to midgreen where plants grow in shady positions. Photograph: Gideon F. Smith.

FIG. 12.8.4   The four-angled stems of *Kalanchoe leblanciae* carry a substantial waxy bloom, a feature shared neither with *K. sexangularis* nor with *K. brachyloba*, with which it has been confused in the past. Photograph: Gideon F. Smith.

FIG. 12.8.5 The leaves of *Kalanchoe leblanciae* are oblong to narrowly oblong and narrow to a subpetiolar base. A few, large, obtuse crenations are evident along the margins. Photograph: Gideon F. Smith.

FIG. 12.8.7 *Kalanchoe* leblanciae carries tall, branched inflorescences in midwinter. Photograph: Gideon F. Smith.

FIG. 12.8.8 Flowers of *Kalanchoe leblanciae* are carried erectly and clustered into dense groups. Photograph: Gideon F. Smith.

FIG. 12.8.6 Small, perfectly formed plantlets that can be removed and grown develop in the axils of the leaves higher up on the peduncle of *Kalanchoe leblanciae*. Photograph: Gideon F. Smith.

flat to slightly lengthwise-folded, fleshy, gracefully recurved to completely down-curved, oblong to oblong-spathulate, *base* attenuate, midgreen, sometimes with a purplish hue during periods of environmental stress, *apex* rounded or obtuse, *margins* dentate, sinuate-crenate, or serrate-lobed, and reddish brown tinted. *Inflorescence* 0.3–1.6 m tall, narrow, much elongated, apically dense, many-flowered corymbose cymes or panicles, branches opposite, erect to patent-erect, subtended by leaflike bracts, with small leafy shoots often produced in axils, and greenish yellow; *pedicels* ± obsolete, 3.0–4.5 mm long when present, and slender. *Flowers* erect; *calyx* consisting of 4 ± separate sepals, basally adnate, *sepals* 1.2–1.8 × 0.8–1.0 mm, triangular-lanceolate, acute, shiny midgreen, and contrasting darker green against corolla tube; *corolla* 8–12 mm long, not twisted apically after anthesis, yellowish green and yellow [reportedly also varying from salmon to bright red, but these colours are not observed], *tube* 6–8 mm long, distinctly 4-angled, box-shaped-square when viewed from below, and yellowish green; *lobes* 2–3 × 1.8–2.0 mm, suboblong to ovate, apex mucronate, rounded or obtuse, bright yellow, and faintly brown-tipped. *Stamens* inserted in or slightly above the middle of the corolla tube, included; *filaments*

FIG. 12.8.9 The flowers of *Kalanchoe leblanciae* have distinctly four-angled corolla tubes that are up to 8 mm long. Photograph: Gideon F. Smith.

± 1.5 mm long, thin, light green; *anthers* ± 0.5 mm long and brownish yellow. *Pistil* consisting of 4 carpels; *carpels* 5–6 mm long and light yellowish green; *styles* 1.0–1.5 mm long; *stigmas* very slightly capitate, light yellow; scales ± 1 mm long and linear. *Follicles* not seen. *Seed* not seen. *Chromosome number*: unknown.

Flowering time:

June–October, peaking in July.

Illustrations:

Pooley (1998: 254, 255, *K. leblanciae* mistakenly figured as *K. brachyloba*); Crouch et al. (2016a: 71); Smith et al. (2017a: 56, plate 2328).

Common names:

Afrikaans: sandwoud-kalanchoe.

English: sand forest kalanchoe.

Geographical distribution:

Although *Kalanchoe leblanciae* was stated by Fernandes (1980: 396–397) to be endemic to Mozambique (one of the countries covered by the *Flora zambesiaca* [FZ] region), specifically to the Manica, Sofala, Sul do Save, and Maputo districts, the species was recorded from Natal [now KwaZulu-Natal], South Africa, by Ross (1972). In Fernandes (1983: 65), the *FZ* treatment of the Crassulaceae, she amended the distribution range and indicated that the species occurred also in South Africa. She gave the distribution in that country as "(Natal, Orange Free State?)". Apart from noting its presence in Mozambique, Descoings (2003) also reported it from "Republic of South Africa (RSA)" (Crouch et al., 2016a). However, *K. leblanciae* was not mentioned in Tölken's (1985) *Flora of Southern African* (*FSA*) treatment of the Crassulaceae. It was therefore omitted from all recent catalogues of southern and

FIG. 12.8.10 *Kalanchoe leblanciae* growing in Licuáti Thicket in the Licuáti Forest Reserve in southern Mozambique. The inflorescence is over 1.5 m tall. The specimen is held aloft by Mr Abilio Manhique, Department of Botany, National Institute of Agronomic Research (INIA), Maputo, Mozambique. Photograph: Abraham E. van Wyk.

South African plants in general (Burgoyne, 2003, 2006) and succulent plants specifically (Dreyer, 1997) and from a preliminary checklist of Mozambican plants (Silva et al. 2004). However, Crouch et al. (2016a) and Smith et al. (2017a) have confirmed that the species is indeed present in KwaZulu-Natal, South Africa; the species is absent from the country's Free State Province. In the north, the *FZ* region is geographically abutted by the *Flora of Tropical East Africa* region, an area from which the species has not been reported (Wickens, 1987) (Fig. 12.8.2).

Distribution by country. Mozambique (Maputo, Gaza, Inhambane, and Sofala provinces) and South Africa (KwaZulu-Natal).

Habitat:

In its natural habitat in Mozambique, *Kalanchoe leblanciae* often occurs in low, forest-like savanna (bushveld)

FIG. 12.8.11 A drought-stressed plant of *Kalanchoe leblanciae* in flower in the Mkhuze Game Reserve, KwaZulu-Natal, South Africa. Photograph: Neil R. Crouch.

with sparse, open canopies; in shrubby grasslands; or in sparse to dense coastal dune scrub established on very sandy soils (Bandeira et al., 2007). The species is an obvious component of sand forest in South Africa (Rutherford et al., 2006), in both the Tembe Elephant Reserve and Ndumo Game Reserve. This forest-type houses not only a number of local Maputaland endemics but also the habitat of various tropical elements that reach their southernmost distribution here. With the exception of the westernmost location recorded for *K. leblanciae*, all collections fall within the Maputaland Centre of Endemism (Van Wyk & Smith, 2001: 87, Map 14).

Conservation status:

In South Africa, the species is regarded as 'Least concern' and not threatened with extinction.

Additional notes and discussion:

Taxonomic history and nomenclature. Kalanchoe leblanciae was described by Raymond-Hamet just over 100 years ago, based on material from "Delagoa Bay", in what is today Maputo, Mozambique (Hamet, 1912a). The first gathering of *K. leblanciae* that we could trace is by Henri-Alexandre Junod, a Swiss Romande missionary, who in October of 1893 pressed what would become the holotype, noting that it was rare. Most other Mozambique collections accessioned in South Africa are from the same general locality as the type. Junod also collected the holotype of *K. luciae* Raym.-Hamet and was commemorated in *K. junodii* Schinz, which is generally considered conspecific with *K. lanceolata* (Descoings, 2003).

Beyond the first description of the species, it was included in the standard regional floristic, monographic, and synoptic works published in the past 40 years, such as Fernandes (1980, 1983) and Descoings (2003). It was not until recently that the presence of the species in South Africa was confirmed (Crouch et al., 2016a; Smith et al., 2017a).

Identity and close allies. There has been some doubt about the taxonomic status of *Kalanchoe leblanciae*. Some

FIG. 12.8.12   Licuáti Thicket, or Sand Thicket, west of Bela Vista, southern Mozambique, is the typical vegetation type in which *Kalanchoe leblanciae* occurs. This type of impenetrable thicket is often associated with so-called sand forest. Photograph: Abraham E. van Wyk.

of this confusion was precipitated at the time that Hamet (1912a: 292) described the species, as he noted that in his earlier monograph on the genus (Hamet, 1907), the species would key out as *K. grandiflora* Wight & Arn., an Indian species. Further taxonomic complications encountered in determining and identifying specimens of *K. leblanciae* resulted from mixed gatherings attached to herbarium sheets (see, e.g. Fernandes, 1980). In the protologue, Hamet (1912a: 292) also noted that *K. leblanciae* could be confused with a number of other *Kalanchoe* species. These include *K. sexangularis*, *K. paniculata*, *K. rotundifolia*, *K. longiflora*, and *K. brachyloba*. *Kalanchoe leblanciae* may be readily distinguished from particularly *K. brachyloba*, *K. hirta*, and *K. longiflora*, all of which have clasping leaf bases. Hamet (1912a) further regarded *K. leblanciae* as similar to *K. rotundifolia*, from which it differs in several leaf and flower characters, some of which were elaborated on by Fernandes (1980: 397). She distinguished *K. leblanciae* from the polymorphic *K. rotundifolia* on account of its thicker, sessile leaves that are of a different shape and various flower characters on which she did not expand. Although the flower colour of *K. leblanciae* has been reported as ranging from "…yellow green, yellow, salmon to bright red…" (Fernandes, 1983; Descoings, 2003; Bandeira et al., 2007), we have yet to encounter specimens with flowers that are not yellow (corolla lobes) and yellowish green (corolla tube). In *K. rotundifolia*, flower colour ranges from salmon, to orange, to bright red, sometimes even bicoloured, but all characteristically show pronounced twisting of the corollas, a character absent from *K. leblanciae*. Fernandes (1983: 42) justifiably distinguished *K. rotundifolia* from the group consisting of *K. brachyloba*, *K. paniculata*, and *K. leblanciae* on the pink to brick flowers of the former species, whereas the latter three consistently have yellow corollas. Confusingly, in the same publication, Fernandes (1983: 65) described *K. leblanciae* as having a corolla "yellow green to yellow or

salmon to bright red, reddish brown throughout or with the lobes darker than the tube on drying". It is not clear whether she had ascertained that nonyellow corollas are observed only once specimens are dry; we note though that one specimen from Inhaca Island in Maputo Bay was documented as having red flowers. Again, *K. rotundifolia*, with red flowers, has been recorded from Inhaca, and it is possible that material of these two species was confused.

Regardless of multiple species having been compared with *K. leblanciae* as possible close allies, we regard *K. sexangularis* as its closest relative. *Kalanchoe leblanciae* is characterised by having oblong to narrowly oblong leaves that remain more or less green and do not take on the same bright red hues that specimens of *K. sexangularis* attain. The margins of the leaves of *K. leblanciae* are crenate with few, large divisions and somewhat similar to those of *K. sexangularis*. However, in the case of *K. leblanciae*, the petiole is very short to hardly distinct from the leaf blade, and the corolla tube is fairly short (7–8 mm long) and nearly Chinese lantern-like square lower down. The leaf axils on the peduncle of *K. leblanciae* almost invariably carry prominent plantlets that will root easily once detached. Such propagules are virtually absent in the case of *K. sexangularis*.

Cultivation. Pooley (1978) recorded that in Ndumo, plants took two to three years to mature and after flowering would die right back before resprouting. In domestic horticulture, plants behave similarly. Tölken (1985) described how in some species this regeneration occurs through the production of slender underground branchlets from the stem base. This would appear to be the case for *K. leblanciae*, as it is with the likes of *K. hirta*. In full sun garden positions, plants sprout profusely from the base and attain a height of about 1 m, whereas in more shaded situations, specimens have fewer stems, are taller, and appear more wispy.

9. KALANCHOE LONGIFLORA SCHLTR. EX J.M.WOOD

Turquoise kalanchoe (Figs. 12.9.1 to 12.9.9)

FIG. 12.9.1   A cluster of low-growing *Kalanchoe longiflora* shrublets. Photograph: Gideon F. Smith.

Kalanchoe longiflora **Schltr. ex J.M.Wood** in *Natal Plants* volume **4**, part **I**: plate 320 (1903). Hamet (1908b: 26); Raymond-Hamet & Marnier-Lapostolle (1964: 79; plate XXVIII, figures 92–93); Jacobsen (1970: 286); Ross (1972: 179) [as *K. longiflora* var. *genuina* Raym.-Hamet; see discussion below]; Tölken (1985: 69); Jacobsen (1986: 618); Descoings (2003: 163); Smith (2005: 145); Smith & Crouch (2009: 85); Court (2010: 91); Smith & Figueiredo (2017b: 124). **Type**: [South Africa.] Natal [KwaZulu-Natal Province].—near the brook Dumbeni, Weenen District [—2830 (Dundee): between Greytown and Weenen (–CD)], 15 April 1891, *J. Medley Wood 4439* (NH NH0005206-0, holo-; K K000232852, iso-).

Synonyms:
None recorded.
Derivation of the scientific name:
From the Latin words *longus* (=long) + *-florus* (=flowered) for the long flowers.
Description:
Perennial, few-leaved, sparsely branched, glabrous, medium-sized, succulent shrub, and 0.2–0.4 m tall. *Stems* few, unbranched or sparsely branched, brittle, erect to leaning to creeping, 4-angled, greenish pink to greenish orange, and surface white flaky. *Leaves* succulent, opposite, erect, subpetiolate to cuneate, light sea-green to turquoise, strongly infused with orange or pink, and

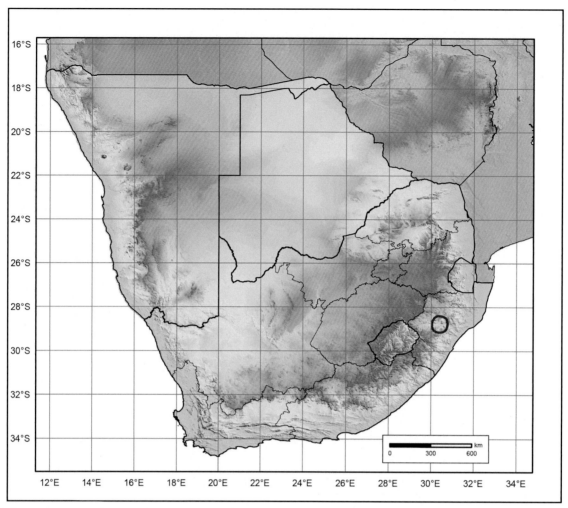

FIG. 12.9.2   Known geographical distribution range of *Kalanchoe longiflora* in South Africa.

FIG. 12.9.3   The stems of *Kalanchoe longiflora* are distinctly four-angled, here distinctly visible in the plant in the centre. Photograph: Gideon F. Smith.

FIG. 12.9.4   The leaves of *Kalanchoe longiflora* are a characteristic sea-green to turquoise, variously infused with pink as here. The leaf margins are dentate with soft crenations, and the leaf blades are not longitudinally folded, that is, they remain flat. Photograph: Gideon F. Smith.

FIG. 12.9.5   In some forms of *Kalanchoe longiflora,* the sea-green to turquoise leaves are orange-infused. Photograph: Gideon F. Smith.

FIG. 12.9.7   The flowers of *Kalanchoe longiflora* have long, shiny, greenish yellow tubes and yellow lobes. Photograph: Gideon F. Smith.

FIG. 12.9.6   The more or less flat-topped inflorescences of *Kalanchoe longiflora* are borne erectly to leaning and are generally sparsely flowered. Photograph: Gideon F. Smith.

FIG. 12.9.8   The follicles (fruit) of *Kalanchoe longiflora* are brittle, grass spikelet-like, and enveloped in the dry, reddish brown remains of the corolla. Photograph: Gideon F. Smith.

never red; *petiole* 2–10mm long when present; *blade* 4–7 × 4–7cm, obovate to nearly orbicular, slightly incurved along margins, and somewhat spathulate; *base* cuneate; *apex* rounded-obtuse; *margins* scalloped into rounded, harmless, and patelliform teeth; *teeth* becoming smaller to obsolescent towards leaf base. *Inflorescence* an erect, sparsely flowered, ± flat-topped thyrse consisting of several open dichasia, to 50cm tall; peduncle often densely covered in large, succulent bracts the same colour as the leaves; *pedicels* 10–18mm long. *Flowers* erect, elongated, greenish yellow (tube), and bright yellow (lobes); *calyx* green, infused with orange towards the tip; *sepals* 3–4mm long, elongated-triangular, acute, and succulent; *corolla* greenish yellow; *tube* 15–18mm long, elongated-urceolate, cylindrical-quadrangular, enlarged lower down, tapering to the tip, distinctly slightly angled,

hardly ever pyramidal, and shiny greenish yellow; *lobes* 4–5 × 4–5mm, broadly obovate to rounded, apiculate, yellow. *Stamens* inserted ¾ of the way up corolla tube, hardly exserted; *filaments* 4–5mm long, thin, greenish yellow; *anthers* 0.4–0.5mm long and yellowish brown. *Pistil* consisting of 4 carpels; *carpels* 10–11mm long and light green; *styles* 5–7mm long; *stigmas* very slightly capitate and yellowish green; *scales* ± 4mm long, linear, and yellowish green. *Follicles* brittle, grass spikelet-like,

FIG. 12.9.9   The leaves of *Kalanchoe longiflora* retain their turquoise colour even when growing in deep shade. Photograph: Gideon F. Smith.

enveloped in dry reddish brown remains of corolla, drying greyish black, and 10–11mm long. *Seeds* 0.50–0.75mm long, brown, and urceolate to banana-shape curved. *Chromosome number*: $2n = 34$ (Baldwin, 1938: 576).

Flowering time:

February–November, peaking in July and August (southern hemisphere).

Illustrations:

Wood (1903: plate 320) [black-&-white drawing prepared by "M.F.", that is, Ms Millicent Flanders nee Franks (06 October 1886–11 February 1961]); Raymond-Hamet & Marnier-Lapostolle (1964: plate XXVIII, figures 92–93); Jacobsen (1986: 618, figure 865); Smith (2005: 145, lower half of page); Pienaar & Smith (2011: 144, 145); Smith & Figueiredo (2017b: 123, 125, 126).

Common names:

Afrikaans: seeplakkie.

English: turquoise kalanchoe.

Geographical distribution:

Kalanchoe longiflora is restricted to the central Tugela River basin of the KwaZulu-Natal Province in southeastern South Africa. The area is host to a range of diverse vegetation types (Edwards, 1967) and is situated in the Maputaland-Pondoland Region of Endemism in South Africa (Van Wyk & Smith, 2001; Steenkamp et al., 2004) (Fig. 12.9.2).

Distribution by country. South Africa (KwaZulu-Natal Province).

Habitat:

The species grows in rocky places in savanna vegetation.

Conservation status:

The species is not threatened.

Additional notes and discussion:

Taxonomic history and nomenclature. Kalanchoe longiflora was first mentioned in print by Wood (1903: Plate 320) after he had collected material in 1891 and subsequently had an undated, black-&-white line drawing prepared for the *Natal Plants* publication series. Despite the species having been illustrated in Raymond-Hamet & Marnier-Lapostolle (1964), its identity remained obscure for several decades, essentially until Ross

(1972: 179) recorded it in his *Flora of Natal* [now the KwaZulu-Natal Province of South Africa]. In the preceding years, it was taxonomically most usually confused with *K. sexangularis*, especially through a doubtfully validly published synonym of the latter, 'K. longiflora Schltr. ex J.M.Wood var. coccinea Marn.-Lap.' (Smith & Figueiredo, 2017b).

In more recent works, such as Newton & Mbugua (1993: 46), the very distinctive *Kalanchoe longiflora* was still considered to consist of two varieties, *K. longiflora* var. *coccinea* (= *K. sexangularis*), and a 'typical' variety, *K. longiflora* Schltr. var. *longiflora*. Newton & Mbugua (1993) regarded both varieties as having a "Tropical Africa[n]" origin, whereas *K. longiflora* is a narrow-range, South African endemic.

Burtt Davy (1926) also speculated that *K. longiflora* might occur within what he defined as the Eastern Mountains 'Botanical Province' of the then Transvaal region he covered. Some 50 years later, Tölken (1978) described a species, *K. rubinea* Toelken, which he stated to have an affinity with *K. longiflora*. *Kalanchoe rubinea*, however, was based on material from the Eastern Transvaal [Mpumalanga] escarpment, stretching from eastern Swaziland northwards to the Soutpansberg, an area much further north than the known distribution range of *K. longiflora*. Five years later, Tölken (1983) synonymised *K. rubinea* with the earlier *K. sexangularis* but was clear in circumscribing *K. longiflora* as an element from the Tugela River basin of KwaZulu-Natal.

The name *Kalanchoe longiflora* is largely devoid of synonyms. This is uncommon in groups of plants, such as succulents, that appeal to collectors. The following names listed in synonymy in some literature are not synonyms:

Reference to the 'name' 'Kalanchoe petitiana var. salmonea Hort.', listed by Jacobsen (1986: 618) under *K. longiflora*, could not be traced. The 'name' 'Kalanchoe petitiana Hort.', also listed by Jacobsen (1986: 618), under *Kalanchoe longiflora* var. *coccinea*, is based on a comment included in Raymond-Hamet & Marnier-Lapostolle (1964) where they refer to a small plant with dense, bright red leaves that was commonly found in florist's shops, either as unnamed specimens, or labelled 'Kalanchoe petitiana'. Raymond-Hamet & Marnier-Lapostolle (1964) concluded that this flower shop-kalanchoe was not at all related to true *K. petitiana* A.Rich., an East African species with a history of taxonomic confusion (see Wickens, 1987: 42), and in fact represented *K. longiflora*. This 'non-name', 'Kalanchoe petitiana Hort.' has become a source of confusion.

Kalanchoe longiflora var. *coccinea*, a name doubtfully validly published, is a synonym of *K. sexangularis* and not of *K. longiflora*.

The name 'Kalanchoe longiflora var. genuina Raym.-Hamet' was not validly published. Infraspecific names such as 'genuinus' that aim to indicate the taxon containing the type of the name of the next higher taxon (in this case a species) are not validly published unless they are autonyms, which *K. longiflora* var. *genuina* is not (McNeill et al., 2012: 57, Article 24.3). 'Kalanchoe longiflora var. genuina' is therefore not a name in the sense of the *International Code of Nomenclature for algae, fungi, and plants* (McNeill et al., 2012: 7, Article 6.3).

Identity and close allies. Tölken (1978: 91) noted that "This species [*K. rubinea*, a synonym of *K. sexangularis*], which is widely used in horticulture, was always confused with *K. longiflora*, an endemic species of central [KwaZulu-]Natal." This is not surprising as the closest relative of *Kalanchoe longiflora* is indeed *K. sexangularis*. Both species are more or less low-growing shrublets with fairly weak, angled stems, and flat leaves with scalloped margins. While *K. sexangularis* is usually slightly growing taller than *K. longiflora*, the latter species is easy to distinguish from *K. sexangularis* given the characteristic sea-green to turquoise leaf colour of *K. longiflora*, hence its common name. In the case of *K. sexangularis*, leaf colour ranges from a red-infused midgreen to intensely bright wine red, especially during times of environmental stress. The leaves of *K. longiflora* are obovate to nearly orbicular, flat, and not folded lengthwise, while those of *K. sexangularis* are broadly elliptic or obovate to oblong, often somewhat folded lengthwise and recurved in the upper half.

As the appropriate specific epithet, 'longiflora', indicates, *Kalanchoe longiflora* indeed has some of the longest flowers in the genus as represented in southern Africa. The flowers have long, shiny, greenish yellow tubes and yellow lobes, a colour combination commonly found among southern African *Kalanchoe* species.

Cultivation. *Kalanchoe longiflora* is easy to grow from stem cuttings that root easily in virtually any soil type. The beautiful orange- or pinkish-infused, sea-green to turquoise leaf colour appears at its best in full sun, but the species will also thrive in semishade and even deep shade. When grown in full sun, the fairly short-stemmed plants remain more compact and when planted in dense groups will rapidly form a striking sea-green groundcover.

In the case of most material of *Kalanchoe longiflora* observed, midwinter to early summer seems to be the typical flowering period. However, some clones may start flowering as early as February (late summer). Interestingly, in the protologue of the name *Kalanchoe longiflora*, Wood (1903) states that plants bear their "...flowers profusely during the summer months". Some variation in flowering period of this species is therefore evident, making it a useful addition to any garden.

10. *KALANCHOE LUCIAE* RAYM.-HAMET

Flipping flapjacks (Figs. 12.10.1 to 12.10.18)

FIG. 12.10.1　The round, soup plate-sized leaves of *Kalanchoe luciae* are most usually heavily red-infused.　Photograph: Gideon F. Smith.

Kalanchoe luciae **Raym.-Hamet** in *Bull. Herb. Boissier*, ser. 2, 8: 256 (1908). Raymond-Hamet & Marnier-Lapostolle (1964: 91, plate XXXIII, figures 112–114); Jacobsen (1970: 287); Ross (1972: 179); Fernandes (1978: 204); Jacobsen (1986: 618); Smith et al. (2016a: 64–72). **Type**: (South Africa) Limpopo Province, Soutpansberg district, Spelonken, *H. Junod s.n.* (G G00418142, holo-; P).

Synonyms:

Kalanchoe aleurodes Stearn in *Gard. Chron.* ser. 3, 89: 475 (1931). **Type**: Southern Rhodesia [Zimbabwe], near Salisbury, raised from seed collected by S.G. Arden and cultivated at Cambridge Botanic Garden, *s.c. s.n.* (CGE holo-, not seen).

Kalanchoe albiflora Forbes in *Bothalia* 4: 37 (1941). **Type**: Swaziland, Ubombo [UBombo] Mountains, Swaziland-Natal [-KwaZulu-Natal], cult. Natal Herbarium 9/1935, *Gerstner sub PRE 26434* (PRE PRE0390335-0, holo-).

Kalanchoe luciae subsp. *luciae*. Fernandes in *Fl. zambes.* 7(1): 57 (1983); Tölken (1985: 71); Descoings (2003: 164). **Type**: See *Kalanchoe luciae* Raym.-Hamet.

Derivation of the scientific name:

The species was named for Mademoiselle Lucy Dufour (fl. 1908), a friend of the French botanist and physician Raymond Hamet [Raymond-Hamet], without further information.

Description:

Perennial or short-lived multiannual, few- to many-leaved, sparsely branched from near the base, glabrous or rarely minutely pubescent, robust succulent, and to 2 m tall. *Stems* arising from a slightly swollen rootstock, erect to curved upwards, horizontally ridged where leaves abscised, and greenish white. *Leaves* erect to patent-erect, succulent, sessile, ± flattened above, slightly convex below, light yellowish green to bluish green, and infused with ruby red especially apically and along margin; *petiole* absent; *blade* 4–16 × 2–9 cm, round to obovate to oblong, and not folded lengthwise; *base* narrow; *apex* rounded-obtuse or truncate; *margins* entire and with a substantial red tint. *Inflorescence* a slender, erect, densely flowered, cylindrical to club-shaped thyrse consisting

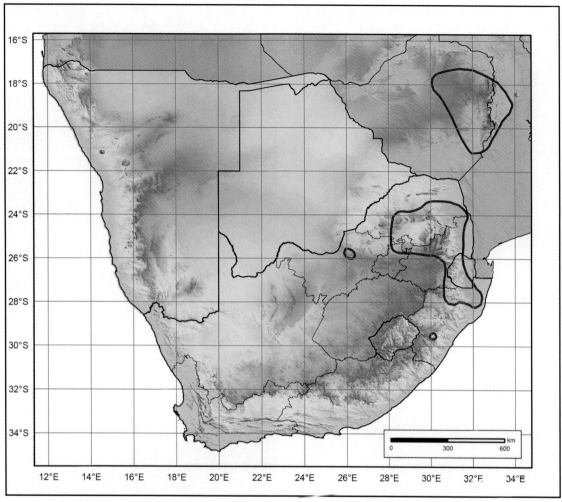

FIG. 12.10.2   Known geographical distribution range of *Kalanchoe luciae* in southern and south-tropical Africa.

FIG. 12.10.3   This form of *Kalanchoe luciae* has light green leaves that are strongly orange-infused. Photograph: Gideon F. Smith.

FIG. 12.10.4   In some instances, the leaves of *Kalanchoe luciae*, especially the outer ones in a rosette, are virtually entirely red. Photograph: Gideon F. Smith.

FIG. 12.10.5   In some forms of *Kalanchoe luciae*, the leaves are more obovate, hardly red-infused, and lack a floury-wax covering during the early developmental stages. Photograph: Gideon F. Smith.

FIG. 12.10.7   Specimens of *Kalanchoe luciae* flowering in the dappled shade of deciduous trees. Photograph: Gideon F. Smith.

FIG. 12.10.6   The leaves of *Kalanchoe luciae* are rarely arranged in distinct vertical rows, as here, when a plant approaches flowering maturity. Photograph; Gideon F. Smith.

of several dichasia terminating in monochasia, to 1.6 m tall; *pedicels* 4–6 mm long. *Flowers* erect to slanted horizontally; usually white; pale greenish white sometimes cream, pinkish, or yellowish; and all parts haphazardly covered with a thin to mostly substantial white waxy bloom; *calyx* light greenish white; *sepals* 3–6 mm long, short-triangular, almost free to the base, and acute; *corolla* light greenish yellow, *tube* 7–12 mm long, more or less quadrangular-urceolate, enlarged in the middle, slightly 4-angled, *lobes* 2.5–4.0 × 3–4 mm, broadly ovate, obtuse to nearly acute, sometimes apiculate, mostly white, and fading pinkish brown when spent. *Stamens* inserted in the middle of the corolla tube, 1–2 mm exserted; *filaments* ± 6 mm long, thin, and light greenish white; *anthers* ± 0.5 mm long and brownish orange. *Pistil* consisting of 4 carpels; *carpels* 6–8 mm long, light yellowish green; *styles* 2–4 mm long; *stigmas* shortly exserted, very slightly capitate, and light yellow; *scales* ± 2 mm long and square to transversely oblong. *Follicles* 9–10 mm long, with the

FIG. 12.10.8   The elongating peduncle of the light green leaf form of *Kalanchoe luciae*. Photograph: Gideon F. Smith.

FIG. 12.10.9   In exposed positions, the leaves on the elongating peduncles of *Kalanchoe luciae* take on a dark shade of red. Photograph: Gideon F. Smith.

appearance of a grass spikelet, surrounded by dark brown dry corolla tube, and dull light green. *Seeds* 1.0–1.5 mm long, oblong, and light brown. *Chromosome number*: unknown.

Flowering time:
April–July, peaking in June (southern hemisphere).

Illustrations:
Raymond-Hamet & Marnier-Lapostolle (1964: 1964: plate XXXIII, figures 112–114); Tölken (1985: figures 9 and 2); Smith & Van Wyk (2008a, b: 154, 155 [incorrectly labelled *K. thyrsiflora*]); Smith et al. (2016a: figures 2, 5, and 7).

Common names:
Afrikaans: sopbordplakkie.
English: flipping flapjacks and paddle plant.
Language unknown: spepejanie, recorded from *W.G. Barnard 328* (PRE).

Geographical distribution:

Kalanchoe luciae has a wide, essentially north-central, northeastern, and eastern range in southern Africa. In Zimbabwe, it may be confused with the endemic *K. wildii* Raym.-Hamet ex R.Fern. (Fernandes, 1983) (Fig. 12.10.2).

Distribution by country. Mozambique, South Africa (North West, Gauteng, Limpopo, Mpumalanga, and northern KwaZulu-Natal provinces), Swaziland, and Zimbabwe.

Habitat:
The species usually grows in grassy patches in savanna vegetation. It often occurs in exposed positions on rocky outcrops.

Conservation status:
Least concern in South Africa. Apparently not assessed elsewhere.

Additional notes and discussion:
Taxonomic history and nomenclature. Following its initial description by Hamet (1908a), *K. luciae* was upheld in most subsequent treatments that provided synopses of the genus (see, e.g. Jacobsen, 1970, 1986 and Descoings, 2003). However, the true identity of the species remained rather obscure, even though it is common in nature and very widely cultivated in domestic and amenity

FIG. 12.10.10   The inflorescence of *Kalanchoe luciae* is densely flowered and cylindrical to club-shaped in outline. Photograph: Gideon F. Smith.

FIG. 12.10.11   The corolla lobes of *Kalanchoe luciae* vary from erectly spreading to recurved. In bud, the sepals of this species lack the reddish brown colouration observed in, for example, *K. montana*. Photograph: Gideon F. Smith.

FIG. 12.10.12   Close-up of the white, urceolate corollas of *Kalanchoe luciae*. Corolla tubes of *K. luciae* are slightly four-angled, and the corolla lobes are most often strongly reflexed. Photograph: Gideon F. Smith.

horticulture in southern Africa. The species was especially confused with *K. thyrsiflora*, the geographical distribution range of which to some extent overlaps with that of *K. luciae*.

Of the species of *Kalanchoe* that are indigenous to southern Africa, at least four bear very distinctive cylindrical inflorescences in which the flowers are densely arranged in a thyrse borne on a robust peduncle. These are *K. luciae*, *K. montana*, *K. thyrsiflora*, and the recently described *K. winteri*. In the *Flora of Southern Africa* treatment of the family Crassulaceae (Tölken, 1985), *K. luciae*, which is the best known of these species (Trager, 2001: 97–98), was interpreted as consisting of two subspecies: the typical one (*K. luciae* subsp. *luciae*) and *K. luciae* subsp. *montana*. This classification coincided with the arrangement proposed earlier by Tölken (1978), in

FIG. 12.10.15   A plantlet emerging at the base of a flowering stem of *Kalanchoe luciae*. Note the prominent leaf scars left by abscised leaves on the stem on the right. Photograph: Gideon F. Smith.

FIG. 12.10.13   Flowers sometimes develop in the axils of the leaflike bracts on the peduncle of *Kalanchoe luciae*. Photograph: Gideon F. Smith.

FIG. 12.10.16   Plantlets forming in the axils of some of the lower leaf-like bracts on the peduncle of *Kalanchoe luciae*. Photograph: Gideon F. Smith.

which he recognised *K. montana* only at subspecific rank. However, Smith et al. (2016a) recently reinstated *K. montana* at species rank, leaving *K. luciae* devoid of infraspecific taxa.

Kalanchoe aleurodes Stearn, a synonym of *K. luciae*, is listed in IPNI as published in "*Journal of Botany* 69: 164. 1931; et in *Gard. Chron.* Ser. III. LXXXIX: 475. 1931". It is likely that the *Journal of Botany* issue was published first and therefore represents the protologue, as a photocopy of that publication is glued to the specimen that was in William Stearn's herbarium and is now at BM (BM000649720). This specimen is annotated as a 'cotype', an obsolete term meaning syntype, or isotype, or even paratype.

Identity and close allies. Kalanchoe luciae has for long been confused with *K. thyrsiflora*, its closest relative.

FIG. 12.10.14   Where a peduncle of *Kalanchoe luciae* was damaged, clusters of flowers sometimes develop. Photograph: Gideon F. Smith.

FIG. 12.10.17   *Kalanchoe luciae* in cultivation at Tshipise, Limpopo province, northern South Africa. Photograph: Gideon F. Smith.

FIG. 12.10.18   *Kalanchoe luciae* in cultivation in Pretoria in the Gauteng province, north-central South Africa. Photograph: Gideon F. Smith.

Morphologically, these two species are indeed very similar, both sporting large, paddle- to soup plate-shaped leaves. However, the leaves of *K. luciae* tend to be larger and more red-infused. In vegetative and reproductive morphology, *K. luciae* differs from *K. thyrsiflora* by having white, pale greenish white, sometimes cream, pinkish, or yellowish flowers, while those of *K. thyrsiflora* are a deep butter yellow.

Cultivation. After *Kalanchoe sexangularis*, *K. luciae* is the second most commonly cultivated species of the southern African kalanchoes. The species grows very easily and will flourish in virtually any type of soil. It will tolerate considerable neglect and still maintain its large, soup plate-sized and soup plate-shaped, rosulately arranged leaves looking firm and healthy. The species is drought-hardy and requires minimum irrigation, looking none the worse for wear. When irrigation is withheld and plants are grown in full sun, the leaves take on a magnificent deep red colour.

Plants grow easily from seed, and whole rosettes that arise from the base of a plant can be removed and planted directly in the spot where they are intended to grow.

11. *KALANCHOE MONTANA* COMPTON

Mountain kalanchoe (Figs. 12.11.1 to 12.11.12)

FIG. 12.11.1   In its natural habitat, as here in western Swaziland, *Kalanchoe montana* often grows wedged between rocks.   Photograph: Gideon F. Smith.

Kalanchoe montana Compton in *J. S. African Bot.* 33: 295 (1967). Compton (1976: 220); Smith et al. (2016a: 69). **Type**: Swaziland, near Devil's Bridge, Pigg's Peak district, Emlembe Mountains, altitude 5500′ [*c.* 1676 m] above sea level, *Compton 29471* (NBG lecto- [Tölken, 1978: 89], NBG0099565-0 designated as lectotype in a second-step lectotypification by Smith et al., 2016a).

Synonym:

Kalanchoe luciae subsp. *montana* (Compton) Toelken in *J. S. African Bot.* 44: 89 (1978). Tölken (1985: 71). **Type**: as above.

Derivation of the scientific name:

From the Latin *montanus* (= mountainous) for the mountainous habitat in Swaziland where the species was first collected.

Description:

Perennial or short-lived multiannual, few- to many-leaved, basally rosulate, solitary or sparsely branched from near the base, often minutely pubescent throughout, sometimes glabrous, robust succulent, and to 1 m tall. *Stems* fleshy, erect to leaning to creeping, and curved upwards. *Leaves* simple, erect to patent-erect, succulent, sessile, ± flattened above, slightly convex below, light yellowish green, sometimes infused with red especially apically and along margin, and very minutely pubescent or glabrous; *petiole* absent; *blade* 4–18 × 2–7 cm, obovate to oblanceolate, rounded-obtuse at apex, and not folded lengthwise; *base* narrow; *apex* rounded-obtuse or truncate; *margins* entire and sometimes somewhat undulate. *Inflorescence* an erect, densely flowered, cylindrical

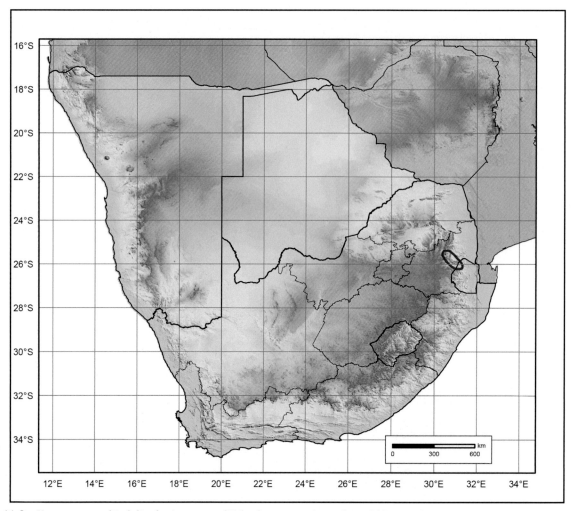

FIG. 12.11.2   Known geographical distribution range of *Kalanchoe montana* in southern Africa.

to club-shaped thyrse consisting of several dichasia terminating in monochasia, to 0.75 m tall; *pedicels* 4.0–6.0 (−10.4) mm long. *Flowers* erect to slanted horizontally, greenish yellow to whitish pink, and all parts sometimes

FIG. 12.11.3   The leaves of *Kalanchoe montana* are often light green and finely pubescent. Photograph: Gideon F. Smith.

FIG. 12.11.4   The leaves of this form of *Kalanchoe montana* are glabrous and light reddish brown-infused. Photograph: Gideon F. Smith.

FIG. 12.11.5 *Kalanchoe montana* growing well camouflaged among grass tufts in northwestern Swaziland. Photograph: Gideon F. Smith.

FIG. 12.11.7 The reddish brown sepals of *Kalanchoe montana* are up to 9 mm long and very conspicuous, especially in the bud stage. Photograph: Gideon F. Smith.

FIG. 12.11.6 The leaves of *Kalanchoe montana* are smaller than those of *K. luciae*, only very lightly red-infused, and all plant parts are minutely pubescent. Photograph: Gideon F. Smith.

minutely pubescent; *calyx* light yellowish green to whitish pink, reddish brown tipped and longitudinally infused, conspicuously so when flowers in bud, and smooth or minutely pubescent; *sepals* 5–8 mm long, lanceolate-elongated, acute, and almost free to the base; *corolla* light greenish yellow to whitish pink; *tube* 13 × 5 mm, more or less pyramidal, broadest and somewhat cigar-shaped-enlarged slightly below the middle, and slightly 4-angled; *lobes* 5 × 2 mm, elongated-triangular, patent-erect to slightly spreading, margins slightly folded in, apiculate, yellow, and fading light brown when spent. *Stamens* 1–2 mm exserted; *filaments* 3–4 mm long, thin, and light greenish yellow; *anthers* ± 0.50–0.75 mm long and purplish brown. *Pistil* consisting

of 4 carpels; *carpels* 5–6 mm long, ovate, and often reddish brown infused; *styles* 5–6 mm long and slender; *stigmas* inserted, minutely capitate, and light yellowish white; *scales* oblong and light green. *Follicles* not seen. *Seeds* not seen. *Chromosome number* unknown.

Flowering time:
February–August, peaking in April.

Illustrations:
Smith et al. (2016a: figures 1, 3, 4, and 6).

Common names:
Afrikaans: bergplakkie.
English: mountain kalanchoe.

Geographical distribution:
It has a narrow geographical distribution range in northwestern Swaziland and southeastern Mpumalanga province of South Africa. *Kalanchoe montana* is restricted to the Barberton Centre of Endemism (Van Wyk & Smith, 2001) (Fig. 12.11.2).

Distribution by country. South Africa and Swaziland.

Habitat:
Found growing exposed on shallow soils amongst rocks in grassland vegetation.

Conservation status:
Least concern in South Africa.

Additional notes and discussion:
Taxonomic history and nomenclature. On retirement, Professor R. Harold Compton, the second Director of the National Botanic Gardens of South Africa, moved to Swaziland where, starting in 1955, he undertook a botanical survey of the country (Rycroft, 1979: 75). The main set of specimens collected under the auspices of the project

FIG. 12.11.9 Corolla tubes of *Kalanchoe montana* are indistinctly pyramidal in shape and the corolla lobes erectly spreading and slightly upturned. Note that the lobes of this form of the species are strongly reddish pink-infused. Photograph: Gideon F. Smith.

FIG. 12.11.8 Close-up of an inflorescence of *Kalanchoe montana*. Photograph: Gideon F. Smith.

consisted of about 8000 numbers (Gunn & Codd, 1981: 122)—11,000 according to Rourke (1976) and Rycroft (1979)—and were kept initially in Mbabane (Smith & Willis, 1997: 31, 34; 1999: 111, 128). Some were eventually transferred to the National Herbarium of South Africa, PRE (Gunn & Codd, 1981: 122), with duplicates also lodged in NBG, the Herbarium in Cape Town named for Compton. After only 13 years, the work on the flora of Swaziland came to initial fruition as an annotated checklist (Compton, 1966), materialising a decade later as a Flora (Compton, 1976; Rourke, 1976).

In the early works on the flora of Swaziland Burtt-Davy & Pott-Leendertz, (1912: 144; Pott, 1920: 126), the only recorded species of *Kalanchoe* with dense, club-shaped inflorescences was *K. thyrsiflora*. During the early 1900s, the most common such *Kalanchoe* material from

FIG. 12.11.10 Dry inflorescence of *Kalanchoe montana*. Photograph: Gideon F. Smith.

FIG. 12.11.11   The rocky montane grassland habitat of *Kalanchoe montana* in the Ngwenya Mountains, Malolotja Nature Reserve, Swaziland. Photograph: Gideon F. Smith.

northern South Africa and Swaziland had not yet been correctly assigned to *K. luciae*, while *K. montana* remained undetected at the time.

Whilst collecting in Swaziland, Compton first encountered specimens of what he would later describe as *Kalanchoe montana*. Compton was evidently not at first convinced that *K. montana* was new, for he identified the early but separate collections of himself and Ben Dlamini as *K. thyrsiflora*. These specimens were collected respectively on 14 March 1957 and 20 February 1961, both from the hills near Mbabane, the capital of Swaziland, at altitudes of *c.* 1500 m above sea level. Both *R.H. Compton 26912 sub PRE0700108-0* and *Ben Dlamini s.n. sub PRE0700312* bear labels written in Compton's hand and are held in PRE, in Pretoria. The labels themselves are headed "Herbarium of the Botanical Survey of Swaziland".

A decade after first encountering specimens of *Kalanchoe montana*, Compton (1967) described the species,

having earlier referred to what would later be the type ["Kalanchoe sp. nov. (C.29471)"] in his aforementioned Swaziland checklist (Compton, 1966: 44, 114).

Two years after *The Flora of Swaziland* (Compton, 1976) appeared, Tölken (1978) reduced *K. montana* to subspecific rank under *K. luciae*. This view is reflected in Tölken (1985: 71), Kemp (1983: 31), and Braun et al. (2004: 37, 95, 102). However, based on the differences between *K. luciae* and *K. montana* as circumscribed by Compton (1967) and as assessed by Smith et al. (2016a, b), *K. montana* was reinstated at species rank.

Identity and close allies. Kalanchoe luciae and *K. montana* have a very similar appearance. However, there are several notable differences on which these two species can be separated. Both species have succulent, discoid leaves, but those of *K. luciae* are soup plate-like rounded, often extensively red-infused on both surfaces and almost always glabrous (devoid of pubescence) with a floury-wax coating on younger leaves. Old mature leaves

FIG. 12.11.12   This small plantlet that developed directly on the basal portion of the stem of *Kalanchoe montana* can be removed and grown on. The part of the plant that flowered will die. The roots of this species are thin and wiry. Photograph: Gideon F. Smith.

lack this bloom. The leaves of *K. montana* are generally light green, usually with very limited red colouring only and smaller than those of *K. luciae*. All plant parts of *K. montana* are nearly consistently pubescent, while those of *K. luciae* are rarely so. Vestiture is an important distinguishing character that has been variously described for *K. montana* in the literature, as "minutely puberulous" in the protologue (Compton, 1967: 295) or as "glandular-downy" (Compton, 1976: 219–220), for example. Tölken (1978) noted that hairiness of *K. montana* was usual but that occasionally glabrous plants are found and perhaps on this basis recognised *K. montana* at only subspecies rank. However, when vestiture is considered in conjunction with the other distinguishing features, *K. montana* is readily separable from the much more commonly encountered *K. luciae* and sufficiently distinct to warrant recognition at species rank.

The flowers of the two species also differ. Those of *Kalanchoe montana* have tubes that are indistinctly pyramidal in shape, whereas in *K. luciae*, these are shorter and urn-shaped, with a more constricted mouth. In the case of *K. montana*, the mouths of the flowers are turned slightly upwards (zygomorphic), while those of *K. luciae* are not. The free portion of the elongated-triangular sepals of *K. montana* is rather long (up to 9 mm), while those of *K. luciae* are short-triangular and only up to 5 mm long. In addition, the sepals of *K. montana* tend to be reddish brown-tipped at bud stage, which is not so for *K. luciae*. The corolla lobes of *K. montana* hardly ever reflex fully, whereas in *K. luciae*, they typically do. The black & white line drawing included in Tölken (1985: 70, figures 3 and 3a, based on Tölken 5571) shows flowers of *K. montana* with slightly reflexed corolla lobes. However, this lobe orientation was evident in only a few specimens that we examined. At anthesis, the anthers and stigmas are much further exserted in the case of *K. luciae*, while they hardly protrude from the mouth of the flowers of *K. montana*.

Kalanchoe montana is thus far the only soup plate-leaved *Kalanchoe* species that has a strongly red- to reddish brown-infused carpels, at least in the case of some populations from Mpumalanga, eastern South Africa.

The peak flowering period of *K. montana* (late summer to autumn) largely precedes that of *K. luciae* (autumn to midwinter).

Cultivation. Kalanchoe montana is virtually unknown in cultivation. Like *K. luciae* and *K. thyrsiflora*, it, however, presents few challenges to ensure healthy growth in a garden setting.

12. *KALANCHOE NEGLECTA* TOELKEN

Umbrella kalanchoe (Figs. 12.12.1 to 12.12.10)

FIG. 12.12.1 *Kalanchoe neglecta* is one of the few southern African species of *Kalanchoe* that has distinctly peltate leaves. Photograph: Gideon F. Smith.

Kalanchoe neglecta **Toelken** in *J. S. Afr. Bot.* 44: 90 (1978). Descoings (2003: 167); Smith & Crouch (2009: 85); Smith et al. (2017b: 311). **Type**: [South Africa], Natal [KwaZulu-Natal], (the) Ubombo (region), Sordwana [Sodwana] Bay, 5 May 1965, *Vahrmeijer & Tölken 835* (PRE PRE0523752-0, holo-).

Synonym:

Kalanchoe rotundifolia (Haw.) Haw. forma *peltata* Raym.-Hamet ex Fernandes in *Bol. Soc. Brot.* Ser. 2, 52: 208 (1978). **Type**: [South Africa], Natal [KwaZulu-Natal], Zululand, Hlabisa, on margin of River forest, 2000 ft., 15 May 1948, *Gerstner 6871* (PRE holo-; K K000232858).

Derivation of the scientific name:

From the Latin 'neglectus' (neglected), in reference to the several decades it took before a name for the species was validly published.

Description:

Perennial, few-leaved, usually unbranched, glabrous, small- to medium-sized succulent, and to 1.2 m tall. *Stems* green, somewhat brittle, few, unbranched, arising from herbaceous base, and erect. *Leaves* opposite-decussate, petiolate, light green, succulent, spreading, and papery on drying; *petiole* 20–80 mm long, not slightly channelled above, and not stem-clasping; *blade* 4–12 × 3–8 cm, broadly elliptic, ovate, cordate or peltate, and saucerlike; *base* deeply cordate in lower leaves, often cuneate higher up; *apex* rounded-obtuse; *margins* entire or coarsely crenate or undulate-crenate into rounded, harmless, crenations, and patelliform. *Inflorescence* to 60 cm tall, mostly erect, apically dense to sparse, many-flowered, flat-topped thyrse with several dense dichasia, rather round in outline when viewed from above, branches opposite,

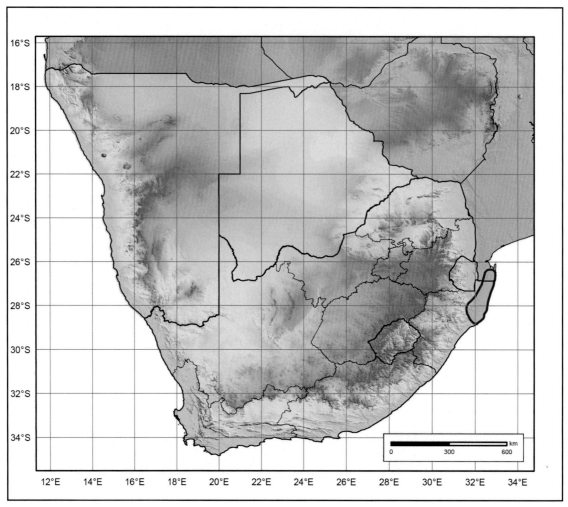

FIG. 12.12.2   Known geographical distribution range of *Kalanchoe neglecta* in southern Africa.

without leafy branchlets in axils, and axis green to bluish green; *pedicels* 4–7 mm long and slender. *Flowers* erect, orange to yellowish orange (upper ½ of tube and lobes), often gradually becoming light green-infused (lower ½ to ⅔ or sometimes as much as ¾ of tube), and lobes same colour as upper ½ of tube; *calyx* light green; *sepals* 2–3 mm long, elongated-triangular, acute, separate, basally slightly fused, curved away from base of corolla tube, and hardly contrasting against green part of corolla tube; *corolla* 6–8 mm long, distinctly enlarged lower down around carpels, distinctly and tightly twisted apically after anthesis, and yellowish orange, gradually green lower down; *tube* 5–7 mm long, globose-rounded, indistinctly 4-angled, not box-shaped-square lower down when viewed from below, longitudinally narrowing above beyond carpels, yellowish orange, and green lower down; *lobes* 2.5–3.0 × 1.5–2 mm, lanceolate, distinctly acute apically, spreading, yellowish orange, and orange more intense towards the lobe margins. *Stamens* included; *filaments* thin; *anthers* 0.5–0.7 mm long. *Pistil* consisting of 4 carpels; *carpels* 6–7 mm long, midgreen;

styles short; *stigmas* very slightly capitate; *scales* ± 2 mm long, narrowly tapering, and linear. *Follicles* enveloped in dry, dark brownish purple remains of corolla, 6–7 mm long; drying dark brown, sharply recurved like a peeled banana at tips. *Seeds* 0.8–1.0 mm long and dark brown. *Chromosome number*: unknown.

Flowering time:

March–July, peaking in June–July (southern hemisphere).

Illustrations:

Van Wyk & Smith (2001: figure 193); Smith & Crouch (2009: 85, bottom); Smith et al. (2017b: 311, bottom).

Common names:

Afrikaans: sambreelkalanchoe.

English: umbrella kalanchoe.

isiZulu: iDlebe lenkau (=monkey's ear).

Geographical distribution:

The species is confined to the Maputaland Centre of Endemism (Van Wyk & Smith, 2001) in northeastern KwaZulu-Natal and southern Mozambique (Fig. 12.12.2).

Distribution by country. Mozambique, South Africa.

FIG. 12.12.3 Plant of *Kalanchoe neglecta*. The leaf blades are often saucerlike folded upwards along the slightly red-infused margins. Photograph: Neil R. Crouch.

FIG. 12.12.5 Adaxial surface of a leaf of *Kalanchoe neglecta*. The leaves of this specimen are cordate and have crenate margins. Photograph: Gideon F. Smith.

FIG. 12.12.4 The stems of *Kalanchoe neglecta* are round in cross-section and remain green and brittle for a long time. Photograph: Gideon F. Smith.

FIG. 12.12.6 Inflorescence of *Kalanchoe neglecta*. Photograph: Neil R. Crouch.

FIG. 12.12.7   Close-up of the orangey yellow flowers of *Kalanchoe neglecta*. Photograph: Neil R. Crouch.

Habitat:

As far as is known, the species is endemic to the north-eastern most corner of KwaZulu-Natal and southern Mozambique (Bela Vista) where it grows in small clusters in sandy soils not far from the Indian Ocean coast.

Conservation status:

The species is not threatened and is included in the Red List category "Least concern'.

Additional notes and discussion:

Taxonomic history and nomenclature. By the time that Fernandes (1978) formally published the combination *Kalanchoe rotundifolia* forma *peltata*, which was proposed by Raymond-Hamet (1960), but that he never validly published, material of the species had already been known for over 50 years, as it had been first collected by Gerstner in 1948.

Tölken (1978) agreed that it was a new entity but preferred to recognise it at the species rank. Since the epithet 'peltata' had already been used in *Kalanchoe* for the Madagascan species *K. peltata* Baill., Tölken (1978) described the species as *K. neglecta*. Shortly afterwards, Fernandes (1978: 208) synonymised *K. neglecta* with her *Kalanchoe rotundifolia* forma *peltata*, a view that she maintained in *Flora zambesiaca* (Fernandes, 1983) but that is not currently accepted.

Identity and close allies. The large, often peltate leaves that somewhat resemble those of the northern hemisphere crassuloid species *Umbilicus rupestris* (Salisb.) Dandy (see Smith & Figueiredo, 2011), commonly known as navelwort, and even the leaves of nasturtiums (*Tropaeolum majus* L., family Tropaeolaceae) separate this species from its closest relative, *K. rotundifolia*. As is the case with the petals of *K. rotundifolia* postanthesis, those of *K. neglecta* also become twisted as they become desiccated.

FIG. 12.12.8   The flowers of this form of *Kalanchoe neglecta* are distinctly orange. Photograph: Gideon F. Smith.

FIG. 12.12.9 The flowers of *Kalanchoe neglecta* are apically twisted in the bud stage and postanthesis. Photograph: Gideon F. Smith.

FIG. 12.12.10 Dark brown, dry inflorescence of *Kalanchoe neglecta* on a specimen growing at Ndumo, KwaZulu-Natal, in eastern South Africa. Photograph: Gideon F. Smith.

Uses. None known.

Cultivation. *Kalanchoe neglecta* is virtually unknown in cultivation. However, it grows easily from seed and, once transplanted when about 5 cm tall, does well in containers and open beds. Plants enjoy dappled shady positions, where they will develop to flowering maturity within one or two seasons. The stems of this species are somewhat brittle and fragile, and care should be taken when handling it when seedlings are transplanted.

Leaf cuttings, with the petiole intact, do not readily strike root; rather, the severed end of the petiole produces callus-like tissue that seems to rather act as a wound sealant, rather than a progenitor of root primordia.

13. *KALANCHOE PANICULATA* HARV.

Rabbit's ear kalanchoe (Figs. 12.13.1 to 12.13.11)

FIG. 12.13.1 In its natural habitat, *Kalanchoe paniculata* often favours vertical cliff faces. Here it grows in association with *Aloe davyana* Schönland (Asphodelaceae). Photograph: Gideon F. Smith.

Kalanchoe paniculata **Harv.** in Harvey & Sonder (eds) *Fl. cap*. 2: 380 (1862) [15–31 October 1862]. Hamet (1908b: 40); Dyer (1947: plate 1007); Raymond-Hamet & Marnier-Lapostolle (1964: 86, plate XXXI, figures 104–106, plate XXXII, figure 107); Compton (1976: 220); Germishuizen & Fabian (1982: 116); Fernandes (1983: 62); Tölken (1985: 66); Hardy & Fabian (1992: 40, Plate 17); Germishuizen & Fabian (1997: 150, plate 67,b); (Van Wyk & Malan, 1997: 126); Retief & Herman (1997: 392); Descoings (2003: 168); Lebrun & Stork (2003: 217); Smith et al. (2003: 22); Smith & Crouch (2009: 86); Court (2010: 92); Gill & Engelbrecht (2012: 112); Smith & Figueiredo (2017d: 73). **Type**: [South Africa] Orange Free State [Free State], Vetrivier, *Zeyher 671* (S S-G-3475, lecto-; BM without barcode, K K000232854, GRA

GRA0001092-0, SAM SAM0036012-1 & SAM0036012-2, isolecto-). Designated by Tölken (1985: 66).

Synonym:

Kalanchoe oblongifolia Harv. in Harvey & Sonder (eds) *Fl. cap*. 2: 379 (1862) [15–31 October 1862]. **Type**: Hopetown District, *Andrew Wyley s.n.* (TCD TCD0001422, holo-).

Derivation of the scientific name:

From the Latin *paniculatus* (= paniculate) for the much-branched inflorescence.

Description:

Multiannual, few-leaved, usually solitary, sparsely branched, glabrous, robust succulent, and to 1.2 m tall; sometimes perennial through basal off sets. *Rootstock* somewhat fattened. *Stems* light green to yellowish

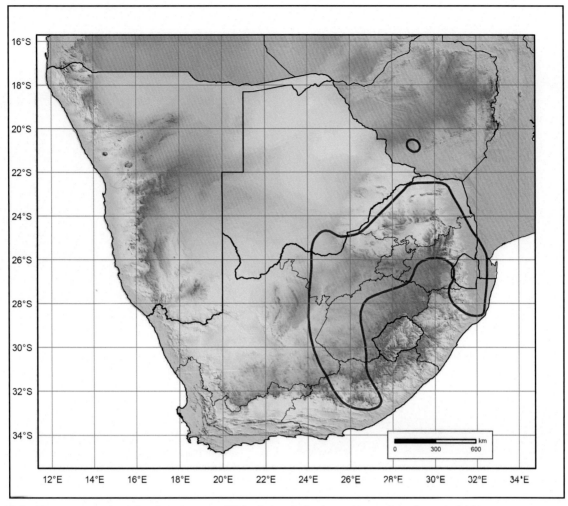

FIG. 12.13.2   Known geographical distribution range of *Kalanchoe paniculata* in southern and south-tropical Africa.

FIG. 12.13.3   Plants of *Kalanchoe paniculata* have an overall yellowish green to light green colour and consistently glabrous leaf surfaces and margins. Photograph: Gideon F. Smith.

FIG. 12.13.4   When young, the leaves of *Kalanchoe paniculata* are booklike-folded, a feature never found in *K. luciae*, a species with which it can be confused. Photograph: Gideon F. Smith.

FIG. 12.13.5   The leaves of *Kalanchoe paniculata* become red-tinged as they mature and lack pubescence throughout. Photograph: Gideon F. Smith.

FIG. 12.13.6   Plants of *Kalanchoe paniculata* sprouting from the base of a spent specimen that has finished flowering. Photograph: Gideon F. Smith.

green, same colour as leaves, robust, usually unbranched, erect to leaning, often with one or more lengthwise running ridges, round to slightly angled, and drying reddish brown. *Leaves* opposite-decussate, sessile lower down, petiolate higher up, light green to yellowish green, sometimes lightly infused with red especially along margin, succulent, spreading, smooth, and coriaceous and papery on drying; *petiole* absent or to 5 cm long, channelled above, and not or scarcely clasping the stem. *Blade* 6–20 × 4–10 cm; oblong to ovate to, more rarely, almost round; often folded lengthwise; sometimes gracefully recurved in upper half; *base* somewhat cuneate to abruptly tapering into petiole; *apex* rounded-obtuse or slightly indented; and *margins* entire. *Inflorescence* to 1.2 m tall, consisting of an erect to leaning, apically dense, many-flowered, flat-topped thyrse with several dichasia, more or less round in outline when viewed from above; branches opposite, erect, straight, slanted upwards at 45 degrees or less, subtended by succulent, boat-shaped, leaflike bracts, without or very rarely with leafy branchlets in axils, and axis light green to yellowish green; *pedicels* 4–5 mm long

FIG. 12.13.7   The inflorescences of *Kalanchoe paniculata* are much-branched to yield thick clusters of flowers. The side branches are held more or less erectly and diverge from the central axis at an angle of less than 45 degrees. In the case of *Kalanchoe paniculata*, flowering is usually initiated early in autumn. Photograph: Gideon F. Smith.

FIG. 12.13.8   In *Kalanchoe paniculata*, the corolla lobes vary slightly in shape, from narrowly-to deltoid-triangular, and are pointed at the apex. Very few flowers in an inflorescence are open simultaneously. Photograph: Gideon F. Smith.

FIG. 12.13.10   In habitat, such as here in sparse bushveld (savanna), *Kalanchoe paniculata* becomes much more conspicuous when in full flower. Photograph: Gideon F. Smith.

FIG. 12.13.9   After flowering, the corollas of *Kalanchoe paniculata* dry to a dark brown colour. Photograph: Gideon F. Smith.

and slender. *Flowers* erect; *calyx* dull yellowish green, *sepals* ± 2 × 1 mm, deltoid-triangular, acute, basally fused for 1 mm, and hardly contrasting against corolla tube; *corolla* 10–12 mm long, somewhat enlarged lower down, very slightly once twisted apically after anthesis, *tube* 9–10 mm long, distinctly 4-angled, box-shaped-square when viewed from below, longitudinally fluted above, yellow to yellowish green, *lobes* 2.0–2.5 × 2.0–2.5 mm, deltoid-triangular to rather narrowly triangular, pointed at apex, apiculate, and bright yellow. *Stamens* inserted in two ranks at about the middle of the corolla tube, included; *filaments* 3–5 mm long,

thin, and yellow; *anthers* 0.3–0.5 mm long and yellowish brown. *Pistil* consisting of 4 carpels; *carpels* 5–6 mm long, light yellowish green, and red-spotted; *styles* 2–3 mm long; *stigmas* very slightly capitate and whitish yellow; *scales* 1.5–2.0 mm long, linear, and light green. *Follicles* brittle, grass spikelet-like, enveloped in dry light to dark brown remains of corolla, dull whitish green, and 8–9 mm long. *Seeds* 0.75–1.00 mm long, reddish brown to dark brown, cylindrical to slightly banana-shape curved. *Chromosome number*: unknown.

Flowering time:

March–May, peaking in April and May (southern hemisphere).

Illustrations:

Raymond-Hamet & Marnier-Lapostolle (1964: plate XXXI, figures 104–106, plate XXXII, figure 107); Germishuizen & Fabian (1982: plate 53); Tölken (1985: 68, figure 8,1); Hardy & Fabian (1992: plate 17); Germishuizen & Fabian (1997: plate 67,b; artwork identical to that included in Germishuizen & Fabian, 1982); (Van Wyk & Malan, 1997: 127, figure 289); Gill & Engelbrecht (2012: 113, top right); Smith et al. (2017b: 312, top); Smith & Figueiredo (2017d: figures 1–12).

FIG. 12.13.11   A rocky outcrop in sparse bushveld (savanna) vegetation in which *Kalanchoe paniculata* occurs on Maanhaarrand, Breedt's Nek, on the Magaliesberg. The silver-leaved *Englerophytum magalismontanum* (Sonder) T.D.Penn. (Sapotaceae) grows between the rocks. Photograph: Gideon F. Smith.

Common names:

Afrikaans: hasie[s]oor (Smith, 1966: 565) and krimpsiektebos[-sie] (Barkhuizen, 1978: 80; Van Wyk & Malan, 1997: 126; Gill & Engelbrecht, 2012: 112).

English: large orange kalanchoe (Long, 2005); rabbit's ear (kalanchoe), and yellow kalanchoe.

siSwati: indabulaluvalo (Long, 2005).

Setswana: segolobe [Kwena dialect], bolatsi [Lete dialect (Selete, southeast, and western Transvaal dialects)] (Cole, 1995: 312), bolatsi, and segolobe (Gill & Engelbrecht, 2012: 112).

Tshivenḓa: tshirndidza (Hardy & Fabian, 1992: 40).

Geographical distribution:

Kalanchoe paniculata is endemic to the eastern half of southern and south-tropical Africa, but is not restricted to a specific Centre of Endemism (Van Wyk & Smith, 2001). The species does not occur in the arid western and west-central karroid parts of the country. It is common over a large part of northeastern and east-central South Africa but has only rarely been recorded from Mozambique to the east (Smith et al., 2003). The geographical range of the species peters out well west of the Drakensberg in central South Africa, stretching to the Hopetown district in the Northern Cape, from where the type specimen of the name *K. oblongifolia* Harv., the only synonym of *K. paniculata*, was collected.

Harvey (1862: 379) admitted that he described *K. oblongifolia* from "…an imperfect specimen".

Unlike most other southern African kalanchoes, *Kalanchoe paniculata* does occur along the western foothills of the climatically severe Drakensberg Mountain in Lesotho (Kobisi, 2005: 37). In the north of South Africa where the climate is milder, the species occurs on both the eastern and western sides of the Drakensberg and in Swaziland and southern Zimbabwe. The distribution of the species in Zimbabwe is, as far as is currently known, restricted to the Matobo [Matopo] Hills in the southwestern parts of the country (Fernandes, 1983: 63). The occurrence of the species in Namibia as reported by Germishuizen & Fabian (1982: 116) and Lebrun & Stork (2003: 218) is an error, as material from this country was subsequently reidentified as belonging to *K. rotundifolia* (Craven et al., 1999: 86). The species occurs in Botswana though (Hargreaves, 1990: 6) (Fig. 12.13.2).

Distribution by country. Botswana, Mozambique, South Africa (Eastern Cape, Free State, Gauteng, KwaZulu-Natal, Limpopo, Mpumalanga, Northern Cape, and North West), Swaziland, and Zimbabwe.

Habitat:

Kalanchoe paniculata is a typical bushveld (savanna) species and quite common in this vegetation type. It only marginally enters into grassland, especially in parts of

South Africa's Gauteng and Free State Provinces and in Lesotho. In the western Free State, it has also been recorded from the eastern extremes of Nama Karoo. Throughout its distribution range, the species favours rocky outcrops, rock faces, and hilly country. In these habitats, the grass and forb cover is often low-growing and comparatively sparse. Much more rarely, the species can also be found in very dense grass-dominated patches, with only the tall inflorescences being visible among the grass culms.

Conservation status:

In South Africa, *Kalanchoe paniculata* is regarded as 'Least concern' and not threatened.

Additional notes and discussion:

Taxonomic history and nomenclature. Kalanchoe paniculata was first described by Harvey (1862: 380) in his treatment of the Crassulaceae for the *Flora capensis* series. In this work, Harvey made extensive use of material collected by both Christian Frederick (Friedrich) Ecklon (1795–1868) and Carl (Karl) Ludwig Philipp Zeyher (1799–1858), two well-known collectors who amassed a considerable set of herbarium accessions while jointly and individually collecting material in South Africa during the first half of the 19th century (Gunn & Codd, 1981). Harvey (1862) based the name *K. paniculata* on a specimen, *Zeyher 671*, which was collected in South Africa's Free State Province [previously the Orange Free State], at Vetrivier. The Vetrivier, which literally translates into English as 'fat river', arises between Marquard and Clocolan in the eastern Free State, near the border with Lesotho, and is a westward-flowing tributary of the substantial Vaalrivier in central South Africa.

Dyer (1947) noted that the specimen (*Zeyher 671*) on which the name *Kalanchoe paniculata* is based was collected en route during an expedition that Zeyher and Joseph Burke (1812–53) undertook by ox wagon from 1839 to 1840, from the Eastern Cape to the Magaliesberg, just north of Pretoria. However, Burke and Zeyher only reached the Vetrivier on 23 February 1841. It was likely around this date that the type specimen was collected. This particular joint expedition that Burke and Zeyher undertook was the first one during which the northern provinces of South Africa (formerly the Transvaal, now the North-West, Gauteng, Limpopo, and Mpumalanga provinces) were extensively collected for botanical and other natural history specimens (Gunn & Codd, 1981: 111).

Less than 10 years after Harvey described *Kalanchoe paniculata* as a new species, *K. brachyloba*, a different species with a wide but slightly more central-southwestern distribution in southern Africa, was described by Britten (1871), based on material from Angola and using a name suggested by Welwitsch (Smith & Figueiredo, 2017a, d). From shortly thereafter, these two species, *K. paniculata* and *K. brachyloba*, became confused in the literature, especially following on from the work of Baker (1899) where he essentially treated *K. brachyloba* under the name *K. paniculata*. Despite the monographic works of Hamet (1907, 1908) on *Kalanchoe*, this confusion persisted for several decades, until Raymond-Hamet & Marnier-Lapostolle (1964: 86), Fernandes (1983: 62), and later Tölken (1985: 66) untangled the two species (see Smith & Figueiredo, 2017a, d). This long-lasting taxonomic confusion inter alia resulted in the synoptic but comprehensive works of Jacobsen (1970, 1986) omitting *K. paniculata* altogether; even the treatment of the species by Fernandes (1983) for the *Flora zambesiaca* project was uncharacteristically devoid of discussion.

Identity and close allies. Kalanchoe paniculata can be easily identified by the overall yellowish green to light green colour of virtually all the plant parts and the lack of pubescence throughout. The species essentially flowers in autumn, being one of the first kalanchoes to produce inflorescences in a season; in some populations, plants will be in full bloom by late March to early April. Although reminiscent of material of *K. thyrsiflora* and to a lesser degree of *K. luciae*, when young, the leaves of *K. paniculata* are oblong to ovate, only rarely round, lengthwise-folded, and gracefully recurved. In terms of floral morphology, *Kalanchoe paniculata* is similar to several other species of *Kalanchoe* in that the corolla tubes are light green, while the free lobes are bright yellow. In shape, the lobes vary slightly, from narrowly to deltoid-triangular, and are pointed at the apex. Very few flowers are open simultaneously, so ensuring multiple visits from pollinators. In time, the flowers dry through light to dark brown.

Uses. Kalanchoe paniculata is widely used in ethnomedicine across its distribution range (Smith et al., 2003). Zulu men scarify their lower foreheads, and the ash of leaves and flowers are rubbed into the wounds. This reputedly attracts the attention of young women they wish to court (Watt & Breyer-Brandwijk, 1962). Long (2005) similarly records the species as being used as a love charm. The Southern Sotho chew fresh, swollen roots as a treatment for colds; alternatively, the roots are dried, pounded, and used as a snuff for the same purpose (Watt & Breyer-Brandwijk, 1962; Germishuizen & Fabian, 1982: 116).

Cultivation. Kalanchoe paniculata has not become popular as a gardening subject and is not commonly cultivated. It is not long-lasting in cultivation and tends to be biannual when grown in gardens. Although when young resembling plants of the horticulturally much more popular *K. luciae* and *K. thyrsiflora*, plants tend to be somewhat more scraggly as they approach maturity, so lacking the appeal of the kalanchoes with soup-plate-sized leaves and club-shaped inflorescences. However, young plants can be easily established in open beds or in pots and will respond well to regular irrigation. Plants at, and near, maturity do not transplant well and tend to die before producing inflorescences.

14. *KALANCHOE ROTUNDIFOLIA* (HAW.) HAW.

Common kalanchoe (Figs. 12.14.1 to 12.14.26)

FIG. 12.14.1   *Kalanchoe rotundifolia* growing near Port Elizabeth. The species was first collected from this general area in South Africa's Eastern Cape province. Photograph: Neil R. Crouch.

Kalanchoe rotundifolia (Haw.) Haw. in *Philos. Mag. J.* 66: 31 (1825). De Candolle (1828: 395); Haworth (1829: 304); Ecklon & Zeyher (1837: 305); Harvey (1862: 379); Balfour (1888: 90); Schönland (1891: 35); Baker (1899: 434); Wood (1899: 76, t. 94); Schinz & Junod (1900: 38) pro parte; Engler (1906: 878); Hamet (1907: 895); Wood (1907: 46); Dinter (1909: 70); Burtt-Davy & Pott-Leendertz, (1912: 144); Hutchinson (1946: 673) [see Spalding (1953: 28)]; Wilman (1946: 79); Mogg (1958: 145); Letty (1962: 149, 152, t. 75,2); Riley (1963: 157); Raymond-Hamet & Marnier-Lapostolle (1964: 87, plate XXXII, figures 108–109); Batten & Bokelmann (1966: 74, and plate 62,6); Edwards (1967: 107, 125, 264); Friedrich (1968: 38); Jacobsen (1970: 289); Lucas & Pike (1971: 39); Van der Schijff & Schoonraad (1971: 489); Venter (1971: 106); Jeppe (1975: 49, plate 28b); Compton (1976: 220); Raadts (1977: 120); Barkhuizen (1978: 80); Gibson (1978: 37, family 79,3 [as *Kalanchoe* sp.]); Fernandes (1980: 405) pro maxima parte (excl. forma *peltata*); Gledhill (1981: 127, plate 28,2); Fernandes (1983: 60) pro maxima parte (excl. forma *peltata*); Kemp (1983: 31); Onderstall (1984: 94); Tölken (1985:62, figure 7); Jacobsen (1986: 627); Wickens (1987:40); Urton (1993: 60, plate 24, figures 1a and b); Germishuizen & Fabian (1997: 150, plate 67a);

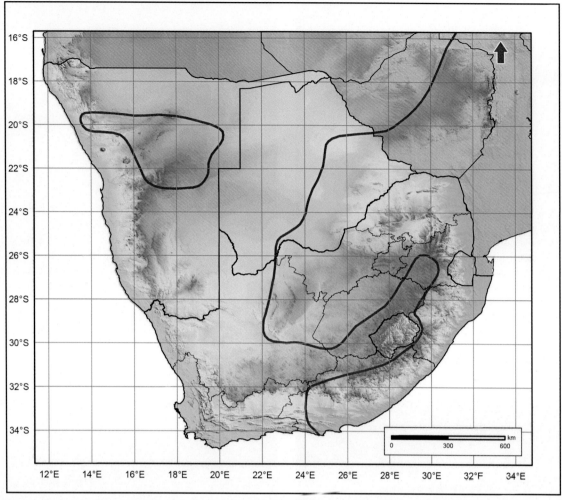

FIG. 12.14.2 Known geographical distribution range of *Kalanchoe rotundifolia* in southern and south-tropical Africa.

FIG. 12.14.3 A dense stand of *Kalanchoe rotundifolia* flowering near Keiskammahoek in the central Eastern Cape Province of South Africa. Photograph: Neil R. Crouch.

FIG. 12.14.4 At Keiskammahoek, flower colour of *Kalanchoe rotundifolia* varies from red to orange. Photograph: Neil R. Crouch.

FIG. 12.14.5   At Bishopstowe, northeast of Pietermaritzburg in KwaZulu-Natal, South Africa, plants of *Kalanchoe rotundifolia* are predominantly quite robust, with fairly tall leafy stems. At this locality, plants have a bluish leaf colour. Photograph: Neil R. Crouch.

FIG. 12.14.7   At Bishopstowe, Pietermaritzburg, in KwaZulu-Natal, the corolla tubes of *Kalanchoe rotundifolia* are reddish orange. Basally, the corolla tubes of *Kalanchoe rotundifolia* are a lighter, green-infused shade of the main colour of the corolla tube and lobes. The sepals are quite short and stubby. Photograph: Neil R. Crouch.

FIG. 12.14.6   Plants of *Kalanchoe rotundifolia* that occur at Bishopstowe, Pietermaritzburg, are variable, with some sporting ovate to obovate leaves. Photograph: Neil R. Crouch.

FIG. 12.14.8   *Kalanchoe rotundifolia* plants usually branch from the base, so producing small shrubby clusters. Photograph taken at Zingela on the banks of the Tugela River, north of Weenen in west-central KwaZulu-Natal, South Africa by Neil R. Crouch.

Retief & Herman (1997: 392); Van Wyk & Malan (1997: 202, 272); Pooley (1998: 54); Vanderplank (1998: 76, plate 33 figure 3); Manning et al. (2001: 154, plate on p. 155, 5); Descoings (2003: 173); Lebrun & Stork (2003: 218); Bandeira et al. (2007: 84, plate 138, figure 78) [as *Kalanchoe rotundifolia* forma *rotundifolia*]; Smith & Van Wyk (2008a, b: 152); Smith & Crouch (2009: 86); Court (2010: 91); Kirby (2013: 231); Figueiredo & Smith (2017b: 111); Smith & Figueiredo (2017e: 63). **Type**: Plate

FIG. 12.14.9 *Kalanchoe rotundifolia* growing at the Hluhluwe-Imfolozi Park, north of the Swart Umfolozi River in northeastern KwaZulu-Natal. Photograph: Neil R. Crouch.

FIG. 12.14.10 At Hluhluwe-Imfolozi in northeastern KwaZulu-Natal, *Kalanchoe rotundifolia* carries small clusters of bright orange flowers. Photograph: Neil R. Crouch.

of *Kalanchoe rotundifolia* dated "1 October 1823" prepared by Thomas Duncanson of sterile material (a stem in leaf). Below the painting is written "Raised from seeds + roots forwarded by Mr [James] Bowie in 1822" (K, neo-). Designated by Tölken (1985: 62), as lectotype; corrected to neotype by Figueiredo & Smith (2017b: 111).

Synonyms:

Crassula rotundifolia Haw. in *Philos. Mag. J.* 64: 188 (1824). De Candolle (1828: 384); Harvey (1862: 365); Schinz & Junod (1900: 38). **Type** as above.

Verea rotundifolia (Haw.) D.Dietr., *Syn. Pl.* 2: 1328 (1840). **Type** as above.

Meristostylus brachycalyx Klotzsch in Peters, *Reise Mossamb., Bot.* 1: 270 (1861). **Type**: Mozambique, Inhambane, *Peters s.n.* (B B_10_0259773, holo-).

Kalanchoe integerrima Lange, *Index seminum hauniense*: 5 (1872): 5 emend. *Bot. Tidsskr.* 10: 139, Table V (1878). **Neotype**: Table V in *Bot. Tidsskr.* 10 (1878). Designated by Smith & Figueiredo (2017e).

Kalanchoe luebbertiana Engl. in Engler & Diels in *Bot. Jahrb. Syst.* 39: 463 (1907). **Type**: [Namibia] Südwestafrika/South West Africa, without precise locality, *Lübbert 67* (B holo- [probably destroyed]; LE LE00013298, PRE PRE0538004-0, iso-).

Kalanchoe seilleana Raym.-Hamet in *J. Bot.* 54 Suppl. 1: 6 (1916). Jacobsen (1970: 289); Jacobsen (1986: 628). **Type**: [South Africa, Northern Cape Province], Prieska, flowered in Mr Armstrong's garden, Port Elizabeth, *s.c. s.n.* (GRA GRA0001088-0, holo-).

Kalanchoe guillauminii Raym.-Hamet in *Bull. Mus. Natl. Hist. Nat.* sér. 2, 20: 467 (1948). **Type**: Cultivated in the greenhouses of the Muséum d'Histoire Naturelle, Paris, from material originating from Port Elizabeth, *Humbert f.353/1933 n°17* (P P03350784, holo-).

Kalanchoe decumbens Compton in *J. S. African Bot.* 33: 294 (1967). **Type**: Swaziland, hill at entrance to Ingwavuma Poort, Lebombo Range, flowered in cultivation at Mbabane, *[R.H.] Compton 29407* (SDNH holo-; NBG NBG0068850-0 and PRE PRE0523106-0, iso-).

Kalanchoe rotundifolia forma *tripartita* Raym.-Hamet ex R.Fern. in *Bol. Soc. Brot.*, Sér. 2, 52: 207 (1978). Ross (1972: 179). **Type**: [South Africa] Natal [KwaZulu-Natal], Muden, *Wylie s.n. sub NH 27707* (NH, holo-).

FIG. 12.14.11   A leafy stem of *Kalanchoe rotundifolia* growing in the Vernon Crookes Nature Reserve, southern KwaZulu-Natal. Photograph: Neil R. Crouch.

FIG. 12.14.12   Near Mbombela, formerly known as Nelspruit, in South Africa's Mpumalanga province, *Kalanchoe rotundifolia* has nearly round, conspicuously red-margined leaves. Photograph: Neil R. Crouch.

Derivation of the scientific name:

A combination of the Latin *rotundus* (= round) + *-folius* (= -leaved), in reference to the somewhat rounded leaves found in some populations of the species.

Description:

Perennial, rarely annual or biennial, few-leaved, sparsely branched, glabrous, low-growing small- to medium-sized, succulent, and 0.2–1.5 m tall. *Stems* green with a slight bloom, few, usually unbranched, thin, erect to leaning to creeping, rooting along the way, sometimes with sterile small-leaved branchlets at the base, usually simple or sometimes branching near the base, and rarely branched higher up. *Leaves* opposite-decussate, petiolate or hardly so, never peltate, mid- to dull-green to pale greyish green, covered in a slight bloom, succulent, erect to spreading, dense low-down on stem and branches, and sparse higher up; *petiole* up to 1cm long, slightly channelled above, and subcylindrical in cross-section; *blade* 1–8 × 1.0–5.5 cm, succulent, oblong, lanceolate, spatulate, narrowly to broadly cymbiform, sometimes

staghorn-like divided, not folded lengthwise, and not recurved in upper half; *base* tapering and narrowly triangular; *apex* rounded-obtuse; *margins* entire and coarsely dentate to crenate. *Inflorescence* 1–40 cm tall, floriferous only at the top, erect to leaning, apically sparsely branched, few-flowered, corymbose cyme, rather round in outline when viewed from above, branches opposite, subtended by small leaflike bracts that dry soon, without leafy branchlets in axils, and axis green to bluish green; *pedicels* 1–8 mm long and slender. *Flowers* erect; usually bicoloured, crimson red, orange, or yellowish orange (upper ½ of tube and lobes); usually gradually becoming green in lower ½ of tube; lobes same colour as upper ½ of tube; *calyx* bluish green; and covered with a slight bloom; *sepals* ± 1.0–2.0 × 1.0–1.5 mm, elongated-triangular, ± separate, basally fused for ± 0.5 mm, acute, hardly contrasting against green part of corolla tube. *corolla* 8–15 mm long; distinctly enlarged lower down around carpels; distinctly and tightly twisted apically after anthesis; crimson red, orange, or yellowish orange; and usually gradually becoming green lower down. *Tube*

FIG. 12.14.13   In the dry season, plants of *Kalanchoe rotundifolia* shrivel and become quite inconspicuous in their natural habitat, as here between Vaalhoek and Caspersnek in Mpumalanga province, South Africa. Photograph: Neil R. Crouch.

FIG. 12.14.14   At Pilgrim's Rest in Mpumalanga, the orangey red corolla lobes of *Kalanchoe rotundifolia* remain somewhat erect. Photograph: Gideon F. Smith.

6–15 mm long; rounded-4-angled; prominently urceolate; strongly globose lower down to somewhat box-shaped-square when viewed from below; longitudinally sharply narrowing beyond carpels above; and crimson red, orange, or yellowish orange. *Lobes* 2.0–4.5 × 2–3 mm; lanceolate to elliptic; distinctly acute apically; subfalcate; spreading but showing diurnal movement (closing at night); and crimson red, orange, or yellowish orange. *Stamens* inserted in several ranks at or just above the middle of the corolla tube, included; *filaments* 3–5 mm long, thin, and yellow-infused to reddish; *anthers* 0.3–0.5 mm long and yellow. *Pistil* consisting of 4 carpels; *carpels* 4–6 mm long and midgreen; *styles* 2–3 mm long; *stigmas* very slightly capitate and yellowish green; *scales* ± 2 mm long, linear, and light yellowish. *Follicles* brittle, grass spikelet-like, enveloped in dry light brownish purple remains of corolla, dull brownish purple, and 4–6 mm long. *Seeds* 0.50–0.75 mm long and brown. *Chromosome number*: $2n = 34$, 68 (Baldwin, 1938; Uhl, 1948; Raadts, 1985, 1989).

Flowering time:

April–November, peaking in June and July (southern hemisphere).

Illustrations:

Wood (1899: 76); Letty (1962: plate 75, figure 2); Raymond-Hamet & Marnier-Lapostolle (1964: 87, plate XXXII, figures 108–109); Batten & Bokelmann (1966: plate 62, figure 6); Gledhill, (1981: 127, plate 28, figure 2); Lucas & Pike (1971: 39, black-and-white line drawing); Jeppe (1975: 49, plate 28b); Gibson (1978: 37, family 79,3 [as *Kalanchoe* sp.]); Turton (1988: plate 36, figure 101); Urton [illustration by Page] (1993: 60, plate 24, figures 1a and 1b); Adams (1976: plate 76); Germishuizen & Fabian (1997: 150, plate 67a); Vanderplank (1998: 76, plate 33, figure 3); Manning et al. (2001: plate on p. 155, 5); Smith & Van Wyk

FIG. 12.14.15   At the Blyderivier Canyon in the Mpumalanga province, the leaf shape of *Kalanchoe rotundifolia* varies from round to more or less cordate. Photograph: Neil R. Crouch.

FIG. 12.14.16   The free corolla lobes of the yellow flowers of the form of *Kalanchoe rotundifolia* growing at the Blyderivier Canyon are held almost horizontally. Photograph: Neil R. Crouch.

(2008a, b: 153); Smith & Crouch (2009: 86, bottom of page); Kirby (2013: 231); Smith et al. (2017b: 314); Smith & Figueiredo (2017e: figures 2–11).

Note. Gibson (1978: 37, family 79,4), which is labelled *Kalanchoe rotundifolia* var. *crenata*, is a rendition of *Kalanchoe hirta*.

Common names:

Afrikaans: nenta(bos) (Smith, 1966: 349, 565; Lucas & Pike, 1971: 39; Urton, 1993: 60; Germishuizen & Fabian, 1997: 150; Van Wyk & Malan, 1997: 202, 272; Pooley, 1998: 54; Vanderplank, 1998: 76; Smith & Crouch, 2009: 86); nentebos (Jeppe, 1975: 49); plakkie (Germishuizen & Fabian, 1997: 150).

English: common kalanchoe (Manning et al., 2001: 154; Smith & Crouch, 2009: 86; Kirby, 2013: 231) and large orange kalanchoe (Long, 2005).

Setswana: bolatsi (Lete dialect [Selete] and Ngwaketse dialect [Sengwakêtse]), moêthimodisô, and serethe (Kirby, 2013: 231; Cole, 1995: 228).

isiXhosa: mfayisele, yasehlatini (Pooley, 1998: 54), umfayisele yasehlatini, and ipewula (Dold & Cocks, 1999).

isiZulu: idambisa, uchane, and umadinsane (Pooley, 1998: 54).

Note. The name 'nenta(bos)' is applied to several different species of southern African Crassulaceae (Powrie, 2004: 51) and refers to poisoning found in small stock, especially sheep and goats, when crassuloid material is consumed.

Geographical distribution:

The southernmost known location of *Kalanchoe rotundifolia* is at Uitenhage and in the adjacent Swartkops River Valley in South Africa's Eastern Cape Province. From there, the species' distribution stretches north-, east-, and northeastwards in a broad sweep through the eastern half of the country, that is, the eastern and central Northern Cape Province, the Free State, KwaZulu-Natal, and northwards through all of South Africa's provinces, except the Western Cape Province. It also occurs further north in south-tropical and eastern Africa, as well as on the Indian Ocean islands of Inhaca [off the coast of Maputo, Mozambique] and Socotra. In southern Africa, the species has been recorded for central and eastern Botswana by Hargreaves (1990: 5–6), Barnes & Turton (1994: 10), and Setshogo (2005: 52) but is apparently absent in western Botswana and the east-central and

FIG. 12.14.17   This form of *Kalanchoe rotundifolia* from south of Burgersfort, Limpopo province, near the border with the Mpumalanga province, South Africa, has deep orange flowers that are carried on peduncles of almost 1 m tall. Photograph: Gideon F. Smith.

FIG. 12.14.19   In the vicinity of Pretoria, *Kalanchoe rotundifolia* has light orange flowers. Photograph: Gideon F. Smith.

FIG. 12.14.20   A low-growing form of *Kalanchoe rotundifolia* with light bluish, ovate to slightly obovate leaves. Photograph taken near Hartbeespoortdam in the North-West province, South Africa, by Gideon F. Smith.

FIG. 12.14.18   *Kalanchoe rotundifolia* growing in deep shade on the southern (cool) slope of a hillock in Pretoria, Gauteng province, South Africa. At this location, the leaves are nearly round. Photograph: Gideon F. Smith.

southern parts of Namibia. The species has also not been recorded from Lesotho nor from the southern and northern massifs of the Drakensberg that stretch south- and northwards beyond Lesotho. It is well-known from the climatically much milder Swaziland (Braun et al., 2004)

FIG. 12.14.21   Close-up of the orange flowers of *Kalanchoe rotundifolia* in its natural habitat near Hartbeespoortdam. Photograph: Gideon F. Smith.

though. In the western parts of south-tropical Africa, the species is absent from Angola (Fernandes, 1982a). Although unconfirmed speculation places the species in Saudi Arabia, it was not included in, for example, Collenette (1985: 188–191) and Lipscombe Vincett (1984). The species was also not included in Edwards (1976: 19–20) in her treatment of some wild flowers from Ethiopia on the Horn of Africa (Fig. 12.14.2).

Distribution by country. Botswana, Mozambique, Namibia, Socotra, South Africa (Eastern Cape, Free State, Gauteng, KwaZulu-Natal, Mpumalanga, North-West, Northern Cape, and Limpopo), Swaziland, Tanzania, and Zimbabwe.

Habitat:

Unlike several other species of *Kalanchoe*, such as *K. winteri* (Crouch et al., 2016b), that are range-restricted and occur in specific Centres of Endemism (Van Wyk & Smith, 2001), *K. rotundifolia* is very much a generalist that occurs over a vast geographical distribution range and grows with great success in multiple habitats. The species has been recorded from both wet and dry habitats in grassland, bushveld (savanna), forests, and thicket vege-tation. It is usually quite common where it occurs and will

FIG. 12.14.22   *Kalanchoe rotundifolia* often grows in the dappled shade cast by bushveld trees, as here near Thabazimbi in the southwestern Limpopo province of South Africa. Photograph: Gideon F. Smith.

FIG. 12.14.23 This form of *Kalanchoe rotundifolia* has narrowly cylindrical leaves that taper at both ends. In this respect, plants resemble some forms of *Cotyledon orbiculata* L. var. *dactylopsis* Toelken (Crassulaceae). Photograph: Gideon F. Smith.

FIG. 12.14.24 A form of *Kalanchoe rotundifolia* with staghorn-like lobed leaf margins. Photograph: Gideon F. Smith.

locally form dense stands of erect, leaning, or creeping stems, often in the shade and within the canopy drip line of trees.

Conservation status:

The species is regarded as being of 'Least concern' and not threatened with extinction.

Additional notes and discussion:

Taxonomic history. At first, the species today known as *Kalanchoe rotundifolia* was described as *Crassula rotundifolia* (Haworth, 1824: 188). Haworth based the description on living material collected in South Africa by James Bowie (*c*. 1789–02 July 1869) who collected plant material for the Royal Botanic Gardens, Kew, the United Kingdom, in the 1810s and 1820s (Smith & Van Wyk, 1989). Bowie's material of *K. rotundifolia* was cultivated at Kew where it flowered and was painted by Thomas Duncanson in October of 1823. Bowie likely collected material of *K. rotundifolia* during a journey he undertook to the eastern region of the Cape Colony (the vicinity of Uitenhage, Port Elizabeth, and Grahamstown) from early 1820 to 29 January 1821 (Figueiredo & Smith, 2017b). By 1822, the material was already in cultivation at Kew and had finished flowering by the time it was described by Haworth (1824).

FIG. 12.14.25 The leaves of *Kalanchoe rotundifolia* are usually to some extent covered in a waxy bloom, as prominently shown here in this form of the species from the Mpumalanga province in eastern South Africa. Photograph: Gideon F. Smith.

FIG. 12.14.26   The upper flower tube and free lobes of *Kalanchoe rotundifolia* become apically twisted postanthesis.   Photograph: Gideon F. Smith.

The often round to obovate, comparatively small cras- suloid leaves of *K. rotundifolia* can easily be taken as belonging to a representative of the widespread *Crassula*. Although Haworth (1824) clearly stated that he had seen the flowers of what he described as *C. rotundifolia*, he did not dissect the flowers, which would have alerted him to the presence of four carpels (not five as in most crassulas). Less than a year after Haworth described the species, he corrected himself and transferred *Crassula rotundifolia* to *Kalanchoe*, as *K. rotundifolia* (Haworth, 1825: 31). Haworth (1824, 1825) did not provide details of the origin of the material of *K. rotundifolia*. Only later did he unam- biguously state the origin of *K. rotundifolia* as "*Habitat* ad Caput Bonae Spei [Cape of Good Hope], ubi legit amicus Dom. Bowie" (Haworth, 1829: 304). From the 17th to 19th centuries, the southwestern tip of South Africa was in Latin often simply given as 'Caput Bonae Spei', some- times abbreviated as 'C. B. S.' or 'C. B. Sp.', or a variation of these abbreviations, when referring to the origin of plant material collected there. This broadly used term variously denotes 'the Cape', 'Cape Town', 'South

Africa', or even 'southern Africa', as the origin of material so referenced.

Kalanchoe rotundifolia was again collected in the Eastern Cape a few years after the species was formally trans- ferred to *Kalanchoe*, when the celebrated plant collectors Christian F. Ecklon [1795–1868] and Carl (Karl) L.P. Zeyher [1799–1858] (Gunn & Codd, 1981: 144–147; 383–387) who were active in South Africa for several decades in the mid-1800s mentioned the species as hav- ing been recorded from near Uitenhage in South Africa's Eastern Cape province (Ecklon & Zeyher, 1837: 305). Some years later, Schonland (1919: 8, 9, 59) agreed with the occurrence of *K. rotundifolia* at Redhouse, between Uitenhage and Port Elizabeth. In this regard, he noted that the species occurs also in Natal [KwaZulu-Natal province] much further east, based on records collected by John Medley Wood who was at the time attached to the present-day KwaZulu-Natal Herbarium in Durban. The only other species of *Kalanchoe* that Schonland (1919) listed for the Uitenhage and Port Elizabeth region was *K. hirta*, a name that we uphold, but that is sometimes included in the synonymy of *K. crenata* (Smith et al. submitted).

Dietrich (1840: 1328) transferred *Kalanchoe rotundifolia* to the genus *Verea*, as *V. rotundifolia* (Haw.) D.Dietr. *Verea* is sometimes attributed to Willdenow (1799: 471), based on the earlier work of Andrews (1798: Plate XXI; where the genus name is spelled *"Vereia"*), but this genus has not been accepted by researchers in the Crassulaceae and is more or less consistently included in the synonymy of *Kalanchoe*, a view with which we agree.

Harvey (1862: 379), in the treatment of *Kalanchoe* for the *Flora capensis* series, included *K. rotundifolia* but expressed some doubt as to whether he was in fact dealing with Haworth's species. However, his description closely coincides with material today referred to as *K. rotundifolia*. As was the case with Ecklon & Zeyher (1837), all the localities that Harvey listed for the species were from the Eastern Cape province, from near Uitenhage, and now also from slightly further east near Grahamstown. Given Harvey's (1862) uncertainty about the identity of the material, he deliberately also catalogued *Crassula rotundifolia*, the basionym of *K. rotundifolia*, in his work (Harvey, 1862: 365).

The first reference to *Kalanchoe rotundifolia* from beyond the borders of South Africa was when Balfour (1888: 90) unambiguously recorded it from Wadi Dilal (600 m above sea level) on the island of Socotra. This is the northernmost extent of the distribution range of the species. At that time, the species was therefore somewhat incongruously known from its southern- and northernmost locations only, with the rest of its distribution range being unrecorded.

The first record of *Kalanchoe rotundifolia* from an African country, other than South Africa, was made by Baker (1899: 434) when he recorded the species from Buluwayo [Bulawayo] in central-southwestern Zimbabwe (then Rhodesia). Seven years later, the occurrence of the species in that country was confirmed by Engler (1906: 878) when he found the species in Matabeleland.

Shortly thereafter, the species was also collected in Mozambique, from "Delagoa Bay" [Maputo Bay] (Schinz & Junod, 1900: 38); Schinz worked from Zürich, Switzerland, while Junod was based in present-day Mozambique. Maputo Bay includes the present-day city of Maputo and all the land that surrounds the Bay (Crouch et al., 2015). Schinz & Junod (1900) noted that their species was the same as the one recorded from the "Kapkolonie" [Cape Colony] by Haworth (1824, 1825, 1829). However, Schinz & Junod (1900) attributed the name to "Harv." and not Haw., even though they clearly linked Haworth (1825) to the plant name in the checklist. The supporting herbarium vouchers that were listed as *H. Junod 284*, *H. Junod 442*, and *H. Junod 443* that consist of a mixture of *Kalanchoe* material, the latter specimen, for example, being the holotype of a separate species,

K. leblanciae (Crouch et al., 2016a; Smith et al., 2017a). Rather confusingly, Schinz & Junod (1900: 38) listed "Crassula rotundifolia Harv." [sic; probably meaning sensu "Harv."] separately from *K. rotundifolia*.

Following on from Harvey's, 1862 treatment of *Kalanchoe* for the *Flora capensis* series, where he mentioned *K. rotundifolia* as also occurring in KwaZulu-Natal, South Africa, Wood (1899: 76) published a drawing of the species, based on material he collected at Inanda, near Durban. Based on his specimens, he noted the leaf margins of his material as being either entire or toothed. The rest of the description also closely coincided with material presently treated as *K. rotundifolia*. Wood was a pioneering plant collector in KwaZulu-Natal (Schrire, 1983; McCracken & McCracken, 1990) and eight years later produced *A Handbook to the flora of Natal* (Wood, 1907), in which he listed *K. rotundifolia*. Like Schinz & Junod (1900), when recording the species for Mozambique seven years earlier, Wood attributed the name *K. rotundifolia* to "Harv." rather than to Haworth. The concept of *K. rotundifolia* as established by Haworth (1825, 1829) was therefore not yet fully understood, at least in southern Africa, as Schinz & Junod and Wood independently interpreted Harvey, in the *Flora capensis* treatment, as having established the name. Balfour (1888: 90) also followed Harvey's concept of the species because he had not seen Haworth's description. Regardless, the early works, such as those of Wood (1907) and later Henkel (1934), on the plants of KwaZulu-Natal greatly facilitated the eventual production of the *Flora of Natal* by Ross (1972). In the latter work, *K. rotundifolia* was treated as Haworth (1824, 1825, 1829) initially interpreted and circumscribed the species.

Hamet (1907) monographed the genus *Kalanchoe* as known at that time. In his treatment, he included *K. rotundifolia* and noted the material first described by Haworth (1824, 1825) as originating from South Africa ("…provenant des déserts de l'Afrique du Sud, …"), although the natural habitats of this species can hardly be described as "deserts" (Hamet, 1907: 895–896). Hamet (1907) also recorded *K. rotundifolia* for Socotra based on Balfour (1888).

In the same year, *Kalanchoe rotundifolia* was for the first time recorded for Namibia (then German South-West Africa) by Engler & Diels (1907: 30) [as *Kalanchoe luebbertiana* Engl.]. Sixteen years later, Dinter (1923b: 77) referred to what was apparently this same Namibian 'species' as '*K. lübberti* Loes', a designation with no status as it is not a name in the sense of the *International Code of Nomenclature for algae, fungi, and plants* (ICN or *Code*). The word "name" as used in the *ICN* (McNeill et al., 2012: 26, Article 6.3) means a name that has been validly published, whether it is legitimate or illegitimate. Further, a name of a taxon has no status under the *Code* unless it is validly published (McNeill et al., 2012: 32, Article 12). The designation '*Kalanchoe lübberti* Loes' is therefore not a synonym

(McNeill et al., 2012: 156, 161, Glossary) of *Kalanchoe rotundifolia*.

In South Africa, Burtt-Davy & Pott-Leendertz, (1912: 144) for the first time recorded *Kalanchoe rotundifolia* from the northern parts of South Africa, by including it in the checklist of the species of what was then known as the Transvaal (this erstwhile South African province is today split into the North-West, Gauteng, Limpopo, and Mpumalanga provinces of the country) and Swaziland. In Retief & Herman's much expanded and updated work on northern South Africa, the occurrence of the species essentially north of the Vaal River in South Africa was repeated and confirmed for the western, southern, northern, eastern, and central parts (Retief & Herman, 1997: 392).

The first synonym of *Kalanchoe rotundifolia* described from South Africa, *K. seilleana* Raym.-Hamet, in which he commemorated his friend Guy Seille, was described based on material from Prieska in the Northern Cape Province (Raymond-Hamet, 1916: 6–9). The holotype of *Kalanchoe seilleana* is a specimen kept at GRA. There is no indication of the collector on its label. It was determined by Raymond-Hamet as "Original der art" and referred to in the protologue as "l'échantillon authentique, conservé dans l'herbier de l'Albany Museum de Grahamstown (colonie du Cap)" [English: the authentic specimen, conserved in the herbarium of the Albany Museum, Grahamstown (Cape Colony)] (Raymond-Hamet, 1916: 8). Seille, who is commemorated in the name, was not the collector of the specimen. Raymond-Hamet (1916) acknowledged the close relationship between *K. seilleana* and *K. rotundifolia* and placed both in the same group (Group 13) he established in his earlier monograph (Hamet, 1907).

A second synonym for *Kalanchoe rotundifolia* from South Africa was described some 32 years later, also by Raymond-Hamet, when he published the name *K. guillauminii* Raym.-Hamet (Raymond-Hamet, 1948: 465). The material on which this name was based originated from Port Elizabeth in the Eastern Cape province, which is the same general location from which James Bowie collected the material on which the name *K. rotundifolia* was originally based by Haworth (1824, 1825). The material was obtained by Humbert during a visit to South Africa in 1933, from Mr Long, director of the gardens of Port Elizabeth (Raymond-Hamet, 1948: 365). It was grown at the greenhouses of the Muséum d'Histoire Naturelle, in Paris, for several years before it flowered. The first specimen to flower was pressed as a Herbarium specimen and examined by Raymond-Hamet. He again recognised the close relationship between *K. guillauminii* and *K. rotundifolia*, noting that the flowers were similar but that the leaves differed.

By the mid-twentieth century, *Kalanchoe rotundifolia* was increasingly recorded in southern Africa, with Wilman (1946: 79), for example, noting it for the northeastern parts of the present-day Northern Cape province, an area at the time, and still often, known as Griqualand West. Wilman was the first Director of the McGregor Memorial Museum in Kimberley, which was included in Griqualand West (Moffett, 2014: 324). Prieska, from where Raymond-Hamet (1916) described *K. seilleana*, fell just beyond the southernmost tip of Griqualand West, so the species was already known from this part of the present-day Northern Cape of South Africa.

In 1967, *Kalanchoe decumbens* was described from Swaziland (Compton, 1967). Compton collected material of several *Kalanchoe* entities in that small country, some of which he later described as species, for example, *K. montana* (Smith et al., 2016a). The type of *K. decumbens* is [R.H.] *Compton 29407*, a collection from a plant that flowered in April 1962 in cultivation at Mbabane, in Swaziland, where Compton was based at the time (Smith et al., 2016a). This specimen, which was collected on 13 November 1959, is kept at SDNH, the National Herbarium of Swaziland, and housed in Mbabane (Smith & Willis, 1997: 34; 1999: 128). The label attached to *Compton 29407* is headed "Herbarium of the Botanical Survey of Swaziland" and carries the following handwritten information: "**Name** Kalanchoe s̶p̶. [i.e. struck through] decumbens Compton. **Loc.** Ingwavuma Hill. 13.11.59. **Dist.** Hlatikulu **Alt.** *c.* 2000. **Notes** Cinnabar red. **Coll.** R.H. Compton No. 29407. **Date** Herb. in Garden. 14.4.62, 18—23—. Det. R.H. Compton". A note added in Compton's hand at the bottom of the label reads "Not matched in PRE", which indicates that Compton compared material he collected with material kept in the National Herbarium of South Africa to determine the species. Two duplicates of this collection exist, at NBG (NBG0068850-0) and at PRE (PRE0523106-0). The NBG specimen is accessible at JSTOR Global Plants, but the PRE specimen is not available. It is, however, listed in PRE's online database. The specimen held in SDNH is the holotype, because Compton clearly stated that "the Swaziland specimens are included in the Herbarium that has been built up during the progress of the survey" (Compton, 1967: 293), adding that duplicates existed at NBG and PRE. Note though that the case of another species described by Compton, *K. montana*, was different in that the so-called duplicate specimens were not duplicates at all, given that they were prepared on different dates, which is why for that species an NBG specimen was designated as type in a second-step lectotypification (Smith et al., 2016a). The two duplicates of *Compton 29407* held at PRE and NBG respectively are therefore duplicates of the holotype, that is, isotypes (McNeill et al., 2012: 17, Article 9.4). Specimens at SDNH and NBG (PRE duplicate could not be examined) have three distinct dates on the label, namely, 14, 18, and 22 (or 23) April 1962. It is likely that those were the dates on which material was collected from the cultivated plant, as it came into flower. However, since the two specimens have a set of three dates, there is no reason why they should not be regarded as duplicates. Before describing *Kalanchoe decumbens*, Compton (1966: 44) listed it as a "sp. nov." under what became the type collection, namely, "*C.29407*". Following description of the species, a "Piece [was] sent to Dr Raymond-Hamet 18.10.1968", according to a note in Compton's handwriting on the type specimen where the piece of material was removed. By 1981, no changes were proposed to the listing of *K. decumbens* in the *Flora of Swaziland*, as it was to be a further four years before Tölken (1985: 62) synonymised this species with *Kalanchoe rotundifolia* (Kemp, 1981: 30).

Kalanchoe rotundifolia forma *tripartita* Raym.-Hamet ex R.Fernandes is the most recently published name that is nowadays included in the synonymy of *Kalanchoe rotundifolia*. Fernandes (1978) validated this name that Raymond-Hamet (1956) proposed (as '*K. rotundifolia* var. *tripartita* Raym.-Hamet') but did not validly publish, for material from South Africa's KwaZulu-Natal Province. Like *K. decumbens*, this forma was synonymised with *K. rotundifolia* by Tölken (1985).

A further entity based on material from South Africa's KwaZulu-Natal Province that Raymond-Hamet (see Raymond-Hamet, 1960) suggested should be recognised in *Kalanchoe rotundifolia* at the rank of varietas was 'Kalanchoe rotundifolia var. peltata Raym.-Hamet'. However, he did not validly publish this combination. Fernandes (1978) eventually validated the name as *K. rotundifolia* forma *peltata* Raym.-Hamet ex R.Fernandes, therefore at the rank of forma rather than varietas. She synonymised *K. neglecta* (Tölken, 1978) with *K. rotundifolia* forma *peltata*. However, we regard the plants from KwaZulu-Natal with peltate, often distinctly cordate, leaves as warranting recognition at the species rank, for which the earliest validly published name at that rank is *K. neglecta*.

Smith & Figueiredo (2017e) have comprehensively reviewed the taxonomic history of *Kalanchoe rotundifolia*.

Identity and close allies. Taxonomically, *Kalanchoe rotundifolia* is one of the most complex of the southern African representatives of the genus, making it challenging to establish diagnostic characters for the species across its inherent variation. The expression of vegetative morphology, especially leaf shape and plant size, is exceedingly variable in this wide-ranging species. Plants are usually rather small-growing, generally remaining less than 0.5 m tall, including the inflorescence. Larger specimens of up to 2 m tall are much rarer, at least in southern Africa, where the species is most diverse. Leaf shape is variable, ranging from nearly perfectly round, through elliptic, elliptic-elongated, and ovate, to obovate. In addition, the leaf margin can be entire, sinuate, dentate, crenate, or lobed. The leaves of *K. rotundifolia* are often to some extent covered in a waxy bloom and range from sessile

to petiolate. However, to date, peltate-leaved forms are not known in the species. The peltate-leaved plants from KwaZulu-Natal and southern Mozambique previously given recognition at the rank of forma under *K. rotundifolia*, as *K. rotundifolia* forma *peltata*, are nowadays recognised as a long-confused species, *K. neglecta*.

The reproductive morphology is more constant in terms of the size and shape of the corolla tube and lobes, as well as in the upper part of the corolla tube and lower part of the corolla lobes invariably becoming twisted at the postanthesis stage. However, flower colour ranges from shades of yellow-infused light orange through pinkish red to scarlet.

The morphological variation found in, and very broad geographical distribution range of, *Kalanchoe rotundifolia* has resulted in eight names appearing in its synonymy (Smith & Figueiredo, 2017e). All eight synonyms are applied to material from the *Flora of Southern Africa* region and Mozambique. Interestingly, no material from further north in Africa has been proposed for recognition at specific or infraspecific rank. Morphologically, *K. rotundifolia* is clearly at its most diverse in a broadly defined southern African, especially along the eastern seaboard (South Africa [Eastern Cape and KwaZulu-Natal], Swaziland, and Mozambique), which has prompted the description of a number of entities at different taxonomic ranks.

Horticulture and cultivation. *Kalanchoe rotundifolia* is exceedingly easy to propagate from stem cuttings and seed. Seed germinates where it falls, and in a garden, plants tend to become weedy. Seedlings transplant with a high degree of success, even when still very small. Cuttings can be taken at any time of year and will rapidly root in open beds or in containers. The leaning, often top-heavy stems of the species will root where they touch the soil. Cultivation is very easy, and plants thrive in any type of soil. The species will flower quickly, often within its first growing season.

Uses. The leaves of *Kalanchoe rotundifolia* are used in a purification ceremony after burial (Riley, 1963: 157). Hutchings (1996: 113) notes that the latter use is by Zulu men. Kirby (2013: 231) additionally records the following uses: as an emetic, and for pain in the solar plexus; the hollow flowering stem is used as a sipping straw, and smoke from burning branches is inhaled to induce sneezing to treat headaches. The dried plant is used like snuff, for headaches and to induce sneezing (Cole, 1995: 228).

Poisonous properties. Plants are known to be poisonous to small stock, causing a chronic form of heart glycoside poisoning (see Onderstall, 1984: 94; Kellerman et al., 2005; Smith & Van Wyk, 2008a, b: 152; Kirby, 2013: 231). This condition is known as nenta (see Afrikaans common names) or krimpsiekte. Goats have been recorded as being especially susceptible to poisoning with *Kalanchoe rotundifolia* (Long, 2005).

Infusions of plants are also used as an emetic (Long, 2005).

15. *KALANCHOE SEXANGULARIS* N.E.BR.

Red-leaved kalanchoe (Figs. 12.15.1 to 12.15.15)

FIG. 12.15.1 *Kalanchoe sexangularis* var. *sexangularis* growing in direct sunlight amongst rocks and boulders in its natural habitat near Mbombela, Mpumalanga, in eastern South Africa. Photograph: Gideon F. Smith.

Kalanchoe sexangularis **N.E.Br.** in *Bull. Misc. Inform. Kew* 1913(3): 120 (1913). Fernandes (1983: 65); Tölken (1983: plate 1878); Tölken (1985: 67); Hardy & Fabian (1992: xviii, 44, plate 19); Germishuizen & Fabian (1997: 154, plate 69d); Retief & Herman (1997: 393); Descoings (2003: 175); Lebrun & Stork (2003: 220); Smith & Van Wyk (2008a, b: 125, 152, 153); Smith & Crouch (2009: 87); Court (2010: 92); Figueiredo et al. (2016: 92). **Type**: [South Africa] from Barberton [annotated in pencil] [Mpumalanga province, 2531 (–CC)], s.d., *Thorncroft s.n.* (K K000232850, holo-).

Synonyms:

Kalanchoe rogersii Raym.-Hamet in *Rec. Albany Mus.* 3: 127 (1915). Compton (1976: 220); Burtt Davy (1926: 144). **Type**: E Transvaal [Mpumalanga], Komati Poort [Komatipoort], [Mpumalanga province, 2531 (–BD)], 14 June 1906, *F.A. Rogers 865* (GRA GRA0001096-0, holo-).

Kalanchoe vatrinii Raym.-Hamet in *J. Bot. (Lond.)* 54, supplement 1: 9 (1916). Jacobsen (1970: 290); Fernandes (1980: 420–426); Jacobsen (1986: 632). **Type**: N.W. Rhodesia [Zambia], Livingstone, N [northern] bank of Zambesi [River], 20 August 1911, *Rogers 7444* (K K000232897, holo-).

Kalanchoe mossambicana Resende ex Resende & Sobr. in *Revista Fac. Ci. Univ. Lisboa, Sér. 2a, C, Ci. Nat.* 2: 199 (1952). Raymond-Hamet & Marnier-Lapostolle (1964: 82); Jacobsen (1970: 287) [as *K. mocambicana*]; Jacobsen (1986: 621) [as *K. mocambicana*]. **Type**: Moçambique, Lourenço Marques, Maputo, Goba, 23-07-1944, *Mendonça 1825* (LISU, holo-; LISC LISC002357 pro maxima parte, iso-).

Kalanchoe rubinea Toelken in *J. S. African Bot.* 44: 90 (1978). Germishuizen & Fabian (1982: 129, plate 55d). **Type**: [South Africa] Transvaal, Soutpansberg, *Galpin 14934* (PRE, holo-; K K000232853, iso-).

'Kalanchoe longiflora Schltr. ex J.M.Wood var. coccinea Marn.-Lap.'

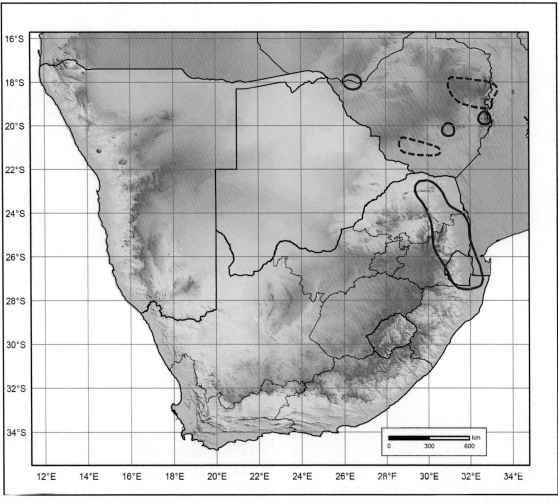

FIG. 12.15.2 Known geographical distribution range of *Kalanchoe sexangularis* var. *sexangularis* (areas outlined in solid red line) and *K. sexangularis* var. *intermedia* (areas outlined in broken red line) in southern and south-tropical Africa.

FIG. 12.15.3 Even when growing in deep shade, as here at the bottom of a cliff on the Soutpansberg, Limpopo province, South Africa, the leaves of *Kalanchoe sexangulairs* var. *sexangularis* retain a light red infusion. Photograph: Gideon F. Smith.

FIG. 12.15.4 The leaves of *Kalanchoe sexangularis* var. *sexangularis* are usually longitudinally folded when growing in direct sunlight. Photograph: Gideon F. Smith.

FIG. 12.15.5   The stems of *Kalanchoe sexangularis* var. *sexangularis* are generally devoid of leaves lower down, and the leaf margins are distinctly crenate, as here in the case of plants growing in Mpumalanga, eastern South Africa. Photographe: Neil R. Crouch.

FIG. 12.15.7   The axils of the bracts on the peduncle of *Kalanchoe sexangularis* var. *sexangularis* very rarely produce leafy branchlets. Photograph: Gideon F. Smith.

FIG. 12.15.6   The leaves of *Kalanchoe sexangularis* var. *sexangularis* are distinctly petiolate. Photograph: Gideon F. Smith.

Derivation of the scientific name:

A combination of the Latin *sex* (= six) + *-angularis* (= angled), in reference to the angled stems of the plant. The stems rarely have six angles though, two and four being more common.

Description:

Perennial, few-leaved, sparsely branched, glabrous, robust succulent, and 0.2–1.0 m tall. *Stems* green to deep wine red, few, unbranched, arising from a brittle, corky base, erect to leaning to creeping, often with several lengthwise running ridges, 4-angled at least on sterile shoots. *Leaves* opposite-decussate, petiolate, green infused with red to a deep crimson red, succulent, spreading or recurved, and coriaceous and rigid on drying; *petiole*

FIG. 12.15.8   The inflorescences of *Kalanchoe sexangularis* var. *sexangularis* can grow over 80 cm tall and often scramble into the lower branches of the trees and shrubs under which these plants are found. Photograph: Gideon F. Smith.

5–50 mm long, channelled above, and lower leaves not clasping the stem but upper leaves distinctly so; *blade* 6–14 × 4–10 cm, broadly elliptic or obovate to oblong, often somewhat folded lengthwise, recurved in upper

FIG. 12.15.9 Lateral view of an inflorescence of *Kalanchoe sexangularis* var. *sexangularis*, showing a few- to many-flowered, flat-topped thyrse with several dichasia. When viewed from above, it appears elliptic in outline. Photograph: Gideon F. Smith.

FIG. 12.15.10 *Kalanchoe* sexangularis var. *sexangularis* flowers have yellowish green corolla tubes. Photograph: Gideon F. Smith.

FIG. 12.15.11 The flowers of *Kalanchoe sexangularis* var. *sexangularis* have bright yellow corolla lobes. Photograph: Gideon F. Smith.

FIG. 12.15.12 The fruit on the dry inflorescences of *Kalanchoe sexangularis* var. *sexangularis* is brittle, grass spikelet-like, and enveloped in the dry light brown remains of the corolla. Photograph: Gideon F. Smith.

half; *base* obcordate in lower leaves and narrowly triangular to cuneate in upper leaves; *apex* rounded-obtuse; *margins* coarsely crenate or undulate-crenate into rounded, harmless, crenations. *Inflorescence* 25–80 cm tall, erect to leaning, apically dense, few- to many-flowered,

flat-topped thyrse with several dichasia, rather ellipsoid in outline when viewed from above, branches opposite, erect to gracefully curved upwards, subtended by leaflike bracts, without or very rarely with leafy branchlets in axils, and axis reddish green to bright crimson red; *pedicels* 4–8 mm long and slender. *Flowers* erect, greenish yellow (tube) to bright yellow (lobes); *calyx* shiny reddish green, strongly infused with small red spots especially towards base, *sepals* ± 2 × 1 mm, triangular-lanceolate, ± separate, basally adnate, acute, and contrasting reddish green against corolla tube; *corolla* 14–17 mm long, somewhat enlarged lower down, not twisted apically after anthesis,

FIG. 12.15.13   The banks of the Crocodile River in Mpumalanga, in eastern South Africa, are typical bushveld habitats of *Kalanchoe sexangularis* var. *sexangularis*. Photograph: Gideon F. Smith.

FIG. 12.15.15   *Kalanchoe sexangularis* var. *sexangularis* grows well in any type of container. If placed in full sun, the leaves turn bright red. Photograph: Gideon F. Smith.

FIG. 12.15.14   The combination of red leaves and yellow flowers has long made *Kalanchoe sexangularis* var. *sexangularis* a favourite accent plant among gardeners. Photograph: Gideon F. Smith.

and yellowish green and yellow; *tube* 13–16 mm long, distinctly 4-angled, box-shaped-square when viewed from below, longitudinally fluted above, and yellow to yellowish green; *lobes* 2.0–3.5 × 2–3 mm, ovate to subcircular to somewhat pyriform, rounded at apex, apiculate, bright yellow, and faintly brown-tipped. *Stamens* inserted well above the middle of the corolla tube and slightly exserted; *filaments* 3–5 mm long, thin, and yellow; *anthers* 0.3–0.5 mm long and greenish yellow. *Pistil* consisting of 4 carpels; *carpels* 7–8 mm long and light yellowish green; *styles* 3–7 mm long; *stigmas* very slightly capitate and whitish yellow; *scales* ± 2 mm long, linear, and reddish brown in upper half. *Follicles* brittle, grass spikelet-like, enveloped in dry light-brown remains of corolla, dull whitish green, and 6–7 mm long. *Seeds* 0.50–0.75 mm long and brown. *Chromosome number*: unknown.

Kalanchoe sexangularis is regarded as consisting of two varieties.

Key to the varieties of *Kalanchoe sexangularis*:

Corolla to 14 mm long, corolla tube ± 13 mm long, leaf margins crenate, petioles 5–50 mm long, and follicles 6–7 mm long. Distribution predominantly southern African..var. *sexangularis*.
Corolla to 17 mm long, corolla tube ± 16 mm long, leaf margins usually lacking crenations, petioles short or absent, and follicles 7–10 mm long. Distribution south-tropical African..var. *intermedia*.

Kalanchoe sexangularis var. **sexangularis**. Fernandes (1983: 66).
   Description:
   The typical variety is in most respects a somewhat smaller plant than var. *intermedia*. Especially, the flowers are smaller, the corolla reaching a length of 14 mm (vs. 17 mm) at maturity, with the follicles 6–7 mm (vs. 7–10 mm) long. The margins of the leaves are consistently crenate and never entire, and the leaves are distinctly petiolate.

Flowering time:

June–November, peaking in August (southern hemisphere).

Illustrations:

Tölken (1983: plate 1878); Germishuizen and Fabian, (1982: 120, 121, plate 55d, as *Kalanchoe rubinea* Toelken); Hardy & Fabian (1992: 44, plate 19); Germishuizen & Fabian (1997: 154, 155, plate 69d, as *Kalanchoe sexangularis*; artwork identical to that included in Germishuizen & Fabian, 1982); Smith & Van Wyk (2008a, b: 125 middle and bottom left, 153 bottom right); Smith & Crouch (2009: 87); Pienaar & Smith (2011: 144); Figueiredo et al. (2016: figures 1–6 and 9).

Common names:

Afrikaans: bosveldplakkie, rooiblaarplakkie, vlamplakkie, and vuurblaar.

English: bushveld kalanchoe and red-leaved kalanchoe.

Geographical distribution:

Kalanchoe sexangularis var. *sexangularis* is an element of a fairly narrow strip of savanna vegetation of the climatically mild eastern parts of southern and south-tropical Africa, essentially the region to the east of the central part and northern extension of the Drakensberg massif. The species does not occur to the west of the Drakensberg, with the low, continental climate-type winter temperatures clearly preventing the colonisation of apparently suitable habitats on the mountain. To the east, it also does not occur in the sandy coastal forests of southern Mozambique and northern KwaZulu-Natal, where it is replaced by its closest and much taller-growing, relative, *K. leblanciae* (Fig. 12.15.2).

Distribution by country. Mozambique, South Africa (northeastern KwaZulu-Natal, Mpumalanga, and Limpopo), Swaziland, Zambia, and Zimbabwe.

Habitat:

Kalanchoe sexangularis var. *sexangularis* is often locally common where it occurs, with a preference to grow on the mid- and upper slopes of rock-strewn koppies in bushveld. It often seems to prefer the dappled semishade provided by tree and shrub canopies.

Conservation status:

In South Africa, the typical variety is regarded as 'Least concern' and not threatened.

Kalanchoe sexangularis var. *intermedia* (R.Fern.) R.Fern. in *Bol. Soc. Brot.* (2.ª série) 55: 99 (1982). Fernandes (1983: 66).

Description:

Kalanchoe sexangularis var. *intermedia* is recognised on the basis of it having longer corollas (up to 17 mm vs up to 14 mm) and carpels [sometimes referred to as 'follicles'] (7–10 mm vs. 6–7 mm) than those of the typical variety, with the corollas additionally having a more attenuate tube. The leaves are slightly smaller with shorter petioles, and leaf margins often lack crenations.

Synonym:

Kalanchoe vatrinii var. *intermedia* R.Fern. in *Bol. Soc. Brot.* (2.ª série) 53: 422 (1980). **Type**: Southern Rhodesia [Zimbabwe], near Salisbury [Harare], 20 September 1936, *Eyles 8846* (K K000232896, holo-).

Derivation of the scientific name:

From the Latin *intermedia* (= intermediate).

Flowering time:

Probably June–September, according to herbarium specimens.

Illustrations:

Unknown.

Common names:

Unknown.

Geographical distribution:

It has a more northerly and restricted distribution than the typical variety (Fig. 12.15.2).

Distribution by country. Mozambique and Zimbabwe.

Habitat.

As in the case of *Kalanchoe sexangularis* var. *sexangularis*, the var. *intermedia* is often locally common in nature, with a preference for the mid- and upper slopes of rock-strewn koppies in bushveld (savanna) in south-tropical Africa.

Conservation status:

Unknown.

Additional notes and discussion on *Kalanchoe sexangularis*:

Taxonomic history and nomenclature. Following its description (Brown, 1913), *Kalanchoe sexangularis* was listed by Pott (1920) and Burtt Davy (1926) in their publications on the flora of the then Transvaal [now North-West, Gauteng, Limpopo, and Mpumalanga Provinces]. Burtt Davy additionally listed *K. rogersii* Raym.-Hamet, which had been described two years after *K. sexangularis*, in 1915. Regarding this name, it has been noted (Figueiredo et al., 2016) that the specimen *Rogers TRV3044* kept at PRE (PRE0523701-0/PRE negative no. 7356) is not an isotype as the collecting number is *867*, not *865* as in the holotype. Burtt Davy also surmised that *K. longiflora* might occur within what he defined as the Eastern Mountains 'Botanical Province' of the region he covered (Burtt Davy, 1926). In 1952, a further species, *Kalanchoe mossambicana*, was described by Resende. Fernandes (1980: 421) corrected the original epithet 'moçambicana' to 'mossambicana' as the letter 'ç' does not exist in the Latin alphabet and needs to be transcribed (see McNeill et al., 2012: Art. 60). She also noted that two existing isotypes were mixed collections: LISC002355 is a specimen of *K. leblanciae* Raym.-Hamet, and LISC002357 consists mostly of *K. mossambicana* but with the small leaf belonging to *K. leblanciae*. Therefore, LISC002355 should not be cited as an isotype.

Some 25 years later, Tölken (1978) described a further species in the group as *K. rubinea* Toelken, which he stated to have an affinity with the turquoise-leaved *K. longiflora* Schltr. ex J.M.Wood, a species restricted to the central Tugela River basin of KwaZulu-Natal in south-eastern South Africa. In the protologue of *K. rubinea*, Tölken (1978) reflected on how true *K. longiflora* was usually confused with the common horticultural subject (*K. sexangularis*), which he was describing as distinct (as *K. rubinea*). Fernandes (1980) suggested that *K. rubinea* was conspecific with *K. vatrinii* Raym.-Hamet, then considered a species distributed in Zimbabwe, Zambia, and Mozambique. Two years later, Fernandes (1982b, 1983) considered *K. vatrinii* to be conspecific with *K. sexangularis*, the latter being an older name and therefore having priority. Tölken (1983) followed this view, and further synonymised *K. rogersii* and his own *K. rubinea* with the earlier *K. sexangularis*.

'Kalanchoe longiflora Schltr. ex J.M.Wood var. coccinea Marn.-Lap.' cited here in synonymy is, according to Descoings (2003: 169), a nom. inval. that dates from 1954. However, we have been unable to trace the source reference but include it in the synonymy of *Kalanchoe sexangularis*, based on the descriptions and identical associated images included in Jacobsen (1970: 287, plate 105/1), and Jacobsen (1986: 618, figure 866). The plant figured by Jacobsen matches *K. sexangularis*. The name applied by Jacobsen was originally cited by Tölken (1978: 90) as "*Kalanchoe longiflora* var. *coccinea* Marn.-Lap. ex Jacobsen in *Handbook of Succulent Plants* 2: 652, figure 866 (1960) and *Sukkulenten Lexikon*: 263, t. 105, 1 (1970)". He determined it to not be validly published and a synonym of *K. rubinea* (= *Kalanchoe sexangularis*). Rather confusingly, it has also been included in the synonymy of the Ethiopian *K. petitiana* A.Rich. (Descoings, 2003: 169).

Kalanchoe sexangularis var. *intermedia* was originally treated as belonging to *Kalanchoe vatrinii* (Fernandes, 1980), which, at the time, was accepted as a species. However, in the early 1980s, *K. vatrinii* was treated as conspecific with *K. sexangularis* (Fernandes, 1982b). *Kalanchoe sexangularis* var. *intermedia* was therefore transferred to *K. sexangularis*, as *K. sexangularis* var. *intermedia* (R.Fern.) R.Fern. This taxon from Zimbabwe and Mozambique was named var. *intermedia* because some specimens by which it is represented had earlier been misidentified as *K. longiflora* and appeared to be intermediate between that species and *K. vatrinii* (= *K. sexangularis*).

Identity and close allies. Although nowadays usually regarded as a distinctive member of *Kalanchoe* in southern

and south-tropical Africa, *K. sexangularis* has in the past been confused with several other species. In its current circumscription, it is well-known as a low-growing to medium-sized, often near-uniformly red-leaved representative of the genus (Figueiredo et al., 2016). At least some of the problems associated with the correct application of the name *K. sexangularis* were related to the interpretation of the type specimen associated with this binomial. Figueiredo et al. (2016) clarified the typification and synonymy of the name *K. sexangularis*. In that interpretation, which we follow here, *K. sexangularis* comprises two varieties, *K. sexangularis* var. *sexangularis* and *K. sexangularis* var. *intermedia*. The latter was not accepted as a good variety by Descoings (2003: 175) who considered it to be a synonym of *Kalanchoe sexangularis*. However, we regard *K. sexangularis* var. *intermedia* as a good variety, being a more northern entity.

Kalanchoe sexangularis is comparatively easy to identify given the tendency of both leaf surfaces to take on a shiny, bright crimson red colour, especially during periods of environmental stress. The closest ally of *K. sexangularis* is *K. leblanciae*, from which it differs by having longer-tubed flowers and vegetative plant parts often prominently red-infused. *Kalanchoe sexangularis* does not, or very rarely, produce leafy shoots from the axils of the bract-like leaves on the inflorescence axis, unlike *K. leblanciae*. In addition, the leaves of *K. sexangularis* are generally distinctly petiolate, while those of *K. leblanciae* are sessile or have pseudo-petiolate attenuate bases. Further, the inflorescences of *K. sexangularis* are broader and more ellipsoid in outline when viewed from above, relative to the much narrower ones of *K. leblanciae*. The flowering shoots of *K. leblanciae* can attain a height of 1.6 m, whilst those of *K. sexangularis* reach only 0.8 m.

Cultivation. Kalanchoe sexangularis is exceedingly easy to propagate from stem cuttings that can be planted directly in the spot where plants are intended to grow. It grows well in virtually any soil type. Leaf cuttings have not been known to strike root. In cultivation, it usually flourishes in full sun but will also thrive in semi-shade given its propensity to grow in association with bushveld trees and shrubs. In more shady situations, the leaves are less red-infused. Shaded plants may exhibit little to no red colouration. Being short-day plants, the species produces its flowers, which have yellowish green tubes and butter-yellow corolla lobes, from midwinter to early summer. In its summer-rainfall habitat, this corresponds to the season when precipitation decreases and temperatures drop, to when these increase again. The combination of red leaves and yellow flowers is visually striking in an otherwise drab, dusty landscape.

16. *KALANCHOE THYRSIFLORA* HARV.

White lady (Figs. 12.16.1 to 12.16.10)

FIG. 12.16.1   A clump of *Kalanchoe thyrsiflora* growing in a shaded position among rocks. Photograph: Gideon F. Smith.

Kalanchoe thyrsiflora **Harv.** in Harvey & Sonder (eds) *Fl. cap.* 2: 380 (1862) [15–31 October 1862]. Hooker (1899: Tab. 7678); Wood (1899: Table 52); Hamet (1907: 894); Wood (1907: 46); Burtt-Davy & Pott-Leendertz, 1912: 144); Pole Evans (1929: plate 341); Berger (1930: 407, figure 196, H–L); Letty (1962: 152, plates 75,3); Raymond-Hamet & Marnier-Lapostolle (1964: 92, plate XXXIV, figures 115–116); Batten & Bokelmann (1966: 73, plate 62,2); Jacobsen (1970: 290); Jacot Guillarmod (1971: 180); Ross (1972: 179); Compton (1976: 221); Gibson (1978: 54); Fernandes (1982b: 99); Germishuizen & Fabian (1982: 118); Fernandes (1983: 59); Tölken (1985: 69); Jacobsen (1986: 630); Killick (1990: 102); Hardy & Fabian (1992: 42, Plate 18); Germishuizen & Fabian (1997: 152, plate 68,a); Retief & Herman (1997: 393); Van Wyk & Malan (1997: 126); Manning et al. (2001: 154); Descoings (2003: 178); Pooley (2003: 128); Smith & Crouch (2009: 88); Court (2010: 91); Gill & Engelbrecht (2012: 112); Smith et al. (2017b: 315). **Type**: [South Africa], Cape, Katrivier near Philipstown, *Ecklon & Zeyher 1953* (S S-G-10758, lecto-; SAM SAM0036022-0 pro parte, isolecto-). Designated by Tölken (1985: 69).

Derivation of scientific name:

From the Latin 'thyrsus', thyrse, and the Latin '-florus', -flowered, alluding to the type of inflorescence.

Description:

Biennial or short-lived multiannual, few- to many-leaved, usually solitary, glabrous throughout, robust succulent, and to 1.5 m tall. *Stem* usually solitary, arising from a slightly swollen rootstock, erect, and slightly horizontally ridged. *Leaves* erect to patent-erect; succulent; sessile; ± flattened above; slightly convex below; light greyish green to bluish green; and infused with red, pink, or orange, especially along margin; *petiole* absent;

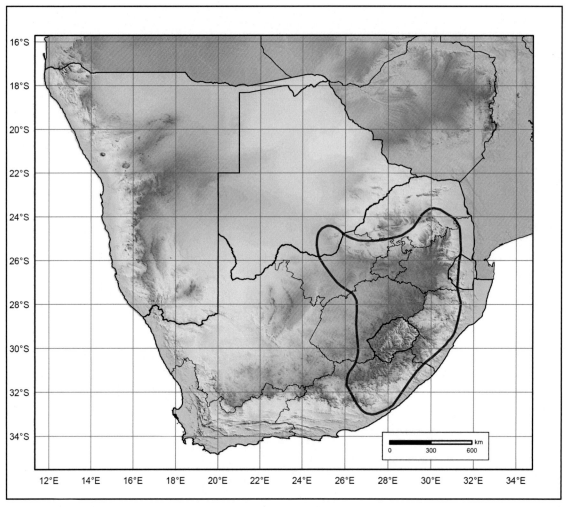

FIG. 12.16.2   Known geographical distribution range of *Kalanchoe thyrsiflora* in southern Africa.

blade 6–14 × 2–9 cm, obovate to oblanceolate, and not folded lengthwise; *base* narrow; *apex* rounded-obtuse; *margins* entire and with a reddish tint. *Inflorescence* a slender, erect, densely flowered, cylindrical to club-shaped thyrse consisting of several dichasia terminating in monochasia, to 1.5 m tall; *pedicels* short. *Flowers* erect to slanted horizontally, usually greyish green, and all parts covered with a substantial white waxy bloom; *calyx* greyish green and hardly contrasting against corolla tube; *sepals* 3.0–4.5 mm long, short-triangular to ovate, almost free to the base, and acute; *corolla* greyish green and densely covered in a waxy bloom; *tube* 12–18 mm long, more or less quadrangular-urceolate, and slightly 4-angled; *lobes* 2–3 × 2–3 mm, somewhat square, recurved, tapering to a point, and bright yellow. *Stamens* included to hardly exserted and visible at mouth of corolla; *filaments* short; *anthers* 1.5–2.0 mm long and yellow. *Pistil* consisting of 4 carpels; *carpels* 10–14 mm long and oblanceolate; *styles* 1.5–3.5 mm long; *stigmas* very slightly capitate; *scales* ± 2.5 mm long, transversely oblong, and truncate. *Follicles* not seen. *Seeds* not seen. *Chromosome number*: unknown.

Flowering time:

April–September, peaking between June and July (southern hemisphere). Some clones can start flowering as early as February.

Illustrations:

Hooker (1899: Tab. 7678); Pole Evans (1929: plate 341); Berger (1930: 407, figure 196, H–L); Letty (1962: plates 75,3); Raymond-Hamet & Marnier-Lapostolle (1964: plate XXXIV, figures 115–116); Batten & Bokelmann (1966: plate 62,2); Jacobsen (1970: plate 106,2); Gibson (1978: 54, plate 54,1); Germishuizen & Fabian (1982: plate 54,a); Tölken (1985: 70, figure 9,1); Jacobsen (1986: figures 892–893); Hardy & Fabian (1992: plate 18); Germishuizen & Fabian (1997: plate 68,a; artwork identical to that included in Germishuizen & Fabian 1982); Van Wyk & Malan (1997: 127, figure 290); Manning et al. (2001: 155, no. 3 on plate); Pooley (2003: 129, bottom centre); Smith & Crouch (2009: 88, top);

FIG. 12.16.3 The leaves of *Kalanchoe thyrsiflora* are generally obovate, rather than soup plate-rounded as in the related *K. luciae*. Photograph: Gideon F. Smith.

FIG. 12.16.5 *Kalanchoe thyrsiflora* growing in a dense, grassy patch in its natural habitat in the Gauteng province, north-central South Africa. The inflorescences of this species can reach a height of 1.5 m. Photograph: Gideon F. Smith.

FIG. 12.16.4 The leaves of *Kalanchoe thyrsiflora* are glabrous and usually not as strongly red-infused and white-waxy as those of *K. luciae*. Photograph: Gideon F. Smith.

Court (2010: 118); Gill & Engelbrecht (2012: 113, top left); Smith et al. (2017b: 315, bottom).

Kalanchoe thyrsiflora was the first species of *Kalanchoe* to be illustrated in *the Flowering Plants of South Africa*.

Common names:

Afrikaans: geelplakkie, plakkie (Germishuizen & Fabian, 1982: 118), meelplakkie, voëlbrandewyn (Pooley, 2003: 128), and plakkie (Batten & Bokelmann, 1966; 73).

English: white lady (Germishuizen & Fabian, 1982: 118).

Southern Sotho [Sesotho]: serelile (Pooley, 2003: 128).

siSwati: utshwala benyoni (Long, 2005).

isiXhosa: utywala bentaka (Manning et al., 2001).

isiZulu: utshwala benyoni (Pooley, 2003: 128).

FIG. 12.16.6 A young specimen of *Kalanchoe thyrsiflora* growing on the banks of the Umzimkulu River, near Harding, KwaZulu-Natal, in eastern South Africa. Photograph: Neil R. Crouch.

FIG. 12.16.7 Close-up of an inflorescence of *Kalanchoe thyrsiflora*. The inflorescences of this species are very densely flowered and cylindrical to club-shaped. Photograph: Abraham E. van Wyk.

FIG. 12.16.8 Flowers of *Kalanchoe thyrsiflora* are usually greyish green; all parts are covered with a substantial white waxy bloom. The corolla tube is more or less quadrangular-urceolate and slightly four-angled, while the bright yellow lobes are somewhat square, recurved, and taper to a point. Photograph: Abraham E. van Wyk.

Geographical distribution:

The type locality of *Kalanchoe thyrsiflora* is recorded as "Cape, Katrivier near Philipstown".

Kalanchoe thyrsiflora does not cross into western south-tropical Africa and has not been recorded for Angola (Fernandes, 1982a) nor for Namibia. Lebrun & Stork (2003: 220) further state that *K. thyrsiflora* is not known from north of the *FSA* region and that this species is restricted to Botswana, Lesotho, and South Africa. It also occurs in Swaziland (Fig. 12.16.2).

Distribution by country. Botswana, Lesotho, South Africa (Northern Cape [possibly], Eastern Cape, Free State, Gauteng, KwaZulu-Natal, Limpopo, and Mpumalanga), and Swaziland.

Habitat:

Kalanchoe thyrsiflora often favours cliffs, foothills, stony outcrops, ridges in grasslands (Van Wyk & Malan, 1997: 126), savannas (Retief & Herman, 1997: 393), and the

FIG. 12.16.9   The perianth lobes of this form of *Kalanchoe thyrsiflora* from KwaZulu-Natal, South Africa, are not as strongly recurved as in the case of plants from further north in southern Africa. Photograph: Gideon F. Smith.

montane belt of the Drakensberg (Killick, 1990: 102). It grows at an altitude of up to 2000 m (Pooley, 2003: 128).

Conservation status:

In South Africa, *Kalanchoe thyrsiflora* is listed as 'least concern'.

Additional notes and discussion:

Taxonomic history and nomenclature. Ecklon & Zeyher (1837: 305) regarded their specimen (*Ecklon & Zeyher 1953*), which became the type of the name *Kalanchoe thyrsiflora*, as representative of *Kalanchoe alternans* Pers. (Persoon, 1805: 446). Persoon (1805) based his *Kalanchoe alternans* on *Cotyledon alternans* Vahl (Vahl, 1791: 51), an altogether different species of *Kalanchoe* from Arabia. Although a few species of *Kalanchoe*, such as *Kalanchoe lanceolata*, do occur from South Africa to the Arabian Peninsula, *K. alternans* has a laxly flowered inflorescence and larger, dull-pinkish flowers. These geographical and morphological differences prompted Harvey (1862: 380) to describe the material collected by Ecklon & Zeyher as a new species, *K. thyrsiflora*.

Some 30 years after *Kalanchoe thyrsiflora* was described, it was introduced into cultivation at La Mortola, Italy. From there, seed was rapidly distributed to various botanical gardens, and by 1891, it was in cultivation at the Royal Gardens, Kew, in the United Kingdom.

From the 1890s onwards, the species started appearing on floristic checklists of the then Natal and Transvaal provinces of South Africa (Wood, 1907: 46; Burtt-Davy & Pott-Leendertz, 1912: 144) and was later also recorded for the neighbouring countries Lesotho (Jacot Guillarmod, 1971) and Swaziland (Compton, 1976).

Identity and close allies. Kalanchoe thyrsiflora has for long been confused with *K. luciae*, its closest relative. The confusion between these two species has reigned in terms of both field identification and in the horticultural trade (Trager, 2001). Vegetative morphologically, *K. thyrsiflora* and *K. luciae* indeed have a similar appearance with both carrying succulent, paddle- to soup plate-shaped and soup plate-sized leaves, initially in a basal cluster. In reproductive morphology, *K. thyrsiflora* differs from *K. luciae* by having bright yellow, recurved corolla lobes

FIG. 12.16.10   Typical bushveld (savanna) habitat of *Kalanchoe thyrsiflora* near Harding in KwaZulu-Natal. Photograph: Neil R. Crouch.

arising from a more cylindrical corolla tube, while the lanceolate corolla lobes of *K. luciae* range from white; to pale greenish white, sometimes cream, or pinkish; to yellowish and arise from a more urceolate corolla tube. The white, floury, waxy bloom that covers most parts of *K. thyrsiflora* is much more pronounced than in *K. luciae*, hence its common name 'white lady' or meelplakkie in Afrikaans (English: 'flour plakkie'). Furthermore, the flowers of *K. thyrsiflora* are sugary scented—with the scent perhaps the strongest in the genus, at least in terms of southern African kalanchoes—while those of *K. luciae* are much more weakly scented.

In habitat, plants of *Kalanchoe thyrsiflora* are usually a dull, pale glaucous green, infused with reddish, orangey, or pinkish hues, with both leaf surfaces uniformly coloured. In the case of *K. luciae*, the red is frequently very pronounced, especially towards the margins of the leaves.

Cultivation. The species grows very easily and will flourish in virtually any type of soil. It will tolerate considerable neglect and still maintain its large, soup plate-sized and soup plate-shaped, oppositely arranged leaves looking firm and healthy. The species is drought-hardy and requires minimum irrigation, looking none the worse for wear. When irrigation is minimised and plants are grown in full sun, the leaves take on a magnificent deep red colour. Plants do particularly well in rockeries, given their natural disposition for growing in rocky places and stony ground.

Plants grow easily from seed, and whole rosettes that arise from the base of a plant can be removed and planted directly in the spot where they are intended to grow.

A plant can mature to flowering age as early as in its second year of growth under favourable conditions, for example, during a good rainy season. In such instances, the species is a true biennial: the plants grow vegetatively in the first year, flower in the second, and then die. In harsher, natural conditions, plants may take anywhere from three to five years to reach flowering age, so responding as true multiannuals.

Poisonous properties. Bullock (1952: 122) recorded *Kalanchoe thyrsiflora* as poisonous, while Batten & Bokelmann (1966: 73) record it as harmful to sheep, and Hardy & Fabian (1992: 42) similarly note that the species is said to be poisonous to stock.

Uses. The roots are used for uneasiness in pregnant women and also in the treatment of colds, intestinal worms, and earaches (Long, 2005). Plants are planted as a charm to smooth away difficulties (Batten & Bokelmann, 1966: 73).

17. *KALANCHOE WINTERI* GIDEON F.SM., N.R.CROUCH & MICH.WALTERS

Wolkberg kalanchoe (Figs. 12.17.1 to 12.17.11)

FIG. 12.17.1 *Kalanchoe winteri* growing in dappled shade among rocks. The leaves of this species are more spreading than in the case of its two large-growing relatives, *K. thyrsiflora* and *K. luciae*, and are less densely packed basally. Plants of *Kalanchoe winteri* resprout from the base annually. Photograph: Gideon F. Smith.

Kalanchoe winteri **Gideon F.Sm., N.R.Crouch & Mich.Walters** in Crouch et al. in *Bradleya* 34: 219 (2016). **Type**: South Africa. Wolkberg, Limpopo province, Thabakgolo Escarpment, Sedibeng sa Lebese Mountain, west of Strasburg, 10 September 2000, *P.J.D. Winter 4430* (PRE holo-; BNRH, PRU, iso-).

Derivation of the scientific name:

The species was named for the collector of the type specimen, Mr Pieter Jacobus de la Rey Winter (1964–), a South African botanist working at the Compton Herbarium, South African National Biodiversity Institute, in Cape Town. Previously, he was the Curator of the L. C. Leach Herbarium of the University of Limpopo, in Polokwane, South Africa.

Description:

Perennial, many-leaved, 1–3 rosettes, sparsely to profusely branched from near the base and higher up, glabrous, waxy, robust succulent, and 0.5(−0.9)m tall in bloom. *Stems* erect to leaning and curved upwards, glabrous, waxy especially at internodes, and light green. *Leaves* opposite, erect to mostly spreading to variously floppy, succulent, sessile, flattened above and below, glabrous, waxy, and light green to bluish green; *axils* often carrying small leafy shoots and short branches that produce flowers in season; *petiole* absent; *blade* 14–16 × 8–14cm, obovate to somewhat oblong, not folded lengthwise, and occasionally light red-infused; *base* narrow and sometimes distinctly auriculate; *apex* rounded-obtuse

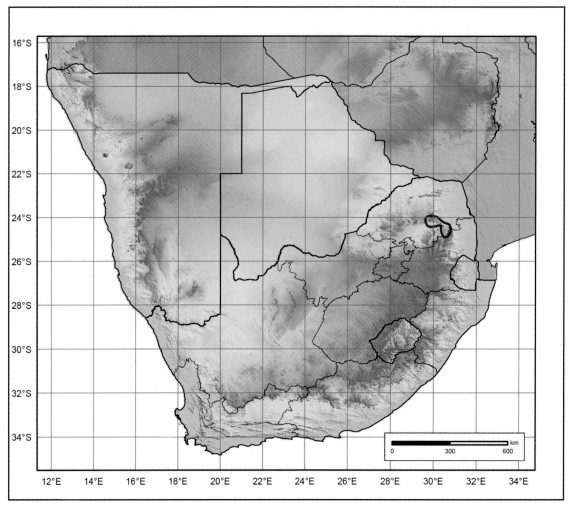

FIG. 12.17.2   Known geographical distribution range of *Kalanchoe winteri* in South Africa.

FIG. 12.17.3   The leaves of *Kalanchoe winteri* are often distinctly auriculate basally. Leaf margins are occasionally red-infused but never the entire lamina. Photograph: Gideon F. Smith.

FIG. 12.17.4   The stems of *Kalanchoe winteri* are brittle and often topple over. Thin, wiry roots that are not at all succulent sprout where the stems touch the ground. Photograph: Gideon F. Smith.

FIG. 12.17.5 The production of lateral shoots (plantlets) in the leaf axils on the peduncle is common in *Kalanchoe winteri*. Photograph: Gideon F. Smith.

FIG. 12.17.6 Close-up of a plantlet that developed in the axil of a leaflike bract on the peduncle of *Kalanchoe winteri*. These can be removed and will easily strike root. Photograph: Gideon F. Smith.

FIG. 12.17.7 Developing inflorescences of *Kalanchoe winteri* are covered in a powdery, somewhat sticky bloom. Photograph: Gideon F. Smith.

or truncate, usually indented; *margins* glabrous, slightly lighter green than blade, and sometimes infused with red. *Inflorescence* a slender, erect, densely flowered,

cylindrical thyrse consisting of several dichasia terminating in monochasia, 0.5(−0.9) m tall; *pedicels* 9–10 mm long. *Flowers* 13–15 mm long, erect to slanted horizontally, pale yellowish green to greenish white (tube) and yellow (lobes), all parts excepting tepal lobes above covered with a substantial white waxy bloom, highly scented, and resinous to the touch; *calyx* midgreen, contrasting against lighter-green corolla tube; *sepals* 3–4 mm long, elongated-triangular, and acute; *corolla* light greenish yellow; *tube* 11–12 mm long, more or less quadrangular, ellipsoid, and distinctly 4-angled; *lobes* 6–8 × 3.5–4.0 mm, triangular, margins slightly to distinctly enrolled, truncated, and bright yellow. *Stamens* 8, inserted just below or in the middle of the corolla tube, and 1–2 mm exserted; *filaments* 3.0–5.5 mm long, thin, and light greenish white; *anthers* 1.4–1.6 mm long and yellow. *Pistil* pyriform and consisting of 4 carpels; *carpels* 9–10 mm long and light green; *styles* ± 4 mm long; *stigmas* very slightly capitate, light yellow, and exserted as far as or slightly less than

FIG. 12.17.8   Inflorescences of *Kalanchoe winteri* are densely flowered. Photograph: Gideon F. Smith.

FIG. 12.17.9   Close-up of an inflorescence of *Kalanchoe winteri*. The
corolla tubes of *Kalanchoe winteri* are four-angled. Unlike those of several
other *Kalanchoe* species, the calyx lobes of *Kalanchoe winteri* are
relatively short. Photograph: Gideon F. Smith.

anthers; *scales* 2.3–2.5 × 1.8–2.1mm, narrowing at the base,
truncate, and repand. *Follicles* brittle, grass spikelet-like,
light to dark brown when dry, enveloped in dry, light
whitish brown remains of corolla, and 6–7mm long. *Seeds*
0.75–1.25mm long and dark brown. *Chromosome number*:
unknown.

Flowering time:
May–September, peaking in July (southern hemisphere).
Illustrations:
Crouch et al. (2016b, figures 1–7 and 12).
Common names:
Afrikaans: Wolkbergkalanchoe.
English: Wolkberg kalanchoe.
Geographical distribution:
Kalanchoe winteri is endemic to a small area in the Lim-
popo Province, in northern South Africa. The Wolkberg
Centre of Endemism from where *K. winteri* was discov-
ered is a recognised area of remarkably high species
diversity (Van Wyk & Smith, 2001) (Fig. 12.17.2).

FIG. 12.17.10   The flowers of *Kalanchoe winteri* have spreading to
reflexed, yellow corolla lobes that are characteristically considerably
longer than broad. The lobe margins are folded inwards and the apex
blunt to very slightly indented. All eight anthers are exserted. Photograph: Gideon F. Smith.

FIG. 12.17.11   Dry inflorescence of *Kalanchoe winteri.* Photograph: Gideon F. Smith.

Distribution by country. South Africa (Limpopo).

Habitat:

Kalanchoe winteri grows on quartzite substrates in grassland vegetation, always in microhabitats on or near rocks where plants are protected from fire. It may be encountered at altitudes of 1370–1750 m above sea level on north, northeastern, eastern, and southwestern aspects, usually in full sun, although at times in the partial shade of shrubs.

It occurs in Northern Escarpment Quartzite Sourveld (Mucina et al., 2006).

Conservation status:

At present, a conservation status of 'rare' should be accorded to *Kalanchoe winteri*. It has a comparatively small known geographical distribution range and is at present known from only three localities within a 50 km range. Furthermore, the species is a habitat specialist.

Additional notes and discussion:

Taxonomic history and nomenclature. The type specimen of *Kalanchoe winteri* was collected in 2000, at which time it was suspected that it was a new species. However, it took a further 16 years before the species was described.

Identity and close allies. Kalanchoe winteri can be confused with *K. thyrsiflora, K. luciae,* and *K. montana,* three species that also have paddle- to soup plate-shaped leaves. However, among the southern African *Kalanchoe* taxa that bear densely flowered, club-shaped to near-cylindrical thyrses, *K. winteri* is the only species with pyriform (pear-shaped) pistils. Furthermore, it may be separated from *K. luciae* by its consistently golden-yellow corolla lobes and ellipsoid corolla; in the case of *K. luciae,* the colour of the corolla lobes varies from whitish, to pale yellowish green, to pale pink. *Kalanchoe winteri* differs from *K. thyrsiflora* in having a less cylindrical and more 4-angled tube that is cigar-shaped enlarged in the middle, oblong rather than square corolla lobes, the lower filament rank inserted deeper in the corolla tube (± ¾ way up the tube) and broader scales. The leaves of *K. winteri* are much less red-infused than those of *K. thyrsiflora* and particularly less so than in *K. luciae.* In this group of species within *Kalanchoe,* the colour of the outer corolla varies with the degree to which a whitish, waxy bloom is present; in *K. montana,* this is frequently absent or obsolescent, whereas in *K. thyrsiflora, K. winteri,* and *K. luciae,* the corolla may appear greyish white when the bloom is intense. *Kalanchoe winteri* is separated from *K. montana* by lacking pubescence, which is to various degrees present in the latter species, which also tends to be smaller-growing. *Kalanchoe winteri* differs from all three of these species in its leaves often being distinctly auriculate basally, while its rosulate leaves are often more spreading, rather than erect. The smaller, oppositely arranged leaves on the peduncle of *K. winteri* are often cup-like clustered. None of the close relatives of *K. winteri* were observed in the immediate vicinity of where it grows. However, *K. luciae* and *K. thyrsiflora* are known to occur on dolomite about 5 km from at least one of the *K. winteri* locations. The flowering periods of these taxa overlap.

Cultivation. Kalanchoe winteri is very easy in cultivation. The non-flowering leafy shoots and short branches that develop in the axils on the peduncle can be placed in a well-drained soil mixture and will easily root. The fine, almost dustlike seed can be sowed in a seedling tray or directly in the spot in a garden where plants are intended to grow. When grown from the peduncular leafy shoots, plants grow very quickly, accumulating significant biomass within a short space of time. It can flower within its first year of growth. In its natural habitat, plants attain a height of about ½ m when flowering; in cultivation, they will grow taller. Carpenter bees, *Xylocopa* cf. *caffra,* and the African honey bee, *Apis mellifera,* visit the flowers of this species.

18. *KALANCHOE BEHARENSIS* DRAKE

Donkey's ear (Figs. 12.18.1 to 12.18.12)

FIG. 12.18.1 *Kalanchoe beharensis* grows as large robust shrubs or as small- to medium-sized, erect to leaning trees that can reach a height of 3 m. Photograph: Gideon F. Smith.

***Kalanchoe beharensis* Drake** in *Bull. Mus. Histoire Natural (Paris)* 9: 41 (1903). Hamet (1908b: 29); Raymond-Hamet & Marnier-Lapostolle (1964: 33; Plate IX, figures 20–22, Plate X, figure 23); Jacobsen (1970: 283); Jacobsen (1986: 607); Boiteau & Allorge-Boiteau (1995: 154); Rauh (1995: 116); Rauh (1998: 300); Descoings (2003: 147); Walters et al. (2011: 257); Smith et al. (2017b: 307). **Type:** Madagascar, Behara, 8 Juillet 1901, *Grandidier s.n.* (P P00374195 & P00374194).

Synonyms:
Kalanchoe vantieghemii Raym.-Hamet in *J. Bot. (Morot)* 20: 109 (1906) 'van Tieghemi'. **Type:** Madagascar, Behara, 8 Juillet 1901, *Grandidier s.n.* (P P00374195 & P00374194).

Derivation of the scientific name: For the occurrence of the species at Behara, Madagascar.

Description:
Perennial shrubs or small trees, sparsely branched, with haphazardly rounded canopies, robust succulent, and to 3.0 m tall. *Stem* simple lower down, branched from ± 1 m above the ground, thick, to 15 cm in diameter, ± straight or variously curved, surface prominently covered by sharp projections left by expanding, and prominent leaf scars; *bark* flimsy and light brownish grey. *Leaves* few to many towards terminal ½ of branches, shed lower down to yield leaf scars, erect to variously curved, succulent, densely pubescent, petiolate, ± flattened above and below, and dull yellowish green to bluish green; *petiole* same colour as leaf blade, succulent, and 4–10 cm long; *blade* 40–160 × 20–90 cm, somewhat elongated-triangular, deltoid to peltate, irregularly folded lengthwise and in width, and succulent; *base* flat and flared

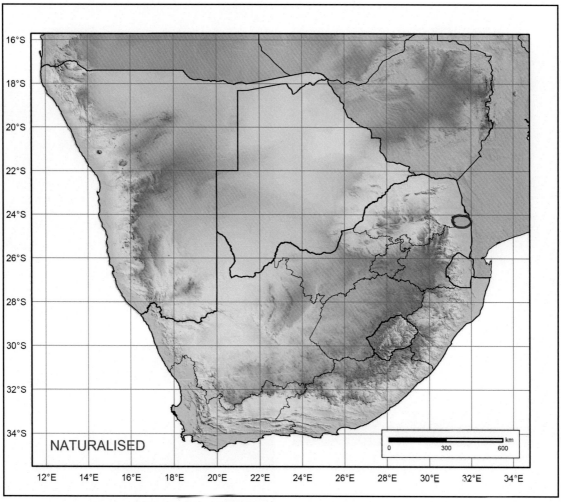

FIG. 12.18.2   Known geographical distribution range of *Kalanchoe beharensis* in South Africa.

downwards beyond point of attachment of petiole and rarely somewhat cuneate; *apex* rounded-obtuse or truncate; *margins* irregularly toothed. *Inflorescence* an axillary, densely branched, many-flowered panicle, up to 40 cm long, rounded, and erect to gracefully curved sideways and upwards; *pedicels* 5–12 mm long and densely pubescent. *Flowers* ± erect or slanted in various directions, densely pubescent, subtended by prominent bracts that are soon shed, pale yellow to yellowish green with longitudinal reddish purple stripes prominent on corolla lobes, less so on corolla tube, and waxy bloom absent; *calyx* light greenish white, very lightly infused with purple especially towards the tip, with or without feint longitudinal and reddish purple stripes; *sepals* 5–7 mm long, elongated-triangular, fused below for ± 2 mm, acute, and obscuring the corolla tube; *corolla* pale yellow to yellowish green with longitudinal reddish-purple stripes; *tube* 6–7 mm long, more or less quadrangular-urceolate, tapering to the mouth, and 4-angled; *lobes* 3 × 4 mm, deltoid, strongly recurved, apiculate, and pale yellow to yellowish green with

prominent longitudinal reddish purple stripes. *Stamens* inserted in the middle of the corolla tube or higher up and 1–2 mm exserted; *filaments* ± 5 mm long, thin, and light greenish white; *anthers* ± 0.5–1.0 mm long, yellow, and exserted. *Pistil* consisting of 4 carpels; *carpels* 3–4 mm long and light yellowish green; *styles* 7–9 mm long; *stigmas* very slightly capitate and light brownish yellow; *scales* rectangular, connate, ± 3 × 1 mm, truncate above, and slightly apiculate. *Follicles* 6–7 mm long, brittle, grass spikelet-like, enveloped in dry, greyish black remains of corolla, dark brownish black. *Seeds* 0.75–1.00 mm long, rectangular to slightly banana-shape curved, tapering at both ends, and dark brown to black. *Chromosome number*: $n = 18$ (http://ccdb.tau.ac.il/search/), Baldwin (1938), Friedmann (1971), and Rabakonandrianina & Carr (1987).

Flowering time:

July–September, peaking in July.

Illustrations:

Raymond-Hamet & Marnier-Lapostolle (1964: Plate IX, figures 20–22, Plate X, figure 23); Jacobsen (1970: Plate 103,

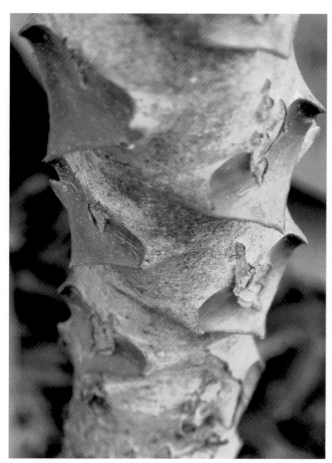

FIG. 12.18.3 Where leaves were shed lower down on a stem or branch of *Kalanchoe beharensis*, the leaf scars are conspicuously edged with sharp points. Photograph: Gideon F. Smith.

FIG. 12.18.4 The form of *Kalanchoe beharensis* most often encountered in South Africa has large, light bluish grey, velvety, boat-shaped leaves. Photograph: Gideon F. Smith.

FIG. 12.18.5 Terminal portion of an inflorescence of *Kalanchoe beharensis*. All the parts are finely tomentose. The recurved corolla lobes are distinctly purple-striped. Photograph: Gideon F. Smith.

FIG. 12.18.6 A cluster of flowers of *Kalanchoe beharensis*. In the bud stage, the corolla is virtually covered by the prominent sepals. Photograph: Gideon F. Smith.

figure 2); Hargreaves & Hargreaves (1972: 7); Jacobsen (1986: 608, figures 847–848), Boiteau & Allorge-Boiteau (1995: 155); Rauh (1998: 300–302, figures 1095–1109); Smith et al. (2017b: 307).

Common names:

Afrikaans: fluweelblaar and donkie-oor.

English: donkey's ear, elephant's ear kalanchoe, and velvet leaf (kalanchoe) (Hargreaves & Hargreaves, 1972: 7).

Geographical distribution:

Kalanchoe beharensis originates from southern, especially southwestern, Madagascar, where it is a component of xerophytic vegetation (Rauh, 1995: 116). It was

FIG. 12.18.7 Plantlets rapidly develop on the cut end of a leaf of *Kalanchoe beharensis* once the wound has sealed. Photograph: Gideon F. Smith.

FIG. 12.18.9 This miniature form of *Kalanchoe beharensis* has fairly narrow, variously curved and curled leaves, and the plants remain small. It has been given the cultivar name 'Nana'. Photograph: Gideon F. Smith.

FIG. 12.18.8 The form of *Kalanchoe beharensis* horticulturally known as either 'var. subnuda' or 'var. aureo-aeneus' has leaves that are sparsely covered in white hairs on the abaxial side and almost glabrous above. This form is here grown in the Jardin Majorelle in Marrakech, Morocco. Photograph taken on 12 April 2017 by Estrela Figueiredo.

introduced to several mild-climate parts of the world and is now widespread in cultivation.

Distribution by country. South Africa (Limpopo and Mpumalanga) (Fig. 12.18.2).

Habitat:

In South Africa, *Kalanchoe beharensis* was recorded from the Kruger National Park (Foxcroft et al., 2008, 2017) and the outskirts of Nelspruit [*Jaca & Cindi* 808 (PRE0988160)]. Plants have escaped from cultivation and became established in bushveld (savanna) vegetation.

Conservation status:

Least concern.

FIG. 12.18.10 Leaves of *Kalanchoe beharensis* 'Curly' are variously twisted and sharply recurved. The leaves are often arranged in four vertical rows. Photograph: Gideon F. Smith.

Additional notes and discussion:

Taxonomic history and nomenclature. Kalanchoe beharensis was described by Drake del Castillo (1903), based on material collected by Guillaume Grandidier (1873–1957) in Madagascar. Three years later, not being aware of this

FIG. 12.18.11   The leaves of *Kalanchoe beharensis* 'Fang' are tomentose, distinctly petiolate, and somewhat elongated-triangular. The leaf blade is deltoid to peltate, irregularly folded lengthwise, and flared downwards beyond the petiole. The margins are irregularly toothed. Both leaf surfaces are adorned with prominent protuberances. Photograph: Gideon F. Smith.

FIG. 12.18.12   The cultivar *Kalanchoe* 'Fern Leaf', also referred to as *K. beharensis* 'Fern Leaf', is a putative hybrid of which the parentage has been suggested to be *K. beharensis* and *K. millotii* Raym.-Hamet & H.Perrier.

name, Hamet (1906) published *Kalanchoe vantieghemii* based on the same type material, so creating a superfluous name. Hamet was soon aware of the illegitimacy of his name and synonymised it with the earlier *K.*

beharensis (Hamet, 1908b). In 1906, Hamet did not describe the leaves of his new species, and later, he noted (Hamet, 1908b) that even though Drake described the species as having ovate-lanceolate glabrous leaves, the remnants of leaves that were part of the collection appeared to have a different shape and there were doubts if they belonged to the same specimen; for this reason, he did not consider that material in his description. However, two years later, Hamet (1910b) revisited the material and concluded that the leaf remains belonged to three taxa *K. verticillata* Scott-Elliot [=*K. tubiflora* (Harv.) Raym.-Hamet], *K. grandidieri* Baill., and *K. beharensis* and was then able to describe the leaves of the latter. The type of *K. beharensis* consists of two specimens, one (sheet P00374194) with inflorescence and the other (sheet P00374195) with inflorescence and separate leaves kept in two envelopes. The former has an original label saying 'feuilles à la page suivante' (leaves on the next page); therefore, the two sheets represent a single gathering. It is assumed that the foreign leaf material noticed by Hamet was removed from the envelopes, but this cannot be verified online, and there are no labels on the specimens with that information.

In 1941, Raymond-Hamet again revisited the species discussing its presence in Tanzania and concluding that it was also indigenous there and not endemic to Madagascar. This was based on information from Ludwig Diels (1874–1945), then Director of the Botanical Museum Berlin-Dahlem. Diels determined a specimen kept in the Herbarium of the Botanical Museum (B. L. Institut Amani, Nr. 3212, Mombo, Steppe, 1 1/2 M. hohes Busch, Gesammelt am 16.8.1910 von Dr. Morstatt) originating from near the Usambaras in Tanzania as *K. beharensis*. Raymond-Hamet could not accept the possibility of it being an escape from cultivation at Amani, because its collection date predates even the introduction of the species into cultivation in France. The Imperial Biological-Agricultural Experimental Station at Amani had been established by the Germans in the then German East Africa at the beginning of the 20th century, and at the time that the plant was collected, the Experimental Station was internationally renowned and visited by many scientists. Exotic species were introduced to the garden of the Station, and many spread and became invasive (Conte, 2004). It is therefore likely that material of *K. beharensis* was also introduced there and escaped. It appears that the specimen that Hermann Morstatt, a zoologist at Amani, collected is no longer extant as most Crassulaceae specimens held at B were among those destroyed during the Second World War, in 1943, two years after Raymond-Hamet's paper was published.

The designations 'Kalanchoe beharensis var. aureoaeneus H.Jacobsen' (1970) and 'Kalanchoe beharensis var. subnuda H.Jacobsen' (1970) were not validly published and are therefore not included in the synonymy of *K. beharensis* above. These two horticulturally desirable leaf forms of the species are popular among collectors and in gardens (see below).

Identity and close allies. Kalanchoe beharensis is one of the largest members of the genus, and one of only a few that will attain tree-size dimensions. Arborescent forms are not known among the southern African species of *Kalanchoe*, and *K. beharensis* cannot be confused with any of the indigenous species. In Madagascar, other tree kalanchoes include *K. arborescens* Humbert (to 4 m tall) and *K. grandidieri* Baill. (to 3 m tall). Neither of the latter two tree-species has entered general horticulture, and both remain largely unknown in collections. Rauh (1998: 298, 305) notes that in cultivation, *K. arborescens* does best when grafted onto *K. beharensis*, while he records *K. grandidieri* as slow-growing.

Along with several other much smaller-growing species, Boiteau & Allorge-Boiteau (1995: 16) included these three tree kalanchoes in their "Groupe X Lanigerae".

Uses. None reported.

Cultivation. Kalanchoe beharensis is very popular in general cultivation as it is very drought hardy, and the furry leaves and 'thorny' stems and branches lend interest and texture to a garden. The species is exceedingly easy in cultivation, grows in virtually any soil type, and thrives on neglect. However, *K. beharensis* cannot tolerate very low temperatures, and during a regular frosty winter's night, the leaves and branch growing tips can be severely damaged. A heavy frost will kill the plant.

A number of forms of *Kalanchoe beharensis* have become very popular in horticulture. Leaves of *K. beharensis* 'var. aureo-aeneus' are densely covered in short, golden-brown hairs on both sides, while those of *K. beharensis* 'var. subnuda' are sparsely covered in white hairs on the abaxial side and almost glabrous adaxially. Rauh (1998: 300) states that the form with nearly glabrous leaves and white hairs may no longer be in cultivation. However, one of us (EF) recently photographed this form in the Jardin Majorelle in Marrakech, Morocco.

Numerous other forms abound in cultivation: one has a small, weakly arborescent habit, and brown-haired, downward curly leaves. It is variously sold in nurseries as cultivar 'Nana' or as cultivar 'Curly'. An interesting and popular form of which the lower sides of the leaves are tuberculate to warty is known as cultivar 'Fang'. It is even available as plastic house plants. Still, other forms have been selected based on interesting leaf shapes and outlines, such as *Kalanchoe beharensis* 'Rose Leaf'.

Propagation is through seed, stem truncheons, branch cuttings, or from severed leaves with the petiole intact.

As is the case with *Kalanchoe grandidieri*, one of the other arborescent Madagascan species, the bark of *K. beharensis*, produces a scented resin (Jadin & Juillet, 1912). For that reason, it burns easily, and plants should not be grown adjacent to barbecue areas.

19. *KALANCHOE DAIGREMONTIANA* RAYM.-HAMET & H.PERRIER

Mother-of-thousands (Figs. 12.19.1 to 12.19.7)

FIG. 12.19.1 A cultivated, drought-stressed plant of *Kalanchoe daigremontiana* with its leaves longitudinally folded upwards. It grows next to the pencil-branched *Euphorbia tirucalli* L. Photograph: Gideon F. Smith.

Kalanchoe daigremontiana **Raym.-Hamet & H.Perrier** in *Ann. Mus. Colon. Marseille*, sér. 3, 2: 128 (1914). Raymond-Hamet & Marnier-Lapostolle (1964: 44, Plate XIII, figures 35–37, Plate XIV, figure 38); Jacobsen (1970: 284); Jacobsen (1986: 612); Boiteau & Allorge-Boiteau (1995: 94); Rauh (1995: 116); Rauh (1998: 130, 320); Descoings (2003: 153). **Type**: Madagascar, Mt Androhiboalava, Marosavoha, Isalo, près Benenitra, Juillet 1910, *Perrier [J.M.H.A. Perrier de la Bâthie] 11798* (P P00374130, holo-).

Synonym:

Bryophyllum daigremontianum (Raym.-Hamet & H.Perrier) A.Berger in Engler & Prantl, *Nat. Pflanzenfam.*, ed. 2 18a: 412 (1930). Walters (2011: 249). **Type**: as for *Kalanchoe daigremontiana*.

Derivation of scientific name:

Named for Monsieur and Madame Daigremont (active *c.* 1914), French Crassulaceae enthusiasts.

Description:

Biennial shrubs, unbranched, glabrous, robust succulent, and to 0.8 m tall. *Stem* simple, erect or leaning and then curved upwards, and brownish. *Leaves* sometimes peltate with blade 'wings' stretching beyond point of petiole attachment; smooth; petiolate; ± flattened above and below; and green and variously spotted with purplish, pinkish, or brownish blotches; *petiole* amplexicaul (stem-clasping), succulent, and 10–40 mm long; *blade* 5–20 × 2–5 cm, ovate, obovate, elongated-triangular, often irregularly folded lengthwise, and succulent; *base* broadened and rounded; *apex* acute; *margins* irregularly

FIG. 12.19.2   The abaxial leaf surfaces of *Kalanchoe daigremontiana* are mottled with irregular, purple blotches that sharply contrast against a light green to creamy green background. Photograph: Estrela Figueiredo.

FIG. 12.19.4   The inflorescence of *Kalanchoe daigremontiana* is apically branched and many-flowered. The peduncle remains straight and erect and is a light greenish purple colour. The pendent flowers are dull light pinkish purple. Photograph: Gideon F. Smith.

FIG. 12.19.3   When grown in deep shade, the leaves of *Kalanchoe daigremontiana* are much greener, especially adaxially. Photograph: Gideon F. Smith.

FIG. 12.19.5   *Kalanchoe daigremontiana* grown as a border plant next to *Portulacaria afra* Jacq. Note the plantlets formed on the leaf margins. Photograph: Gideon F. Smith.

crenate with plantlets developing on the marginal crenations. *Inflorescence* a terminal, apically branched, many-flowered head-shaped corymb; up to 20 cm tall; and erect; peduncle straight and light greenish purple; *pedicels* 5–10 mm long and glabrous. *Flowers* pendent, glabrous, subtended by small bracts that soon shrivel, dull light pinkish purple, longitudinally infused with yellow in centre of petal, waxy bloom absent, and papery when dry; *calyx* tubular for ± ⅔, shiny light green infused with purple, purple more prominent towards calyx base and sepal tips, and purple arranged in feint longitudinal lines; *sepals* 7–9 mm long, free portion short-triangular to deltoid, fused for ± 4–6 mm, acute, and obscuring ¼ of the corolla tube; *corolla* dull light pinkish purple and longitudinally infused with yellow in centre of petal; *tube* 15–19 mm long, more or less cylindrical to somewhat campanulate, and flared at the mouth; *lobes* ± 7 × 4–5 mm, obovate, apically rounded, and apiculate. *Stamens* inserted below middle of corolla tube ± upper level of carpels and included or hardly exserted; *filaments* ± 12 mm long, thin, and light purplish red; *anthers* ± 1.0–1.5 mm long, black, and hardly exserted. *Pistil* consisting of 4 carpels; *carpels* ± 6 mm long and shiny midgreen; *styles* ± 5 mm long; *stigmas* very slightly capitate, green; *scales* ± square, free, ± 1.5 × 1.5 mm, and slightly indented above. *Follicles* not seen. *Seeds* not seen. *Chromosome number*: $n = 30$ (http://ccdb.tau.ac.il/search/Kalanchoe%20daigremontiana/).

FIG. 12.19.6 Potted specimens of *Kalanchoe daigremontiana* (centre and back) offered for sale in the Djema el-Fna Square, Marrakesh, Morocco. Photograph: Estrela Figueiredo.

Flowering time:

July–September (southern hemisphere).

Illustrations:

Raymond-Hamet & Marnier-Lapostolle (1964: Plate XIII, figures 35–37, and Plate XIV, figure 38); Jacobsen (1970: Plate 103, Fig 3), Jacobsen (1986: 611, figures 852–853), Boiteau & Allorge-Boiteau (1995: 93); Rauh (1995: 117, figure 287); Rauh (1998: 317–319, figures 1183–1189), and Walters (2011: 250–251, figures 286–287).

Common names:

English: alligator plant, devil's backbone, maternity plant, mother of thousands, and Mexican hat plant.

Geographical distribution:

Southeastern Madagascar (see Boiteau & Allorge-Boiteau, 1995: 199, Map 8). It was introduced elsewhere as an ornamental plant, in many instances eventually becoming weedy and naturalised. It is considered an invasive species in several countries. In Africa, it is considered invasive in South Africa (Foxcroft et al., 2008).

Distribution by country. Madagascar (native). Introduced in several countries, such as China, Mexico, USA, Bahamas, Cuba, Haiti, Dominican Republic, Puerto Rico,

Venezuela, Italy, the Madeira archipelago (Portugal), the Canary Islands and Balearic archipelago (Spain), Australia, New Caledonia, and New Zealand.

Habitat:

It occurs in dry zones of southwestern Madagascar. Jacobsen (1986: 612) records the species as found "at the Mont Androhibolave (Onilahy), in the gneiss of the Marosavoha, [and] in the sandstone of the Isalo and Makay, Fiherena [Fiherenana]".

In southern Africa, the species has been noted as commonly cultivated around homesteads in the greater Durban area in KwaZulu-Natal, in mild-climate eastern South Africa, and in the climatically more severe Gauteng province. As *Kalanchoe daigremontiana* has yet to be recorded in natural vegetation in the subcontinent, a distribution map is not provided for the species.

Conservation status:

Not applicable.

Additional notes and discussion:

Taxonomic history and nomenclature. This is one of the few *Kalanchoe* species that is not burdened with a vast synonymy. The only recorded synonym is *Bryophyllum*

FIG. 12.19.7   *Kalanchoe ×houghtonii*, a hybrid between *K. daigremontiana* and *K. tubiflora*, cultivated in a pot in São Brás Alportel, in the district of Faro, in the Algarve, Portugal's southernmost province. Photograph: Gideon F. Smith.

daigremontianum, which was published when Berger (1930: 412) transferred the species to that genus, a classificatory view that we do not share in this work.

Identity and close allies. Although a very variable species, *Kalanchoe daigremontiana* can be generally distinguished on the basis of its elongated-triangular leaves that are often folded longitudinally upwards; the irregularly crenate leaf margins that develop a large number of plantlets; the variously purplish-, pinkish-, or brownish-spotted leaf blades; and its dull light pinkish purple flowers.

Along with *Kalanchoe tubiflora*, *K. daigremontiana* was included in Groupe VI Bulbilliferae by Boiteau & Allorge-Boiteau (1995: 16).

Kalanchoe daigremontiana hybridises easily with other species of *Kalanchoe* (Resende, 1956), including *K. tubiflora* and *K. rosei* Raym.-Hamet & H.Perrier, and especially the hybrid with *K. tubiflora* has become a weed in its own right (Silva et al., 2015: 74, figures 39a–39c, 77). A selection of this hybrid has been named *K. ×houghtonii*, which in itself has been further classified into a number of cultivars, such as *K. ×houghtonii* 'Garbí', based on

vegetative or reproductive characters (Guillot Ortiz et al., 2014: 104–105). While *K. ×houghtonii* has the general appearance of *K. daigremontiana*, it tends to be a slightly smaller plant in all respects, and the leaves while well-spotted with dark purplish brown blotches, in particular, are much narrower. *Kalanchoe ×houghtonii* has not yet been recorded for southern Africa.

Reproductive strategies. *Kalanchoe daigremontiana* propagates in a number of different ways. It produces copious amounts of seed and numerous small plantlets on the margins of its leaves. While some plants will die completely after flowering, others will sprout from the base.

Cultivation. *Kalanchoe daigremontiana* is one of the easiest species of *Kalanchoe* to grow. Its multiple reproductive strategies make it very proliferous in cultivation, and plants very quickly become weedy in gardens and invasive in the wild. This species should not be grown in mild-climate areas given its invasive tendencies.

In the past *Kalanchoe daigremontiana* was extensively used in the production of a range of interspecific hybrids, several of which have been made available commercially (Shaw, 2008).

20. KALANCHOE FEDTSCHENKOI RAYM.-HAMET & H.PERRIER

Purple scallops (Figs. 12.20.1 to 12.20.10)

FIG. 12.20.1 The regular and variegated forms of *Kalanchoe fedtschenkoi* are here grown together in a bed in the Royal Botanic Garden Sydney, New South Wales, Australia. The variegated form of this species has white to yellowish white, variegated leaves. Photograph: Gideon F. Smith.

Kalanchoe fedtschenkoi **Raym.-Hamet & H.Perrier** in *Ann. Inst. Bot.-Géol. Colon. Marseille*, série 3, 3: 75–80 (1915). Raymond-Hamet & Marnier-Lapostolle (1964: 45; Plate XIV, figures 39–40); Jacobsen (1970: 284); Jacobsen (1986: 613); Boiteau & Allorge-Boiteau (1995: 104); Rauh (1995: 116, 149); Rauh (1998: 320, 374); Descoings (2003: 156); Smith & Figueiredo (2017c: 83). **Type**: Madagascar, Fianarantsoa, Ihosy, Mt Tsilongaba-lala, bassin de Mangoky, September 1911, *H. Perrier de la Bâthie 11797* (P P00431060, image not available online).

Synonyms:

Bryophyllum fedtschenkoi (Raym.-Hamet & H.Perrier) Lauz.-March. in *Comptes R. Hebd. Séances Acad. Sci., D (Paris)* 278(20): 2508 (1974). Walters (2011: 251). **Type**: as above.

Kalanchoe fedtschenkoi var. *isalensis* Boiteau & Mannoni in *Cactus (Paris)* 21: 73 (1949). Jacobsen (1970: 284); Jacobsen (1986: 613); Boiteau & Allorge-Boiteau (1995: 105). **Type**: Madagascar, Fianarantsoa, Plateau de l'Isalo, 30 July 1928, *Humbert & Swingle 4954 bis* (P P00431063

image not available online, lecto-). Designated by Boiteau & Allorge-Boiteau (1995: 105).

'Kalanchoe fedtschenkoi Raym.-Hamet & H.Perrier var. typica'.

Derivation of the scientific name:

Kalanchoe fedtschenkoi was named after Boris Alexje-witsch (Alexeevich) Fedtschenko (1872–1947), curator of the Imperial Botanical Garden in St Petersburg, Russia. Today, the Garden is also known as the Botanic Gardens of the Komarov Botanical Institute or the Komarov Botanical Garden.

Description:

Perennial, spreading, robust, tuft-like, glabrous, brittle shrubs, sparsely to densely branched, with haphazard untidy canopies, succulent, to 75 cm tall, and annually increasing in size. *Stems* thin, green to greenish purple when young, becoming brown, variously erect or pros-trate (creeping), rooting along the way, leaning branches developing long near-woody stilt-like roots; *bark* peeling to flaking with age, flimsy, and brown. *Leaves* many,

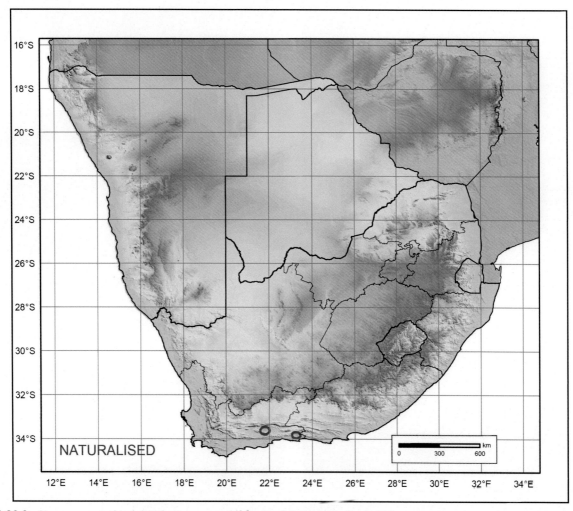

FIG. 12.20.2   Known geographical distribution range of *Kalanchoe fedtschenkoi* in South Africa.

FIG. 12.20.3 *Kalanchoe fedtschenkoi* is a succulent shrub with haphazard, multi-branched canopies. This specimen is cultivated in the Botanical Garden of the University of Porto, in northern Portugal. Photograph: Gideon F. Smith.

FIG. 12.20.4   With age, the stems of *Kalanchoe fedtschenkoi* turn brown with a flaky bark and develop stilt-like roots. Photograph: Gideon F. Smith.

FIG. 12.20.5 The obovate to nearly round leaves of *Kalanchoe fedtschenkoi* have glabrous surfaces, and the leaf margins are irregularly indented with coarse crenations, especially in the upper half. Photograph: Gideon F. Smith.

FIG. 12.20.7 Inflorescences of *Kalanchoe fedtschenkoi* are apically much-branched, head-shaped, and many-flowered. Some inflorescence branches terminate in sterile, curved, pedicel-like branchlets. The calyx obscures up to three-quarters of the corolla tube. Photograph: Gideon F. Smith.

FIG. 12.20.6 In the bud stage, the flowers of *Kalanchoe fedtschenkoi* are completely enveloped in the prominent calyx tube. Photograph: Gideon F. Smith.

FIG. 12.20.8 The drooping, orange flowers of *Kalanchoe fedtschenkoi* are longitudinally yellow-infused. Photograph: Gideon F. Smith.

densely packed in young plants, more widely dispersed on older stems, erect to variously slanted away from branches, succulent, glabrous, petiolate, flattened above and below, often slightly thickened below in line with petiole, bluish to purplish to brownish green, distinctly waxy, and wax easily rubs off; *petiole* thin, same colour as leaf blade, succulent, and 1–10 mm long; *blade* 20–70 × 15–40 mm, obovate to nearly circular, and succulent; *base* narrow and cuneate; *apex* rounded-obtuse; *margins* irregularly toothed with coarse, acute or subacute crenations in the upper half, depressions between teeth sometimes dark brown purple, somewhat angled, and forming bulbils (adventitious buds) in the crenations especially once

leaves are detached; lower half of leaves sometimes with a few small, irregular, and hardly noticeable crenations only. *Inflorescence* a terminal, apically branched, many-flowered head-shaped corymb; up to 200 mm tall; and

FIG. 12.20.9   The regular form of *Kalanchoe fedtschenkoi* cultivated in a plant container at a craft market in northwestern Swaziland. Photograph: Gideon F. Smith.

FIG. 12.20.10   The variegated-leaved form of *Kalanchoe fedtschenkoi* grown in a clay pot in Cullinan, Gauteng, in north-central South Africa. Photograph: Gideon F. Smith.

erect; some corymb branches terminating in sterile, curved, pedicel-like branchlets; peduncle straight and light brownish purple; *pedicels* 6–10mm long and glabrous. *Flowers* pendent, glabrous, subtended by small bracts that soon shrivel, orange, longitudinally infused with yellow in centre of petal, waxy bloom absent, papery when dry and dries same purplish colour as calyx; *calyx* tubular for ± ¾, light greenish purple, purple more prominent towards sepal tip, and purple arranged in feint longitudinal lines; *sepals* 15–20mm long, free portion elongated-triangular, fused for ± 10–15mm, acute, and obscuring from ⅔ to ¾ of the corolla tube; *corolla* orange to pinkish orange and longitudinally infused with yellow in centre of petal; *tube* 18–21mm long, more or less cylindrical to somewhat campanulate, and flared at the mouth; *lobes* ± 5×5mm, spathulate to obovate, and apically obtuse-flattened with a slight indentation. *Stamens* inserted very low-down in corolla tube at ± upper end of carpels and included or hardly exserted; *filaments* ± 20mm long, thin, and light purplish red; *anthers* ± 0.5–1.0mm long, black, and hardly exserted; *pollen* greyish yellow. *Pistil* consisting of 4 carpels; *carpels* 6–8mm long and light shiny green; *styles* 18–20mm long; *stigmas* very slightly capitate, green, and later slightly exserted; *scales* rectangular to square, free, ± 1×1mm, and slightly indented above. *Follicles* not seen. *Seeds* not seen. *Chromosome number*: $2n = 34$ (Boiteau & Allorge-Boiteau, 1995: 104).

Flowering time:

June–September, peaking in August (southern hemisphere).

Illustrations:

Raymond-Hamet & Marnier-Lapostolle (1964: Plate XIV, figures 39–40); Jacobsen (1970: Plate 103, figure 4); Jacobsen (1986: figure 855); Rauh (1995: 151, figures 387–389); Boiteau & Allorge-Boiteau (1995: 103); Rauh (1998: 374, picture on the left, excluding insert); Walters (2011: figures 288–290), and Smith & Figueiredo (2017c: 81–84).

Common names (local):

Afrikaans: perskalanchoe and persplakkie (a reference to the overall bluish purple colour of the plants).

English: purple scallops.

Geographical distribution:

It originates from central and south central Madagascar. Rauh (1995: 149) treats *Kalanchoe fedtschenkoi* as a "Montane (cloud) forest succulent". Reference to the species additionally occurring naturally in southeastern parts of the island (Walters, 2011: 252) could not be confirmed (see Boiteau & Allorge-Boiteau, 1995: 200, Map 10). It was introduced in several parts of the world, and it is now widespread.

Distribution by country. Madagascar (native). Introduced in Australia, Cameroon, Dominican Republic, Florida (USA), Haiti, Hawaii, Madeira [Island] (Portugal), Mexico, and Puerto Rico.

Habitat:

In South Africa, the species was recorded from the Western Cape Province in the Klein Karoo (Fig. 12.20.2). A population was found in natural vegetation north of Calitzdorp, with plants growing in rich, humic soils. Calitzdorp falls in the Little Karoo Centre of Endemism that is exceedingly rich in indigenous and endemic succulents (Van Wyk & Smith, 2001).

Conservation status:

Not applicable.

Additional notes and discussion:

Taxonomic history and nomenclature. The species was collected for the first time by H. Perrier de la Bâthie in September 1911 at Mont Tsitongabalaa, near Ihosv (Bassin du Mangoky) in Madagascar. This collection would be the type of the species that was described a few years later by Raymond-Hamet & Perrier de la Bâthie (1915). A variety was described in 1949, as *Kalanchoe fedtschenkoi* var. *isalensis* Boiteau & Mannoni. Originally, Boiteau & Mannoni (1949: 74) did not designate a type for this name, only citing the following collections: "*Centre*: HUMBERT 4.954 bis, plateau de l'Isalo, sur grès, vers 800–1.000 m.; BOITEAU 2.038, même plante cultivée à Tananarive, floraison juillet" [English: "*Center*: HUMBERT 4.954 bis, Isalo plateau, on sandstone, from 800–1.000 m.; BOITEAU 2.038, same plant cultivated in Tananarive, flowering July"]. The variety was lectotypified by Boiteau & Allorge-Boiteau (1995). It was considered a synonym by Descoings (2003).

The designation 'Kalanchoe fedtschenkoi Raym.-Hamet & H.Perrier var. typica' (Boiteau & Mannoni, 1949) was not validly published, as infraspecific names such as 'typica' that aim to indicate the taxon containing the type of the name of the next higher taxon (in this case a species) are not validly published unless they are autonyms, which 'K. fedtschenkoi var. typica' is not (McNeill et al., 2012: 57, Article 24.3). 'Kalanchoe fedtschenkoi var. typica' is therefore not a name in the sense of the *International Code of Nomenclature for algae, fungi, and plants* (McNeill et al., 2012: 7, Article 6.3).

Identity and close allies. Kalanchoe fedtschenkoi is a perennial, low-growing, branched, succulent shrub, with haphazard, untidy canopies. The stems are thin, green to greenish purple when young, becoming brown with a flaky bark, and variously leaning when older, rooting along the way. The branches develop long, very tough stilt-like roots. The leaves are glabrous, flattened but succulent, and distinctly waxy; the wax easily rubs off. Leaves are obovate to nearly round in outline with a narrowed base. Leaf margins are irregularly toothed with coarse, acute, or subacute crenations in the upper half but sometimes also lower down. The drooping, orange flowers are yellow-infused longitudinally and carried in dense, apically branched inflorescences. Some inflorescence branches terminate in sterile, curved, pedicel-like branchlets. The calyx consists of four fused sepals that obscure up to three-quarters of the corolla tube.

The recognition of a single variety, *Kalanchoe fedtschenkoi* var. *isalensis* Boiteau & Mannoni (1949: 73–74) in *K.*

fedtschenkoi, has been proposed. This variety was differentiated from the 'typical' form of the species on the following differences: leaves have only two crenations at the apex, leaf blades are narrower, inflorescences are longer, flowers are longer and pink infused with yellow ("deeper red" sensu Jacobsen, 1970: 284; "more red" sensu Rauh, 1995: 149), branches are violet-coloured (Rauh, 1995: 149), and the waxy coat on the leaves is less conspicuous, thus giving plants a less violet appearance (Boiteau & Allorge-Boiteau, 1995: 105). These authors stated that *K. fedtschenkoi* var. *isalensis* does not differ sufficiently to warrant its recognition at species rank; they do regard the differences as listed above as deserving of recognition, at varietal rank, to draw attention to variation in *K. fedtschenkoi*. However, the variety was not upheld by Descoings (2003: 156).

Boiteau & Allorge-Boiteau (1995: 16) included *Kalanchoe fedtschenkoi* in Group VII Suffrutescentes. The other five Madagascan species, *K. laxiflora* Baker, *K. marnieriana* H.Jacobsen, *K. rosei* Raym.-Hamet & H.Perrier, *K. serrata* Mannoni & Boiteau, and *K. waldheimii* Raym.-Hamet & H. Perrier, included in the group are virtually unknown in cultivation in South Africa. Note that Descoings (2003: 144) regarded the groups tabulated by Boiteau & Allorge-Boiteau (1995: 16) as "…too heterogeneous and artificial" and that they "…cannot be used for a better understanding of the genus".

Reproductive strategy. In South Africa, the most significant reproductive strategy of *Kalanchoe fedtschenkoi* is the formation of bulbils along the leaf margins, especially in the depressions between the crenations. Such plantlets typically form when the leaves are deliberately or accidentally detached from a plant or when leaves attached to the plants are damaged. Plants suffering from environmental stress have also been known to form leaf marginal bulbils at a higher rate. Healthy leaves that are attached to a plant hardly ever form bulbils.

The cultivar generally referred to as *Kalanchoe fedtschenkoi* 'Variegata' is very widely cultivated in gardens. It has attractive yellowish white sections on the leaves, especially along the leaf margins. These whitish blotches contrast with the predominant bluish, purplish, or brownish green colour of the leaves and dark brown purple notches between the crenations on the leaf margins.

Cultivation. Kalanchoe fedtschenkoi is exceedingly easy in cultivation. Given its ability to become naturalised in places very remote from its natural geographical distribution range in Madagascar, the species should not be grown in southern Africa.

21. KALANCHOE GASTONIS-BONNIERI RAYM.-HAMET & H.PERRIER

Life plant (Figs. 12.21.1 to 12.21.5)

FIG. 12.21.1 *Kalanchoe gastonis-bonnieri* cultivated in a garden near Durban in South Africa's KwaZulu-Natal province. Note the plantlets forming at the leaf tips. Photograph: Neil R. Crouch.

Kalanchoe gastonis-bonnieri **Raym.-Hamet & H.Perrier** in *Ann. Sci. Nat. Bot.*, sér. 9, 16: 364 (1912). Raymond-Hamet & Marnier-Lapostolle (1964: 47; Plate XV, figures 41–42); Jacobsen (1970: 285); Jacobsen (1986: 614); Sajeva & Costanzo (1994: 152); Boiteau & Allorge-Boiteau (1995: 116); Rauh (1995: 92, 116, 331–332); Sajeva & Costanzo (2000: 173); Descoings (2003: 156). **Type**: Madagascar, Tampoketsa, Bemarivo, 8.1905, *H. Perrier 11831* (P P00374106, lecto-). Designated by Boiteau & Allorge-Boiteau (1995: 116).

Synonyms:

Kalanchoe adolphi-engleri Raym.-Hamet in *Bull. Soc. Bot. France* 102: 239 (1955). **Type**: not designated.

Bryophyllum gastonis-bonnieri (Raym.-Hamet & H.Perrier) Lauz.-March in *Comptes R. Hebd. Séances Acad. Sci.,*

D (Paris) 278(20): 2508 (1974). Walters et al. (2011: 254). **Type**: as for *Kalanchoe gastonis-bonnieri*.

Kalanchoe gastonis-bonnieri var. *ankaizinensis* Boiteau ex L.Allorge in Boiteau & Allorge-Boiteau, *Kalanchoe de Madagascar*: 119 (1995). **Type**: "Boiteau s. n. (Dufournet coll.) (matériel en alcool, P, photos)" (Boiteau & Allorge-Boiteau, 1995: 119).

Derivation of the scientific name:

Named for Prof. Dr Gaston Eugène Marie Bonnier (1853 [1851 according to Boiteau & Allorge-Boiteau (1995: 116)]–1922), botanist attached to the Université Paris-Sorbonne, France (Eggli & Newton, 2004: 92). He was very committed to his teaching role, and his *Traité de botanique* became a classic textbook. Bonnier is also known for making botany accessible to the public, for

FIG. 12.21.2 *Kalanchoe gastonis-bonnieri* in full flower. Photograph: Neil R. Crouch.

FIG. 12.21.4 The pendulous flowers of *Kalanchoe gastonis-bonnieri* are glabrous and yellowish green infused with reddish purple veins. The prominent calyx is tubular, urceolate, and pale green with reddish purple veins. Photograph: Neil R. Crouch.

example, through publishing several Floras such as *Flore du nord de la France et de la Belgique* and *Flore des environs de Paris*. He was also a coauthor of the *La Flore complète de France, Suisse et Belgique*.

Description:

Perennial, sometimes biennial and monocarpic, few-leaved, few-branched, glabrous, small- to medium-sized, terrestrial, succulent, to 70 cm tall, and forming dense groups. *Stems* short, sometimes branched low-down, and erect. *Leaves* basally laxly rosulate to subrosulate, opposite-decussate, petiolate, white-pruinose especially above, green or bluish green with irregular brownish-green markings, succulent, and spreading to erectly spreading; *petioles* short or prominent, broad, and often not very distinct; *blade* 12–55 × 5–10 cm, ovate-lanceolate, and folded lengthwise; *base* cuneate; *apex* harmlessly acute, forming plantlets, these developing roots while still attached to the leaf apex; *margins* sinuate to coarsely crenate. *Inflorescence* 40–55 × 30–60 cm, erect, apically dense, many-flowered, lax cyme, and corymbose; *pedicels* 5–15 mm long and slender. *Flowers* pendulous to slightly spreading, yellowish green infused with reddish purple

veins, and glabrous; *calyx* 20–25 mm long, tubular, urceolate, prominent, and yellowish to pale green with reddish purple veins; *sepals* fused for ¾ of their length into lobes 5–6 × 4–5 mm, ± deltoid, acute, yellowish to greenish white pruinose, and glabrous; *corolla* 4–5 cm long, cylindrical, and yellowish green infused with reddish purple veins; *tube* 3.0–4.0 cm long, cylindrical, and yellowish green infused with reddish purple veins; *lobes* ± 10 × ± 6 mm, spreading to slightly recurved, triangular-ovate, and apically acuminate. *Stamens* inserted below the middle of the corolla tube and included; *filaments* short; *anthers* ± 3 mm long and reniform. *Pistil* consisting of 4 carpels; *carpels* 8–11 mm long; *styles* prominent; *stigmas* capitate; *scales* half-round to square. *Follicles* not seen. *Seeds* not seen. *Chromosome number*: $2n = 34$ (Baldwin, 1938: 576).

Flowering time:

In winter in the southern hemisphere.

Illustrations:

Raymond-Hamet & Marnier-Lapostolle (1964: 47; Plate XV, figures 41–42); Jacobsen (1986: 614, figure 857); Rauh (1992: Plate 2051); Sajeva & Costanzo (1994: 152); Boiteau

FIG. 12.21.3   The apically dense, many-flowered inflorescence of *Kalanchoe gastonis-bonnieri* is carried erectly. Photograph: Neil R. Crouch.

& Allorge-Boiteau (1995: 117); Rauh (1995: 331, figures 999–1000, 332, figures 1001–1005); Sajeva & Costanzo (2000: 173); Walters et al. (2011: figures 291–293).

Common names:

Afrikaans: spookplakkie and witplakkie.

English: life plant and palm beach bells.

Geographical distribution:

The species occurs in mountainous parts in central-northwestern Madagascar (see Map 12 in Boiteau & Allorge-Boiteau, 1995; Rauh, 1995: 331).

Distribution by country. Madagascar.

Habitat:

The natural habitat of *Kalanchoe gastonis-bonnieri* is "…lava fields near Ankerika, south of Antsonihy, in Tampoketsa in [the] Bemarivo valley, and in Ampasima-tera" in central-northwestern Madagascar (Rauh, 1995: 331).

In southern Africa, the species has been noted as commonly cultivated around homesteads in parts of east-central KwaZulu-Natal, in mild-climate eastern South Africa. As *Kalanchoe gastonis-bonnieri* has yet to be recorded from natural vegetation in the subcontinent, a distribution map is not provided for the species.

Conservation status:

Not applicable.

Additional notes and discussion:

Taxonomic history and nomenclature. In 1912, Hamet (not yet as Raymond-Hamet) and Perrier de la Bâthie collaborated on a paper on Madagascan kalanchoes, in which they described six species as new:

1. *K. rolandi-bonapartei* Raym.-Hamet & H.Perrier
2. *K. gastonis-bonnieri* Raym.-Hamet & H.Perrier
3. *K. bouvieri* Raym.-Hamet & H.Perrier
4. *K. guignardi* Raym.-Hamet & H.Perrier (the epithet to be corrected to *guignardii*)
5. *K. mangini* Raym.-Hamet & H.Perrier (the epithet to be corrected to *manginii*)

FIG. 12.21.5 Plantlets that form at the tips of the leaves of *Kalanchoe gastonis-bonnieri* are often a ghostly white colour, so giving rise to the Afrikaans common name, 'spookplakkie'. Photograph: Neil R. Crouch.

6. *K. milloti* Raym.-Hamet & H.Perrier (the epithet to be corrected to *millotii*)

The taxonomic status of *Kalanchoe bouvieri* and *K. guignardii* is contentious. The other species are widely accepted.

Lauzac-Marchal (1974: 2508) transferred *Kalanchoe gastonis-bonnieri* to the genus *Bryophyllum*, but Descoings (2003: 156) again treated it as a species of *Kalanchoe*.

In the protologue of *Kalanchoe gastonis-bonnieri*, two specimens collected by Perrier (Tampoketsa, Bemarivo, August 1905, and Ampasimentera, August 1906) are cited, without collection numbers. These correspond to the specimens *Perrier 11831* (P) and *11796* (P), respectively. These specimens are syntypes. Boiteau & Allorge-Boiteau (1995) designated as lectotype *Perrier 11831* (P) when they referred to that specimen as "TYPE". According to the *ICN* (McNeill et al., 2012), the phrase "designated here" was only a requirement after 1 January 2001 (McNeill et al., 2012, Art. 7.10.), and the use of the word "TYPE" instead of lectotype is treated as an error to be corrected (McNeill et al., 2012, Art. 9.9).

Boiteau & Allorge-Boiteau (1995: 119) stated that the variety *Kalanchoe gastonis-bonnieri* var. *ankaizinensis* was filed in the herbarium under the name *K. ankaizinensis* Boiteau, based on material cultivated in Tsimbazaza. ["Cette variété est en fait séparée dans l'herbier sous le nom de *K. ankaizinensis* Boiteau et est cultivée à Tsimbazaza (photos, Allorge 1994)".] This does not constitute valid publication of the designation 'Kalanchoe ankaizinensis'. Further, as far as we could ascertain, "Allorge 1994" is not a literature reference, but rather an indication of the date that photographs were taken. The specimen and photos cited as type for *Kalanchoe gastonis-bonnieri* var. *ankaizinensis* Boiteau ex L.Allorge are not listed in the online catalogue of the Paris Herbarium.

There is no type designated for the name *Kalanchoe adolphi-engleri* Raym.-Hamet. No collections are cited in the protologue of the name, which only mentions that the species is from southeastern Madagascar.

Identity and close allies. *Kalanchoe gastonis-bonnieri* can be easily identified based on its often nearly white leaf colour, elongated-deltoid leaves that can be heavily pruinose, and the basally rosulate arrangement of the leaves. This is clearly a species that belongs in *K.* subg. *Bryophyllum* (bulbiliferous leaves and pendulous flowers, among other characters), even though it has in the past been referred to the typical subgenus of *Kalanchoe*. For example, Rauh (1992: first text page accompanying Plate 2051) unambiguously states that *K. gastonis-bonnieri* belongs to "section *Bryophyllum*", while Rauh (1995: 116), probably in error, places the species in *K.* subg. *Kalanchoe*. In the infrageneric classification of Smith & Figueiredo (2018a), *K. gastonis-bonnieri* belongs in *K.* subg. *Bryophyllum*.

Boiteau & Allorge-Boiteau (1995: 16) placed *Kalanchoe gastonis-bonnieri* in their Group IX Proliferae, the group in which *K. pinnata* was also included.

Invasiveness. Especially mature plants of *Kalanchoe gastonis-bonnieri* form adventitious plantlets at the leaf apex. These easily become severed from the leaves and will strike root and grow where they are dropped. In this way, plants rapidly form small, spreading colonies.

Cultivation. The species is very easy in cultivation—too easy in fact, as it will soon overstay its welcome and become weedy. Such material that is removed from gardens is often disposed of in an irresponsible way, so leading to the establishment of populations of exotic species in natural vegetation (Smith & Figueiredo, 2017c). *Kalanchoe gastonis-bonnieri* should therefore ideally not be cultivated in southern Africa (Walters et al., 2011: 254).

22. KALANCHOE PINNATA (LAM.) PERS.

Cathedral bells (Figs. 12.22.1 to 12.22.7)

FIG. 12.22.1 Cultivated plants of *Kalanchoe pinnata* growing in a shaded bed in a garden in Tshipise, Limpopo province, northern South Africa. These plants grow aggressively and will soon take over the entire bed. Photograph: Gideon F. Smith.

Kalanchoe pinnata **(Lam.) Pers.**, *Syn. pl.* 1: 446. (1805). Hamet (1908b: 21); Raymond-Hamet & Marnier-Lapostolle (1964: 55); Jacobsen (1970: 288); Jacobsen (1986: 623); Boiteau & Allorge-Boiteau (1995: 128); Rauh (1995: 116); Sajeva & Costanzo (2000: 174); Descoings (2003: 169); Smith et al. (2017b: 312); Smith & Figueiredo (2018b: 220). **Type**: Isle de France [Mauritius] *Sonnerat s.n.* (P P00297646, holo-); see Wickens (1987: 27).

Synonyms:

Cotyledon pinnata Lam., *Encycl.* 2(1): 141 (1786). **Type**: Isle de France [Mauritius] *Sonnerat s.n.* (P P00297646, holo-); see Wickens (1987: 27).

Bryophyllum calycinum Salisb., *Parad.* 1: T. 3 (1805). De Candolle (1828: 396); Schönland (1891: 34). **Type**: not designated.

Cotyledon calycina (Salisb.) Roth, *Nov. pl. sp.*: 217 (1821). **Type**: not designated.

Verea pinnata (Lam.) Spreng., *Syst. veg.*, ed. 16, 2: 260 (1825). **Type**: as for *Cotyledon pinnata* Lam.

Cotyledon rhizophylla Roxb., [*Hort. Bengal.*: 34 (1814), nomen], *Fl. ind.* 2: 456 (1832). **Syntypes**: India, *Roxbugh s.n.* (BM BM000649752), *Roxbugh s.n.* (BM without barcode), and *Roxbugh s.n.* (K K000838478).

Bryophyllum pinnatum (Lam.) Oken in *Allg. Naturgesch.* 3(3): 1966 (1841). Fernandes (1982a: 33); Fernandes (1983: 68); Wickens (1987: 28); Walters et al. (2011: 242). **Type**: as for *Cotyledon pinnata* Lam.

Bryophyllum germinans Blanco, *Fl. Filip.*, ed. 2 [F.M. Blanco]: 220 (1845). [Cited as "*Bryophyllum germinans* Blanco, Fl. Filip. 2: 47, t. 147 (1878)" by Fernandes (1983: 70)]. **Type**: "from the Philippines?" vide Fernandes (1983: 70).

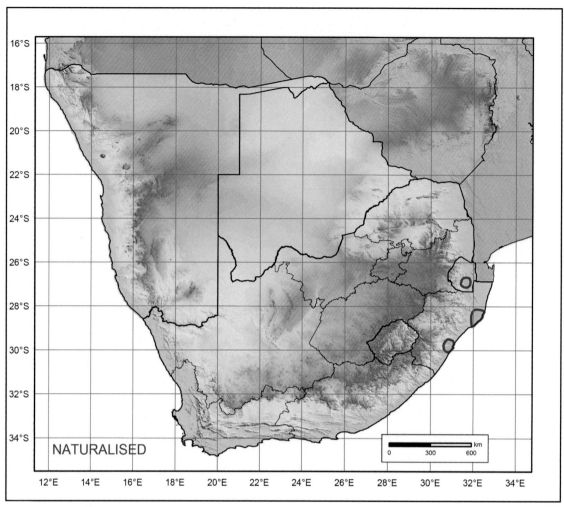

NATURALISED

FIG. 12.22.2   Known geographical distribution range of *Kalanchoe pinnata* in southern Africa.

FIG. 12.22.3   The leaves of *Kalanchoe pinnata* are compound, consisting of three to five round to oblong leaflets. Sometimes, a compound leaf is reduced to a terminal leaflet only. The terminal leaflet is usually the largest of the leaflets. Photograph: Gideon F. Smith.

FIG. 12.22.4   The basal portions of the corolla tubes of the large flowers of *Kalanchoe pinnata* are characteristically obscured by the calyces, more so when the flowers are developing, as here. Photograph: Gideon F. Smith.

FIG. 12.22.5 In *Kalanchoe pinnata*, the colour of the smooth, shiny, inflated, broadly cylindrical calyx varies from green with purple markings to purple with green markings. Where enveloped by the calyx, the corolla tube is greenish white, while the exposed part varies from red, to purplish, to greenish purple. Photograph: Gideon F. Smith.

FIG. 12.22.7 Plant of *Kalanchoe pinnata*. The margins of the leaflets of the pinnately compound leaves are crenate-dentate. Photograph: Neil R. Crouch.

FIG. 12.22.6 Flowers of *Kalanchoe pinnata* are pendulous; the corollas are greenish white basally where enveloped by the calyx but red, purplish, or greenish purple above where exposed beyond the calyx. Photograph: Neil R. Crouch.

Bryophyllum pinnatum (Lam.) Kurz in *J. Asiat. Soc. Bengal, Pt. 2, Nat. Hist.* 40(2): 52 (1871). nom. illeg. **Type**: as for *Cotyledon pinnata* Lam.

Bryophyllum pinnatum (Lam.) Oken [as "S. Kurz 1876"] var. [β] *simplicifolium* Kuntze, *Rev. gen. pl.* 1: 228 (1891). **Type**: not designated.

Kalanchoe pinnata var. *brevicalyx* Raym.-Hamet & H. Perrier in *Ann. Inst. Bot.-Géol. Colon. Marseille*, série 3, 3: 88 (1915). **Type**: Madagascar, Haut-Bemarivo, *H.Perrier 11286* (P P00374120, holo-).

Kalanchoe pinnata var. *calcicola* H. Perrier in *Arch. Bot. Bull. Mens.* 2(2): 21 (1928). Jacobsen (1970: 288). **Type**: Madagascar, dans mon jardin de Tananarive, Diego-Suarez, 3.1926, *Perrier 17637* (P P00444062, holo-).

Kalanchoe calcicola (H. Perrier) Boiteau ex L.Allorge, Boiteau & Allorge-Boiteau, Kalanchoe *de Madagascar*: 132 (1995). **Type**: as for *Kalanchoe pinnata* var. *calcicola*.

Bryophyllum calcicola (H.Perrier) V.V.Byalt in *Novosti Sist. Vyssh. Rast.* 32: 51 (2000). **Type**: as for *Kalanchoe pinnata* var. *calcicola*.

FIG. 12.22.8   The petioles of *Kalanchoe pinnata* are prominent and strongly purple-infused. Photograph: Gideon F. Smith.

Nomenclatural notes.

The homotypic synonymy and typification of *Kalanchoe pinnata* (Lam.) Pers. were comprehensively addressed by Smith & Figueiredo (2018b: 220–223). The heterotypic synonyms and some of the homotypic ones that we are aware of are discussed below.

'Sedum madagascariense Clus.' is a pre-Linnean designation with no nomenclatural standing.

Kalanchoe pinnata (Lam.) Pers. var. *floripendula* Pers., *Syn. pl.* 1: 446 (1805). Apart from making the combination *K. pinnata*, Persoon (1805: 446) additionally distinguished one variety under *Kalanchoe pinnata*, as "β. *floripendula*, fol. subconcavis crenis nudis. Lam. l. c. Crassuvia floripendia. Commers. Lam. l. c.". This therefore seems to be a proposed combination based on 'Crassuvia floripendula' [note spelling of the epithet], a designation that, as far as we could ascertain, was not validly published as a name in the sense of the *International Code of Nomenclature for algae, fungi, and plants* (McNeill et al., 2012).

Bryophyllum calycinum Salisb. (see "**Synonyms**" above) was described from a plant cultivated in the Calcutta [Kolkata] Garden in Kolkata, the capital of the West Bengal state of India. No specimens that can be unambiguously linked to the name were found at K or BM.

Boiteau & Allorge-Boiteau (1995: 128) listed "*Vereia pinnata* (Lamarck) Andrews" as having been published [by Andrews] in "Bot. Repos.: 21, 1797.", but as far as we could ascertain, there is no mention of the combination "*Vereia pinnata*" in that work of Andrews. Andrews (1798: 21) is a plate of *Vereia crenata* (=*Kalanchoe crenata*).

Boiteau & Allorge-Boiteau (1995: 128) listed "*Verea pinnata* (Lamarck) Willdenow" as having been published [by Willdenow] in "Spec.: 471, 1799, Pl II", but as far as we could ascertain, there is no mention of the combination "*Verea pinnata*" in that work of Willdenow. Willdenow (1799: 471) is the text for *Verea crenata* (=*Kalanchoe crenata*).

'Kalanchoe brevicalyx (Raym.-Hamet & H. Perrier) Boiteau' in P. Boiteau & L. Allorge-Boiteau, Kalanchoe de Madagascar: 133 (1995). This designation was not validly published as a reference to the place of publication of the basionym is not cited.

'Kalanchoe pinnata var. genuina'. This designation appears in Raymond-Hamet & H. Perrier (*Ann. Inst. Bot.-Géol. Colon. Marseille*, série 3, 3: 85. 1915), but was not validly published. Infraspecific names such as *genuina* that aim to indicate the taxon containing the type of the name of the next higher taxon (in this case a species) are not validly published unless they are autonyms.

'Cotyledon calyculata'. This designation is listed in IPNI as having been authored by "Sol. ex Sims—Bot. Mag. 34: sub t. 1409. 1811", but in that publication, this designation appears as a synonym of *Bryophyllum calycinum*, and it is therefore not validly published. In the African Plant Checklist and Database (APCD; Genève), it is listed as having been authored by "Sol. in DC, *Prodr.* 3: 396 (1828)", but in that publication, it is also cited as a synonym of *B. calycinum* and was therefore not validly published there either.

In the treatment of *Bryophyllum* in the *Flora of China* project, the combination *Bryophyllum pinnatum* is attributed to "(Linnaeus f.) Oken" (Fu & Gilbert, 2001: 204), with the place where this combination was made [by Oken] given as "Allg. Naturgesch. 3(3): 1966. 1841". The basionym is cited as "*Crassula pinnata* Linnaeus f., Suppl. Pl. 191. 1782". However, Oken (1841: 1966) did not reference *Crassula pinnata* nor Linnaeus fil. The citation "*Bryophyllum pinnatum* (Linnaeus f.) Oken" given by Fu & Gilbert (2001: 204) is therefore incorrect. If it is interpreted as a new combination, as *Bryophyllum pinnatum* (L.f.) K.T.Fu & M.G.Gilbert, it would be a nom. illeg. as it will be a later homonym of Oken's earlier, validly published combination. "*Kalanchoe pinnata* (L.f.) Persoon" is also cited by Fu & Gilbert (2001) but as it is only listed in the synonymy of *Bryophyllum pinnatum*, along with *Bryophyllum calycinum* Salisb., the combination is not validly published.

Derivation of the scientific name:
From the Latin 'pinnatus', 'winged', 'finned', or 'feathered', for the deeply dissected leaf blades.

Description:
Perennial, sometimes monocarpic, many-leaved, unbranched but sprouting from the base, glabrous throughout, medium-sized to large succulent, and to 2m tall. *Stems* robust, erect to leaning-erect, and somewhat woody below. *Leaves* to 20cm long, opposite-decussate, leaf pairs widely spaced, petiolate, green, sometimes with purplish lines, somewhat shiny, softly leathery-succulent to succulent, and spreading to slightly down-curved; *petiole* to 10cm long, subterete, and amplexicaul to half-clasping the stem; *blade* 5–20 × 4–10cm, first leaves simple, 3- to 5-foliolate higher up, sometimes reduced to terminal leaflet, and pinnate; *blade segments* sometimes slightly asymmetrical, circular to oblong to oblong-oval to ovate-elongate, often lengthwise slightly boat shape-folded upwards, and terminal leaflet the largest; *apex* obtuse-rounded; *margins* broadly crenate to harmlessly dentate, reddish purple, and with bulbils produced in the notches between the crenations. *Inflorescence* a very large, erect to leaning, branched, paniculate cyme, and up to 80cm tall; *pedicels* thin, 10–25mm long. *Flowers* pendulous; and greenish white below where enveloped by calyx; red, purplish, or greenish purple above where exposed beyond calyx; *calyx* inflated, broadly cylindrical to campanulate-tubular, thinly succulent, hardly depressed at the base, tube 20–30mm long, ± cylindrical, green with purple markings to purple with green markings, smooth, and shiny; *sepals* 4, fused, apically acuminate-cuspidate, free portions (lobes) softly acute-tipped, 5–10 × 5–11mm, contrasting against corolla tube, and clasping the corolla tube; *corolla* somewhat balloon-like inflated; longer than calyx tube; greenish white below where enveloped by calyx; and red, purplish, or greenish purple above; *corolla tube* 25–40mm long; tubular; greenish white below; and merging into red, purplish, or greenish purple; *lobes* 9–14 × 4–7mm, deltoid-cuspidate, tapering to an acute apex, and somewhat spreading. *Stamens* inserted below middle of corolla tube and very slightly exserted; *filaments* long; *anthers* ± 3.0 × 1.5–2.0mm and ovate. *Pistil* consisting of 4 carpels; carpels basally connate, 12–14mm long, tapering into styles, and styles 20–30mm long; *scales* 1.6–2.6 × 1.5–2.0mm, ± rectangular. *Follicles* not seen. *Seeds* not seen. *Chromosome number*: $2n = 36$ (Friedmann, 1971: 105).

Flowering time:
Midwinter to spring (July–September) in the southern hemisphere.

Illustrations:
Raymond-Hamet & Marnier-Lapostolle (1964: Plate XIX, figures 57–58, and Plate XX, figure 59); Fernandes (1983: 68, Tab. 7); Jacobsen (1986: 623–624, figures 876–878); Wickens (1987: 29, figure 5); Sajeva & Costanzo (2000: 174); Boiteau & Allorge-Boiteau (1995: 129); Walters et al. (2011: 243–245, figures 275–279); Smith et al. (2017b: 312, bottom); Smith & Figueiredo (2018b: 221).

Common names:
Afrikaans: makplakkie.
English: air plant, Canterbury bells, cathedral bells, and curtain plant (Walters et al., 2011: 242).

Geographical distribution:
The species is indigenous to Madagascar but has become extensively naturalised in mild-climate parts of southern Europe; Africa; Asia; Australia; and North, Central, and South America.

Distribution by country. Madagascar. It has been introduced to most tropical parts of the world, often becoming naturalised.

Habitat:

In South Africa, the species was recorded from the subtropical south coast of the KwaZulu-Natal Province north and south of the city of Durban. It has also become established in south central Swaziland (Fig. 12.22.2).

In the mid-1800s, the indefatigable botanical collector in Angola, Friedrich Welwitsch, collected the species in Sierra Leone, when he was travelling to Angola, and sent it to Portugal in 1854 as part of a shipment of living material to be grown in botanical gardens and nurseries. He recorded it as "Bryophyllum spec. (Aff. Br. Calycino)" (Welwitsch, 1856: 251). *Bryophyllum calycinum* is a synonym of *Kalanchoe pinnata* (Lam.) Pers. *Kalanchoe pinnata* is recorded for Sierra Leone in the *Flora of Tropical Africa* (Britten, 1871: 390) and in the *Flora of West Tropical Africa* (Hutchinson & Dalziel, 1954: 116), with the earliest collection cited being a specimen collected by Charles Barter in 1857, that is, a few years after Welwitsch visited the region at Regent, 10 km (6 miles) east of Freetown, during the ill-fated 1857–59 Niger Expedition. Welwitsch took the plant to Luanda and grew it in his garden. Fernandes (1982a: 34) noted the species as occurring in Cuanza Norte, Luanda, and Huíla in Angola and also that it had been found in Malawi, Mozambique, and Zimbabwe in the adjacent *Flora zambesiaca* region.

Conservation status:

Not applicable.

Additional notes and discussion:

Taxonomic history and nomenclature. Four years before Lamarck (1786) published the name *Cotyledon pinnata*, Linnaeus [filius] (1782: 191) published a description for a plant he named *Crassula pinnata*. However, the description that the younger Linnaeus fil. (1782: 191) provided for his *Crassula pinnata* cannot be unambiguously applied to plants today known as *Kalanchoe pinnata*. Although the leaves that are described as "pinnatis" and the flower colour that is given as "rubris" indicate that Linnaeus fil. might have had material of *K. pinnata* at hand, the margins of the leaflets are clearly described as entire. The margins of the leaflets of *K. pinnata* are in fact distinctly crenate-dentate. We therefore interpret the name *Crassula pinnata* L.f., *Suppl. pl.*: 191 (1782 [1781 published in April 1782]) as being of unresolved application (Smith & Figueiredo, 2018b: 222).

Four years after Linnaeus (1782) published the name *Crassula pinnata*, Lamarck (1786: 141) published the name *Cotyledon pinnata*. When, just over 20 years later, Persoon (1805: 446) published the combination *Kalanchoe pinnata*, he clearly cited "Lam. enc." [= Lam., *Encycl.* 2(1): 141 (1786).], as the source of the epithet 'pinnata' and not Linnaeus fil. (1782: 191).

De Candolle (1828: 395) treated *Kalanchoe alternans*, one of the species recognised by Persoon (1805), as a species of *Kalanchoe*. De Candolle unambiguously linked this name to Persoon (1805) by citing the latter as "ench.", which is also a widely used abbreviation for that work. Notably, De Candolle (1828: 396) placed *Bryophyllum calycinum* Salisb., one of the synonyms of *K. pinnata*, in the genus *Bryophyllum*. One year later, Haworth (1829) did not include *Kalanchoe pinnata* in his monograph of *Kalanchoe*, so perhaps tacitly accepting *Bryophyllum* as a genus separate from *Kalanchoe*. He was though aware of De Candolle's work of 1828, citing it as "*De Cand.* Prod. Syst. Veg. 3. 395." in the references he listed for the family Crassulaceae.

Following on from these works that date from the first few decades of the 1800s, *Kalanchoe pinnata* was included in all subsequent works that dealt with the genus. Although the species is reasonably stable in the expression of its characters, a number of names that are nowadays included in the synonymy of *K. pinnata* were published over the next 100 years (see **Synonyms**, above).

Identity and close allies. Globally, *Kalanchoe pinnata* is arguably the best known of all the kalanchoes. It is easy to identify as it has bright green, compound leaves adorned with softly scalloped, purplish red marginal indentations that are often referred to as 'dentate-crenate'. The lantern-shaped flowers are quite large and decorative. Flower colour is light yellow (lower down) to greenish with reddish purple longitudinal lines that become more conspicuous as the flowers mature.

Boiteau & Allorge-Boiteau (1995: 16) included the species in their "Groupe IX Proliferae". Two other kalanchoes that are naturalised in southern Africa, *K. gastonis-bonnieri* and *K. prolifera*, are included in this group.

Reproductive strategy. In South Africa, the most significant reproductive strategy of *Kalanchoe pinnata* is the formation of bulbils in the notches between the leaf marginal crenations. Such plantlet formation also occurs on the margins of severed leaves.

Cultivation. The species is exceedingly easy in cultivation and will grow in virtually any soil-based medium in the tropics and subtropics.

Invasiveness. Kalanchoe pinnata spreads and establishes itself easily through the plantlets formed on the leaf margins, especially in subtropical and tropical parts of the world. This species should not be cultivated in such climatic zones.

Uses. Numerous uses have been recorded for *Kalanchoe pinnata*, in places as widely separated as Africa, India, Brazil, and the West Indies (see Walters, 2011: 243 for references). However, in southern Africa, the species apparently does not feature in traditional medicinal and other uses.

23. *KALANCHOE PROLIFERA* (BOWIE EX HOOK.) RAYM.-HAMET

Blooming boxes (Figs. 12.23.1 to 12.23.7)

FIG. 12.23.1   A naturalised specimen of *Kalanchoe prolifera* growing near the city of Durban on the subtropical south coast, KwaZulu-Natal Province, South Africa. Photograph: Neil R. Crouch.

Kalanchoe prolifera (Bowie ex Hook.) Raym.-Hamet in *Bull. Herb. Boissier*, Ser. II. 8: 19 (1908b). Raymond-Hamet & Marnier-Lapostolle (1964: 56; Plate XX, figures 60–62); Jacobsen (1970: 288); Jacobsen (1986: 625); Boiteau & Allorge-Boiteau (1995: 124); Rauh (1995: 116, 156); Descoings (2003: 170); Smith et al. (2017b: 313); Figueiredo & Smith (2018: 22). **Type**: Hort. Kew, 1857, Madagascar, *s.c. s.n.* (K K000232802, lecto-, designated by Figueiredo & Smith, 2018: 24).

Synonym:

Bryophyllum proliferum Bowie ex Hook. in *Curtis's Bot. Mag.* 85: t. 5147 (1859); Baker (1883: 139); Berger (1930: 410, figure 197 A–C); Fernandes (1983: 68); Crouch & Smith (2007: 206–208); Walters (2011: 245). **Type**: as above.

Nomenclatural note. In the protologue [Hook. in *Curtis Bot. Mag.* 85: t. 5147 (1859)] of the name *Bryophyllum*

proliferum, a single collection was cited, without the word 'type' used, along with an illustration, t. 5147, which also forms part of the protologue. Since no type was indicated by Hooker, a lectotype therefore had to be designated from these two elements (the specimen and the illustration). Figueiredo & Smith (2018: 24) designated the specimen as the lectotype (McNeill et al., 2012: 16, Article 9). This matter is discussed more fully below.

Derivation of the scientific name: from the Latin 'prolifer', proliferating, for the numerous bulbils produced on the inflorescence.

Description:

Perennial, many-leaved, unbranched but sprouting from the base, glabrous throughout, large succulent, and exceptionally to 2–3 m tall. *Stems* robust, erect to leaning, somewhat woody below, ± 4-angled, raised scars of

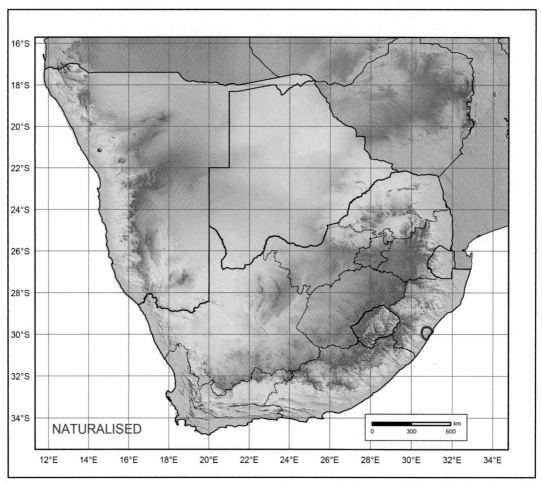

FIG. 12.23.2   Known geographical distribution range of *Kalanchoe prolifera* in South Africa.

abscised leaves conspicuous. *Leaves* to 30 cm long, opposite-decussate, leaf pairs widely spaced, petiolate, green, somewhat shiny, succulent, and spreading to down-curved; *petiole* to 16 cm long, broadened at the base, clasping the stem, and purplish red; *blade* pinnatisect or pinnate and rarely undivided; *blade segments* 7–15 × 1–5 cm, sometimes asymmetrical, decurrent at the base, oblong to lanceolate to ovate-elongate, and often lengthwise boat-shape-folded upwards; *apex* rounded; *margins* crenate to harmlessly dentate and reddish purple. *Inflorescence* 35–90 × 20–45 cm, a very large, erect to leaning panicle, often with aborted flowers and bulbils; *pedicels* 8–15 mm long and thin. *Flowers* pendulous, green below where enveloped by calyx, red above where exposed beyond calyx, and sometimes light green to yellowish green throughout; *calyx* inflated, tubular, tube 12–16 mm long, campanulate, 4-angled, green, and slightly scabrid; *sepals* 15–20 mm long, fused, apically acuminate-cuspidate, free portions (lobes) 2.5–4.0 × 2.5–4.0(−7) mm, softly acute, contrasting against red corolla tube, and clasping the corolla tube; *corolla* somewhat inflated and enlarged lower down, light green below

where obscured by calyx, and red above where exposed; *tube* 15–25 mm long, tubular, slightly constricted above carpels, suburceolate above constriction, light green below, merging into red above, and sometimes light green to yellowish green throughout; *lobes* 2–3 mm, acuminate-cuspidate, and tapering to an acute apex. *Stamens* inserted below middle of corolla tube and exserted; *filaments* long; *anthers* 2.0 × 1.5 mm, ovate, and visible beyond the mouth of the corolla tube. *Pistil* consisting of 4 carpels; carpels basally connate, 7–8 mm long, and styles 1.5–2.0 mm long; *scales* 1.4–1.6 × 2.0–2.5 mm and ± orbicular. *Follicles* not seen. *Seeds* not seen. *Chromosome number*: 2n = 34 (Friedmann, 1971).

Flowering time:

August–September (southern hemisphere).

Illustrations:

Hooker (1859: t. 5147); Berger (1930: 409, figure 197 A–C) [as *Bryophyllum proliferum*]; Raymond-Hamet & Marnier-Lapostolle (1964: Plate XX, figures 60–62); Jacobsen (1970: Plate 105, figure 4); Jacobsen (1986: 624–625, figures 879–880); Rauh (1995: 156, figures 407–409); Boiteau & Allorge-Boiteau (1995: 123); Walters

FIG. 12.23.4 The sparsely branched inflorescence of *Kalanchoe prolifera* can reach a height of over 1 m. Photograph: Neil R. Crouch.

FIG. 12.23.3 Inflorescences of *Kalanchoe prolifera* are borne in an erect to leaning position, often with aborted flowers and bulbils attached to the side branches. Photograph: Neil R. Crouch.

et al. (2011: 247–248, figures 281–285); Smith et al. (2017b: 313); Figueiredo & Smith (2018: 23, figures 1–3).

Common names:

Afrikaans: boksplakkie and lanternplakkie.

English: blooming boxes, Chinese lantern (kalanchoe), and green mother of millions (Walters et al., 2011: 245).

Geographical distribution:

The species occurs on the Central Plateau of Madagascar (Rauh, 1995). It has been introduced to most tropical parts of the world, often becoming naturalised.

Distribution by country. Madagascar.

Habitat:

In South Africa, the species was recorded from the subtropical south coast of the KwaZulu-Natal Province near the city of Durban (Fig. 12.23.2). Fernandes (1983: 68) noted the species as a garden escape at Harare, Zimbabwe.

Conservation status:

Not applicable.

Additional notes and discussion:

Taxonomic history and nomenclature. The name *Bryophyllum prolifera* was given to a plant that could have been brought from Madagascar to Cape Town by James Bowie. However, we were unable to ascertain whether Bowie visited Madagascar during his exploratory botanical travels in the first half of the 19th century. Material of the species could well already have been in cultivation in Cape Town, where Bowie was based from 1827 until his death on 2 July 1869 (Smith & Van Wyk, 1989). Indeed, Hooker (1859: text accompanying Tab. 5147) states that "Our plants were raised from cuttings, sent from the Cape of Good Hope, and which he [presumably Bowie or 'he' might have been a typographical error for 'we'] received as dried specimens for the herbarium, by Mr. Bowie". Hooker further commented that the species was unlike anything described until then in *Bryophyllum* or *Kalanchoe*, even though there was already a *Kalanchoe delagoensis* described from Delagoa Bay [Mozambique], but evidently, little was known about the latter species. *Kalanchoe delagoensis* was a Madagascan, not Mozambican, species that Ecklon & Zeyher (1837) apparently described from cultivated, or naturalised, material

FIG. 12.23.5   Close-up of the flowers of *Kalanchoe prolifera*. The exposed part of the corolla tube of this form is red. Photograph: Neil R. Crouch.

FIG. 12.23.6   This form of *Kalanchoe prolifera* has a greenish corolla tube. Note the formation of plantlets on the inflorescence. Photograph: Neil R. Crouch.

from Delagoa Bay, Mozambique (Figueiredo & Smith, 2017c; also see under *K. tubiflora*).

Preserved specimens that Bowie sent to the United Kingdom are kept at both K and BM (Smith & Van Wyk, 1989). No material of *Kalanchoe prolifera* sent to the United Kingdom by Bowie was found at BM. When Hamet (1908b) published the combination *K. prolifera*, he cited the type as follows: "Bowie—échantillon authentique de *Bryophyllum proliferum*". Descoings (2003: 170) followed Hamet's view; however, he was uncertain whether the material was indeed kept, or existed, at K

by querying at least the herbarium where the material might be kept: "**T**: Madagascar Bowie s.n. [K?]". In the Kew Catalogue, there is a specimen annotated 'Bryophyllum Hort. Kew 1857 Madagascar' filed in *Bryophyllum* that is not labelled as type (K K000232802). This specimen was designated as the lectotype. Note, however, that the specimen is not attributed to Bowie as collector.

Hamet (1908b) also cited two additional collections by Baron ("nos. 1270! et 1475"). Boiteau & Allorge-Boiteau (1995) designated as lectotype *Baron 1270* (P). That

FIG. 12.23.7   A multitude of plantlets developing on the peduncle of a spent inflorescence of *Kalanchoe prolifera*.   Photograph: Neil R. Crouch.

specimen is not in the P catalogue online, but a duplicate is held by BM (BM000645892). However, the Baron specimens cannot be part of Hooker's original material that dates from 1859, as Baron only travelled to Madagascar in 1872. In 1883, Baker (1883: 139) mentioned that Baron had "rediscovered *B. proliferum* Bowie (Baron 1270, 1465), figured in Bot, Mag. t. 5847". Hamet (1908b) likely cited the Baron collections based on Baker's comment, as he only marked as examined no. 1270 and not no. 1465. The lectotypification proposed by Boiteau & Allorge-Boiteau (1995) was not effective as a lectotype should have been selected from among the two elements that are undoubtedly the only original material cited in the protologue of the name *Bryophyllum proliferum* (Hooker, 1859). Note though that material mentioned in the protologue of *B. proliferum* is devoid of a collector's name, collecting number, and date.

Figueiredo & Smith (2018: 22–24) clarified the typification of the name *Kalanchoe prolifera* and its basionym *Bryophyllum proliferum*.

Along with nine other species, Boiteau & Allorge-Boiteau (1995: 16) included *Kalanchoe prolifera* in their "Groupe IX Proliferae". Two of the other species that

they included in that group, *K. gastonis-bonnieri* and *K. pinnata*, have also become naturalised in South Africa.

Wickens (1987: 27) noted that *Kalanchoe prolifera* (as its synonym, *Bryophyllum proliferum*) had become naturalised in parts of tropical east Africa, but he did not include a treatment of the species in that floristic work. At the time that Fernandes (1982a) published her work on the flora of Angola, the species was not known for that country, but the following year, she noted it as escaping from gardens in Salisbury [Harare], Zimbabwe (Fernandes, 1983: 68).

Identity and close allies. Kalanchoe prolifera is a large, fast-growing leaf succulent that can reach a height of about 3 m under ideal growing conditions. It is one of the largest nonarborescent *Kalanchoe* species from Madagascar. Stems are mostly unbranched and can grow quite long; at a distance, young, erect plants resemble small palm trees or even the tropical perennial *Zamioculcas zamiifolia* (Lodd.) Engl. (family Araceae). The pinnately compound light to midgreen, succulent leaves of *K. prolifera* are carried in opposite pairs along the stems; a leaf pair is arranged along the stem at right angles to the next. The leaf margins and petioles are reddish purple, especially when plants are grown under strong light conditions.

At flowering maturity, which plants can take a few years to reach, a candelabrum-like inflorescence of close to 1 m tall overtops the plants. In the case of material naturalised in southern Africa, the flowers are green below and reddish above, with the green portion almost entirely obscured by the prominent calyx tube. Note though that some forms of the species can have light green to yellowish green corollas throughout.

The common name of the species, blooming boxes, refers to the Chinese lantern-like shape of the green calyces of the flowers that well hide the reddish or light green to yellowish green corolla.

Reproductive strategy. In South Africa, the most significant reproductive strategy of *Kalanchoe prolifera* is the formation of bulbils on the inflorescence and on the margins of severed leaves.

Cultivation. The species is very easy—too easy in fact—in cultivation. It spreads primarily from bulbils that form especially on the inflorescence and detached leaves and from seed. This species should ideally not be grown in a garden, given that it has a tendency to become invasive. After flowering, plants wilt and die but may sucker, in some cases quite slowly, from the base. Bulbils can also form higher up along the stems.

Uses. Kalanchoe prolifera is reputedly used against rheumatism in Madagascar (Githens, 1949; Descoings, 2003: 171). In southern Africa, no medicinal uses have been recorded for the species. Plants are however grown around homesteads as a protective charm.

24. *KALANCHOE TUBIFLORA* (HARV.) RAYM.-HAMET

Chandelier plant (Figs. 12.24.1 to 12.24.9)

FIG. 12.24.1   A dense infestation of *Kalanchoe tubiflora* at Praia da Baía de Porto Covo, along the Atlantic coast of Portugal, about 170 km south of Lisbon. The exotic *Opuntia ficus-indica* (L.) Mill. grows to the left of *K. tubiflora*. Photograph: Gideon F. Smith.

Kalanchoe tubiflora **(Harv.) Raym.-Hamet** in *Beih. Bot. Centralbl.*, (Abt. 2, 29: 41–44. 1912). Raymond-Hamet & Marnier-Lapostolle (1964: 60; Plates XXII–XXIII, figures 70–71); Jacobsen (1970: 290); Jacobsen (1986: 631); Boiteau & Allorge-Boiteau (1995: 90); Rauh (1995: 116); Rauh (1998: 206, 320); Figueiredo & Smith (2017c: 771). **Type**: Delagoa Bay, *s.c.* (Forbes) *s.n.* (S S-G-10717).

Synonyms:

Kalanchoe delagoensis Eckl. & Zeyh., *Enum. pl. afric. Austral.*: 305 (1837), nom. nud. Descoings (2003: 153); Smith & Van Wyk (2008a: 150); Smith et al. (2017b: 308).

Bryophyllum tubiflorum Harv. in Harvey & Sonder, *Fl. cap.* 2: 380 (1862); Fernandes (1983: 67). **Type** as for *Kalanchoe tubiflora*.

Kalanchoe verticillata Scott-Elliot in *J. Linn. Soc., Bot.* 29: 14 (1891) **Type**: Madagascar, Fort Dauphin, *Scott-Elliot 2983* (K K000232791 & K000232792).

Bryophyllum delagoense (Eckl. & Zeyh.) Schinz in *Mém. Herb. Boiss.* 10: 38 (1900). Tölken (1985: 73); Tölken & Leistner (1986: Plate 1938); Walters (2011: 237). Combination not validly published because the basionym is not validly published.

Geaya purpurea Costantin & Poiss. in *Compt. Rend. Acad. Sci. Paris* 147: 636 (1908). **Syntypes**: Madagascar, Nord du Cap Ste Marie, *Geay 6372* (P P00374200), Plateau, au nord du Cap Ste Marie *Geay 6335* (P P00444108), Nord du faux Cap, *Geay 6336* (P P00444109).

Bryophyllum verticillatum (Scott-Elliot) A.Berger in Engler & Prantl, *Nat. Pflanzenfam.*, ed. 2 18a: 411 (1930).

Derivation of the scientific name:

From the Latin *tubus*, tube, and Latin *florus*, flowered, referring to the tubular flowers, a character shared by most other species of *Kalanchoe*.

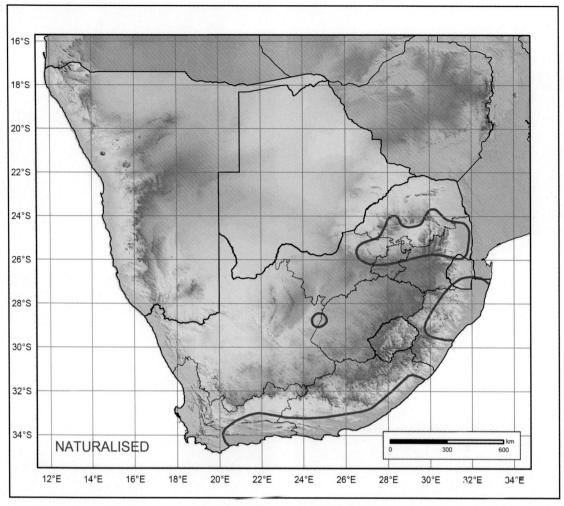

FIG. 12.24.2 Known geographical distribution range of *Kalanchoe tubiflora* in southern Africa.

FIG. 12.24.3 The thin, erect stems of *Kalanchoe tubiflora* can reach a height of about 2 m but usually remain much shorter. Photograph: Gideon F. Smith.

FIG. 12.24.4 Leaves of *Kalanchoe tubiflora* are evenly spaced along the stems, slanted away from the branches at ± 45 degrees, succulent, glabrous, sessile, terete, cylindrical to narrowly oblong, and bluish to purplish to brownish green. Photograph: Gideon F. Smith.

FIG. 12.24.5   A multitude of plantlets (bulbils) are formed especially at the tips of the leaves of *Kalanchoe tubiflora*. These easily strike root where they fall. Photograph: Gideon F. Smith.

FIG. 12.24.7   Some forms of *Kalanchoe tubiflora* have crimson red flowers. Photograph: Gideon F. Smith.

FIG. 12.24.6   The light purple calyx of *Kalanchoe tubiflora* is deeply incised to yield a short tube and fairly long lobes. Photograph: Gideon F. Smith.

FIG. 12.24.8   Orange-red-flowered form of *Kalanchoe tubiflora*. The flowers are arranged in dense, head-shaped inflorescences. Photograph: Gideon F. Smith.

Description:

Biennial or semiperennial, erect to leaning to procumbent, glabrous, brittle shrubs, sparsely branched or suckering near the base, with canopies terminating in an inflorescence, succulent, and to 2m tall. *Stems* thin to medium-sized in diameter, light yellowish brown to yellowish grey, and usually erect or toppling over under the weight of the inflorescence. *Leaves* many, sparsely arranged and evenly spaced throughout, slanted away from branches at ± 45 degrees, succulent, glabrous, sessile, terete, cylindrical to narrowly oblong, bluish to purplish to brownish green to grey-green with irregular dark green or bluish purple spots and somewhat waxy; *petiole* absent; *blade* 3–12 cm × 2–5 mm, cylindrical to narrowly oblong, and succulent; *base* narrow; *apex* with 2–9 small conical teeth at apex, usually with bulbils (adventitious buds) borne in their axils, also once leaves are detached. *Inflorescence* a terminal, apically branched, many-flowered, head-shaped to rounded thyrse; up to 20 cm in diameter; erect; peduncle long, straight, and light yellowish grey; *pedicels* 6–12 mm long and glabrous. *Flowers* pendent, glabrous, subtended by small bracts that soon shrivel, corolla much exceeding the calyx at maturity, buds yellowish, mature flowers various shades of red, from crimson to deep orange, basally longitudinally

FIG. 12.24.9   A dense stand of *Kalanchoe* ×*houghtonii*, a hybrid between *K. tubiflora* and *K. daigremontiana*, naturalised in Porto Covo, Portugal. Photograph: Gideon F. Smith.

infused with yellow along tube angles, waxy bloom ± absent, papery when dry, and drying purplish brown; *calyx* campanulate, tubular for ± ⅓, light greenish red, red-infusion more prominent towards sepal tip, purplish when young, and red arranged in feint longitudinal lines; *sepals* 10–11mm long, fused for ± 3–4mm, free portion elongated-triangular, acute, and obscuring lower ± ⅓ of the corolla tube; *corolla* various shades of red, from crimson to deep orange, and basally longitudinally infused with yellow along tube angles; *tube* 25–30mm long, cylindrical to distinctly campanulate, bulging in the middle, and slightly flared at the mouth; *lobes* ± 15 × 5mm, spatulate to obovate, apically obtuse-rounded, and minutely apiculate. *Stamens* inserted very low-down in corolla tube at ± same level as carpels and included to hardly exserted; *filaments* 25–28mm long, thin, light greenish red, and basally light green; *anthers* ± 1mm long, black, and included to hardly exserted; *pollen* greyish yellow. *Pistil* consisting of 4 carpels; *carpels* 5–8mm long and light shiny green; *styles* 20–25mm long; *stigmas* very slightly capitate, green, inserted, and later hardly exserted; *scales* rectangular, free, ± 0.5 × 1.0mm, and flat or slightly indented above. *Follicles* not seen. *Seeds* not seen. *Chromosome number*: $2n = 68$ (Boiteau & Allorge-Boiteau, 1995: 92).

Flowering time:

June–September, peaking from late June to August (southern hemisphere).

Illustrations:

Scott-Elliot (1891: Plate 3); Raymond-Hamet & Marnier-Lapostolle (1964: Plates XXII–XXIII, figures 70–71); Jacobsen (1970: Plate 103, figure 3, plant on the right); Hargreaves & Hargreaves (1972: 7); Tölken (1985: 74, figure 10); Jacobsen (1986: figures 895–896); Tölken & Leistner (1986: Plate 1938); Boiteau & Allorge-Boiteau (1995: 91); Rauh (1998: 375); Smith & Van Wyk (2008a, b: 151, bottom of page); Walters et al. (2011: figures 269–274).

Local common names:

Afrikaans: kandelaarplant (Walters et al., 2011: 237).

English: chandelier plant, mother of millions, and pregnant plant (Walters et al., 2011: 237).

isiZulu: indunjane (Walters et al., 2011: 237).

Origin:

East-central and southern Madagascar (see Boiteau & Allorge-Boiteau, 1995: 199, Map 8).

Naturalised distribution range in South Africa:

The species has become naturalised along much of South Africa's coast, from the southern Cape, eastwards to the Mozambican border (Fig. 12.24.2). It further occurs in all the provinces of South Africa, including especially in the eastern (Mpumalanga province) and north-central (North-West, Gauteng, and Limpopo provinces) parts of the country; centrally, it is also recorded from the Free State. It also occurs in Swaziland and quite possibly in all the southern and south-tropical African countries, including the somewhat more arid Botswana and Namibia. Although therefore showing a preference for mild-climate parts of the subcontinent, it is the only naturalised *Kalanchoe* species that also tolerates the much more severe climate above the Great Escarpment.

Conservation status:

Not applicable.

Additional notes and discussion:

Taxonomic history and nomenclature. Originally, *Kalanchoe tubiflora* was described under the name *Kalanchoe delagoensis* by Ecklon & Zeyher (1837) based on a plant fragment collected at "Delagoa Bay" by John Forbes. Forbes was part of an expedition led by W.F. Owen to survey the east coast of Africa. Delagoa Bay was the name used for present-day Maputo (Mozambique) and surrounding areas, where the expedition stopped on their way to Madagascar. It is likely that the material originated from Madagascar and not Mozambique. Nevertheless, the name *Kalanchoe delagoensis* was not validly published as it is considered a nomen nudum (see Figueiredo & Smith, 2017c).

The next name to be created for this plant was *Bryophyllum tubiflorum* Harv. Based on the same material collected by Forbes, Harvey (1862) provided a description of the species, which he however regarded as a bryophyllum, and he listed *Kalanchoe delagoensis* as a synonym. If *Kalanchoe delagoensis* was validly published, Harvey's name would be illegitimate because he listed the former (*Bryophyllum tubiflorum*) as a synonym. When treating the plant as a bryophyllum, Harvey should have used the epithet *"delagoense"*. For this reason, the combination *Bryophyllum delagoense* (Eckl. & Zeyh.) Schinz was later made (Schinz & Junod, 1900). However, as *Kalanchoe delagoensis* is considered as not having been validly published, the correct epithet for the taxon is *"tubiflora"*, and accordingly, a new combination under *Kalanchoe* was made by Hamet (1912b).

In 1891, the plant was redescribed (as *Kalanchoe verticillata*) and illustrated for the first time, from new material collected in Madagascar by George F. Scott-Elliot (1862–1934). There are two sheets of that collection in the Herbarium at the Royal Botanic Gardens, Kew, and neither are marked as holotype; therefore, they are syntypes of the name *K. verticillata*.

In 1908, a further name, *Geaya purpurea*, was published by Constantin and Poisson for this plant, which they wrongly interpreted as a representative of the Heath family, the Ericaceae. To these authors, the plants looked like a saprophyte or parasite, and they even suggested that it could represent a new family. In the protologue of *Geaya purpurea*, the original material cited is "6372-6335, Nord du Cap-Sainte-Marie; n° 6336, Nord du Faux Cap". These three collections are syntypes, even though 6372 has a recent label saying holotype. Boiteau & Allorge-Boiteau (1995) cited the 6372 specimen as holotype (meaning lectotype).

Identity and close allies. Kalanchoe tubiflora is a short-lived perennial that can be easily recognised in both the vegetative and reproductive phases. It has linear-tubular leaves that can reach a length of over 10 cm but usually remain much shorter. It carries a magnificent, densely clustered inflorescence of bright orange to red, tubular-to bell-shaped flowers in winter. *Kalanchoe tubiflora* and *K. daigremontiana* were the only two species included in the "Groupe VI Bulbilliferae" by Allorge & Allorge-Boiteau (1995: 16 [unnumbered]).

Kalanchoe tubiflora easily hybridises with other kalanchoes that used to be recognised in *Bryophyllum*. One such entity, *K. ×houghtonii* D.B.Ward, recorded by Silva et al. (2015) as occurring at Cascais and Estoril on the west coast of Portugal north of Lisbon, is often misidentified as one of its parents, *K. daigremontiana*. The other parent of this hybrid is *K. tubiflora*; both parents originate from Madagascar and are widely cultivated globally, including in Portugal. *Kalanchoe ×houghtonii* has known invasive tendencies (Guillot Ortiz et al., 2014) and in Portugal, as elsewhere, is showing signs of moving from domestic gardens to the natural vegetation. Although the hybrid is similar in appearance to *K. daigremontiana*, it is a smaller plant in all respects and has narrower leaves, and the flowers are a pinkish rather than a more purplish colour (Guillot Ortiz et al., 2014). *Kalanchoe ×houghtonii* has fortunately not yet been introduced to South Africa, as far as we could determine.

Reproductive strategy. The species produces a multitude of bulbils (plantlets) on its leaves and copious amounts of seed. Plants also sprout from the base.

Given how easily *Kalanchoe tubiflora* becomes naturalised and invasive in mild-climate countries, including in South Africa, its possible biocontrol has received attention (Witt, 2004; Witt et al., 2004). However, the two stem-boring weevils, *Alcidodes sedi* and *Osphilia tenuipes*, investigated are not host-specific and can also complete their life cycles on other indigenous Crassulaceae.

Uses. It is widely planted as a protective charm around homesteads, given its tolerance of horticultural neglect and hardiness under a wide range of environmental conditions. Given that it is therefore 'difficult to kill', it is believed that it will provide protection to dwellings where it is planted (Von Ahlefeldt et al., 2003).

Although *Kalanchoe tubiflora* is poisonous to livestock and humans (Kellerman et al., 2005), no stock losses have been reported from southern Africa as a result of the consumption of plants of this species.

Cultivation. Kalanchoe tubiflora is exceedingly easy in cultivation. Given its ability to become naturalised in places very remote from its natural geographical distribution range in Madagascar, the species should not be grown in southern Africa. Tölken & Leistner (1986) aptly noted that "…there can be little doubt that it [*Kalanchoe tubiflora*, as *Bryophyllum delagoense*] would be a collectors" item if it was rare and difficult to grow'. While it is therefore indeed at first sight a highly desirable species to have in domestic horticulture, it will soon become unpleasantly weedy.

25. *KALANCHOE BLOSSFELDIANA* POELLN.

Flaming Katy (Figs. 12.25.1 to 12.25.15)

FIG. 12.25.1   Two colour variants, bright pink and orangey yellow, of *Kalanchoe blossfeldiana* flowering in a windowsill box in the garden of Weavind Place Guest House, Weavind Park, Pretoria, Gauteng, north-central South Africa. Photograph: Gideon F. Smith.

Kalanchoe blossfeldiana **Poelln.** in *Repert. spec. nov. regni veg.* 35: 159 (1934). Blossfeld (1934: 36); Jacobsen (1970: 283); Jacobsen (1986: 608); Rauh (1995: 116, 117, 148, 149); Rauh (1998: 300); Descoings (2003: 148); Smith & Van Wyk (2008b: 150). **Lectotype**: Madagascar, de mon jardin de Tananarive, provenance Massif Tsaratanana, *H. Perrier 17883* (P P00374114). Designated by Van Voorst & Arends, (1982: 575).

Synonyms:

Kalanchoe globulifera H.Perrier var. *coccinea* H.Perrier in *Arch. de Bot. II, Bull. Mens.* 2: 26 (1928). **Type**: same as for *Kalanchoe blossfeldiana* Poelln.

Derivation of the scientific name:

For Robert Blossfeld (1882–1945), German horticulturalist employed at Pottsdam, near Berlin, and later in Lübeck.

Description:

Perennial, usually many-leaved, densely branched, glabrous throughout, small- to medium-sized, rounded, succulent shrublets, and 0.2–0.4 m tall. *Stems* green to reddish green, a few arising from a sometimes brittle, corky base, erect to slightly leaning, round in cross-section, and rarely with one or two lengthwise running ridges. *Leaves* opposite-decussate, petiolate, mid- to dark green, reddish-infused if grown in exposed positions, shiny, somewhat succulent, spreading to slightly erect, and papery on drying; *petiole* 5–30 mm long, channelled above, and leaves not clasping the stem; *blade* 4–7 × 2.5–4.0 cm, ovate-oblong to oblong, and flat; *base* rounded to somewhat cuneate; *apex* rounded-obtuse; *margins* coarsely crenate or undulate-crenate into

FIG. 12.25.2   The bright yellow flower selection of *Kalanchoe bloss-feldiana*. Photograph: Gideon F. Smith.

FIG. 12.25.5   Orange- and yellow-flowered selections of *Kalanchoe blossfeldiana* grown along the sidewalk fence of The Blackstone Restaurant in Chicago, Illinois, USA. Photograph: Gideon F. Smith.

FIG. 12.25.3   The pure white flower selection of *Kalanchoe blossfeldiana*. Photograph: Gideon F. Smith.

FIG. 12.25.6   *Kalanchoe blossfeldiana* cultivars offered for sale in the Golegã town square, central Portugal. Photograph: Gideon F. Smith.

FIG. 12.25.4   The bright red, single-flowered form of *Kalanchoe blossfeldiana* in full flower in Lubango, southern Angola. Photograph: Gideon F. Smith.

FIG. 12.25.7   Potted specimens of *Kalanchoe blossfeldiana* sold in the Safari Nursery, Lynnwood Road, Pretoria, South Africa.

FIG. 12.25.8 *Kalanchoe blossfeldiana* grown in a terracotta pot at the entrance to the Volta & Meia Restaurant in Figueira da Foz, north-coastal Portugal. Photograph: Gideon F. Smith.

FIG. 12.25.10 Selections of *Kalanchoe blossfeldiana* sold in the shop of the Missouri Botanical Garden, St Louis, Missouri, USA. Photograph: Gideon F. Smith.

FIG. 12.25.9 This 'Rosebud', or double-flower, selection of *Kalanchoe blossfeldiana* has bright pink flowers. Photograph: Gideon F. Smith.

FIG. 12.25.11 Miniature specimens of *Kalanchoe blossfeldiana* that look like rooted plant tips in full flower are sold as "Kalanchoe Mini" by Lidl in Alcanena, central Portugal. Photograph: Gideon F. Smith.

rounded, harmless teeth especially above the middle, often curved upwards sometimes, and red. *Inflorescence* 15–20 cm tall, erect, apically dense, many-flowered, flat-topped thyrse, branches opposite, erect, subtended by leaflike bracts, without leafy branchlets in axils, and axis green to reddish green to brown; *pedicels* 3–4 mm long, and slender. *Flowers* erect, crimson red to scarlet (selected colour forms include virtually all colours, except black and blue), light green lower down, and scarlet higher up (tube) to bright scarlet (lobes); *calyx* shiny reddish green, strongly infused with small red spots especially towards tips and margins; *sepals* 4–6 mm long, fused for ± 0.5–1.0 mm, triangular-lanceolate, ± separate, basally hardly adnate for ± 0.5 mm, acute, flared away from tube, and hardly contrasting against corolla tube; *corolla*

12–13 mm long, somewhat enlarged lower down, not twisted apically after anthesis, and light green and crimson red to scarlet; *tube* 7–10 mm long, 4-angled, rounded to slightly box-shaped-square when viewed from below, slightly longitudinally fluted above, and light green and crimson red to scarlet; *lobes* 5–6 × 3–4 mm, ovate to oblong-ovate to oblong-deltoid, rounded-acute at apex, and mucronate, bright crimson to scarlet. *Stamens* inserted well above the middle of the corolla tube and included; *filaments* 1.0–1.5 mm long, thin, and yellow; *anthers* 0.30–0.75 mm long, rounded to slightly elongated, and greenish yellow turning brown. *Pistil* consisting of 4 carpels; *carpels* 5–6 mm long and bright light green; *styles* 2–3 mm long; *stigmas* capitate, whitish yellow, and visible at mouth of corolla tube; *scales* ± 2 mm long, linear, and

FIG. 12.25.12   Potted specimens of *Kalanchoe blossfeldiana* available in the Floral Rua da Saudade, Alcanena, central Portugal. Photograph: Gideon F. Smith.

FIG. 12.25.13   A mustard yellow, double-flowered form of *Kalanchoe blossfeldiana* growing in a trough outside the Palace El Badi in Marrakech, Morocco. Photograph: Estrela Figueiredo.

light yellowish green. *Follicles* brittle, grass spikelet-like, enveloped in dry, light-brown remains of corolla, dull whitish green, and 5–6 mm long. *Seeds* 0.3–0.4 mm long and light brown. *Chromosome number*: variable, $2n = 34$ (Baldwin, 1938: 576); $2n = 68$, and aneuploids known (see Rowley, 2010: 33, Rowley, 2017: 139).

Flowering time:

July–September, peaking in July; manipulated to flower at any time of year.

Illustrations:

Jacobsen (1986: 608, figure 849); Boiteau & Allorge-Boiteau (1995: 189, as 'Kalanchoe coccinea var. blossfeldiana'); Rauh (1995: 148, figures 380–382); Smith & Van Wyk (2008b: 151, top right, middle left).

Common names:

Afrikaans: blomkalanchoe.

English: Flaming Katy and Madagascar widow's thrill.

Geographical distribution:

Kalanchoe blossfeldiana originates from northeastern Madagascar. The species was discovered by Perrier de la Bâthie at Mount Tsaratanana at an altitude of 2000–2200 m, and some years later, it was collected again by Humbert near the springs of the Sambirano River (Van Voorst & Arends, 1982; Rauh, 1995: 148), from where it was introduced into cultivation all over the world, especially as a long-flowering houseplant. Although it is widespread in cultivation, it is not invasive.

Distribution by country. Madagascar. Plants have not escaped from cultivation in southern Africa.

Habitat:

In Madagascar, *Kalanchoe blossfeldiana* grows at altitudes of 1600 m and higher in humid cloud forests in the Tsaratanana Mountains. In this high-mountain location, the midsummer temperature often does not go very high.

Conservation status:

Not threatened.

Additional notes and discussion:

Taxonomic history and nomenclature. Material today known as *Kalanchoe blossfeldiana* was originally described as *Kalanchoe globulifera* var. *coccinea* by the French botanist Perrier de la Bâthie (1928: 26). Six years later, Von Poellnitz (1934: 159) elevated this variety to species rank, and since the epithet *coccinea* was earlier used for an African species of *Kalanchoe* by Britten in his treatment of the Crassulaceae for the *Flora of Tropical Africa* (Britten, 1871: 395), Von Poellnitz named the species *Kalanchoe blossfeldiana*, for Robert Blossfeld. The Madagascan species *Kalanchoe globulifera* is today still recognised as a species in its own right.

Later attempts by Boiteau and Allorge-Boiteau to separate *Kalanchoe blossfeldiana* and *K. globulifera* var. *coccinea* as varieties of the same species ('K. coccinea') resulted in two invalid designations. Both 'Kalanchoe coccinea (H. Perrier) Boiteau var. blossfeldiana (Poelln.) Boiteau' in Boiteau & Allorge-Boiteau (1995: 190) and 'Kalanchoe coccinea (H.Perrier) Boiteau' in Boiteau & Allorge-Boiteau (1995: 188) are not validly published. The reason is that when two or more different names based on the same type are proposed simultaneously for the same taxon by the same author (so-called alternative names), none of them is validly published (McNeill et al., 2012: 76, Article 36.2). Furthermore, for the first designation Boiteau indicated as basionym 'Begonia blossfeldiana' (instead of *Kalanchoe blossfeldiana*). This mistake has been transposed to specimen labels and online databases.

Identity and close allies. *Kalanchoe blossfeldiana* is arguably the best known of all the kalanchoes. This Madagascan species is exceedingly common in cultivation, predominantly as a houseplant. Selections with exquisitely beautiful flowers have been produced on an

FIG. 12.25.14 A single-flowered form of *Kalanchoe blossfeldiana* with very large, purplish pink blooms cultivated in a yellow, urn-shaped pot in the Jardin Majorelle in Marrakech, Morocco. Note how the margins of the leaves curve upwards yielding somewhat spoon-shaped leaves. Photograph taken on 12 April 2017 by Estrela Figueiredo.

enormous scale and are sold in millions through florist shops and plant nurseries.

Plants are small- to medium-sized, rounded shrublets that carry a few, erect branches. The leaves are shiny mid- to dark green, dentate-margined, and quite flat but distinctly succulent. Inflorescences are densely flowered and flat-topped, with the flowers borne erectly, rather than pendulous, as is the case with many Madagascan kalanchoes.

Kalanchoe blossfeldiana is included in *Kalanchoe* subg. *Kalanchoe*, which includes species with pendulous or erect flowers.

Although not always an indication of a close relationship, *Kalanchoe blossfeldiana* has been known to hybridise with a number of other *Kalanchoe* species, including the Madagascan *Kalanchoe daigremontiana* and the African and Arabian *K. flammea* Stapf (Jacobsen, 1986: 609). *Kalanchoe flammea* is today often regarded as a synonym of *K. glaucescens* Britten.

Uses. By far, the most important use of *Kalanchoe blossfeldiana* is in the horticultural trade. Flowers are exceptionally long-lasting, and a plant can easily produce blooms for 2 months or even longer. The species has

benefitted from extensive selection of desirable strains. As long ago as the 1930s, shortly after the species was described, Blossfeld, for whom the species was named, had already started selecting superior material for the horticultural market (Blossfeld, 1934).

Cultivation. Kalanchoe blossfeldiana is very popular in general indoor cultivation as plants grow well in small pots that take up little space, and their root systems are rather shallow. The growing medium can be very light and well aerated with lots of decomposed compost and gritty sand added, which make the potted plants conveniently easy to transport. Given the apparent preference of the species for temperate, rather than desert-like climates, it is perfectly suited for growing in areas with a continental-type climate where it will thrive if kept indoors or under glass. In greenhouse conditions, plants should be placed in airy, well-ventilated areas. Likely, given their usually smooth, shiny leaf surfaces, plants should ideally not be watered from above with 'heavy' water, as white, crystalloid deposits easily stain the leaves. Plants need high-light levels and manipulation of the day and night temperatures to stimulate them to grow as chunky, potted specimens. Under low-light

FIG. 12.25.15   What the original *Kalanchoe blossfeldiana* likely looked like: a specimen with single, red flowers. Photograph: Gideon F. Smith.

conditions, plants tend to become etiolated. A good temperature range to maintain to ensure that the species thrives is in the mid-teens (*c.* 15°C).

Untold quantities of potted kalanchoes, especially of *Kalanchoe blossfeldiana* and hybrids of which it is one parent, are sold each year. In *c.* 1990, the small European country of Belgium alone had 12 producers that distributed such material worldwide (Van Nieuwerburgh et al., n.d.: 391; Shaw, 2008: 28). Rowley (2017: 137) states that 65 million plants of the double-flowered cultivars of *Kalanchoe blossfeldiana* were produced in Holland annually. Somewhat confusingly, hybrids of *Kalanchoe blossfeldiana* are often simply sold under the name of the species, that is, with no indication that it contains genetic material derived from a range of other, mostly Madagascan, kalanchoes.

Kalanchoe blossfeldiana can be grown from seed that should ideally be sown shortly after it was harvested. However, this propagation method is hardly ever used by the plant-purchasing public. Nowadays, plants in full glorious flower can be obtained at any time of the year from florist shops and plant nurseries. Cuttings can be taken after plants have flowered, and while these will strike root readily, such plants cannot compete with the vigour of the large-scale-produced material offered for sale in the horticultural trade.

From a horticultural point of view, *Kalanchoe blosssfeldiana* can be described as the ultimate short-day plant, and it therefore naturally flowers in the winter months. In the case of *K. blossfeldiana*, 'short-day' equates to a day length limited to 12½ h (see Jacobsen, 1986: 608–609 for a useful discussion on the cultivation and propagation of *K. blossfeldiana* and hybrids of which it is one parent). Artificially, introducing such short days is a widely used mechanism to stimulate plants of *Kalanchoe*, and many other unrelated species, to flower, so ensuring that flowering material can be made available for sale at any time of the year.

A range of double-flowered cultivars of *Kalanchoe blossfeldiana* with very large blooms have been produced (see Rowley, 2010, 2017: 137–139 for a brief history of this double-flowered group). In the trade, these fantastically flowered forms are known variously as Rosebud kalanchoes, Rosebud Flaming Katys, or simply as belonging to the Double Flaming Katy Group.

26. *KALANCHOE HUMILIS* BRITTEN

Zebra kalanchoe (Figs. 12.26.1 to 12.26.6)

FIG. 12.26.1 *Kalanchoe humilis* generally remains quite small and low-growing. Photograph: Gideon F. Smith.

Kalanchoe humilis **Britten** in Oliver (ed.), *Fl. Trop. Afr.* 2: 397 (1871). Hamet (1908b: 39); Hamet (1910a: 23); Berger (1930: 405); Jacobsen (1970: 285); Fernandes (1980: 366); Fernandes (1983: 56); Jacobsen (1986: 615); Wickens (1987: 55–56); Descoings (2003: 159). **Type**: Mozamb. Distr. Zambesi land [Mozambique], Moramballa [Morrumbala], *H. Waller s.n.* (K K000232904, holo-).

Synonyms:

Kalanchoe prasina N.E.Br. in *Gard. Chron.* ser. 3, 35: 211 (1904). **Type**: Nyasaland [Malawi], cultivated in Kew Gardens, *Brown s.n.* (K K000232903, holo-).

Kalanchoe figueiredoi Croizat in *Bull. Jard. Bot. État Bruxelles* 14: 366 (1937), as 'figuereidoi'; corrected to "figueiredoi" by Jacobsen (1970: 284). Raymond-Hamet & Marnier-Lapostolle (1964: 102; Plate XXXV, figure 122;

Plate XXXVI, figure 123); Raadts (1977: 149). **Type**: Cultivated plant at Jardin Botanique Bruxelles, from rhizomes originating from Metonia (Niassa, Mozambique), *Gomes e Sousa s.n.* (BR, holo-; A A00042502, K K000232905, iso-).

Derivation of the scientific name:

From the Latin *humilis* (= humble, low, and modest), for the comparatively low-growing habit of the plants.

Description:

Perennial, low-growing, glabrous, sparsely branched, tuft-like, rosettes, with small rounded canopies, succulent, and to 80 cm tall. *Stems* thin, erect, leaning, creeping or trailing, weak, purplish green, and arising from a thickened rhizomatous base. *Leaves* few to many, sparsely to densely (especially lower down) arranged along branches, erect to patent-erect, succulent, glabrous, shortly petiolate through subsessile to sessile, light green to bluish green,

FIG. 12.26.2   The leaves of *Kalanchoe humilis* can be densely purple-mottled, especially when grown in direct sunlight. Note that spotted and unspotted leaves can occur on the same plant. Photograph: Gideon F. Smith.

FIG. 12.26.3   Some forms of *Kalanchoe humilis* can develop comparatively elongated, robust stems. Photograph: Gideon F. Smith.

FIG. 12.26.4   Inflorescences of *Kalanchoe humilis* are sparsely flowered. Photograph: Gideon F. Smith.

FIG. 12.26.5   The small, bluish purple flowers of *Kalanchoe humilis* are shaped like tiny lanterns. Photograph: Gideon F. Smith.

densely and irregularly reddish purple-spotted depending on age and exposure to insolation, and bloom not waxy; *petiole* same colour as leaf blade, succulent, and absent or to 5 mm long; *blade* 2–10 × 1–3 cm, upper surface flat or convex, lower surface convex, obovate to spathulate, and succulent; *base* narrow and cuneate; *apex* ending in a rounded tooth; *margins* upper ½ irregularly toothed with small crenations, and reddish purple-infused. *Inflorescence* a diffuse, terminal, branched, few- to many-flowered head-shaped cyme, dichasial branches subtended by fleshy, carpel-shaped bracts, 10–40 cm tall, erect to leaning to one side; peduncle ± straight and light green infused with purple; *pedicels* 5–15 mm long and glabrous.

FIG. 12.26.6 *Kalanchoe humilis* plants (horizontal row in the centre) offered for sale in a nursery in Phoenix, Arizona, USA. Photograph: Gideon F. Smith.

Flowers 5–7 mm long, erect to horizontal, glabrous or very finely pubescent, subtended by small succulent bracts that are soon shed, light green (base of tube) to very light purple (upper part of tube) with a network of distinct, purple veins, and waxy bloom absent; *calyx* purple; *sepals* 1.0–1.5 mm long, free portion deltoid- to elongated-triangular, basally fused for <1 mm, acute, apically slightly curved away from the corolla tube, and glabrous or rarely very finely pubescent; *corolla* light green (base of tube) to very light purple (upper part of tube) with a network of distinct, purple veins, and waxy bloom absent; *tube* 4–6 mm long, 4-angled, basally slightly broadened, box kite-shaped above, and very slightly flared at the mouth; *lobes* ± 1.0–2.5 × 1–2 mm, erect to hardly flared, oblong, and apiculate. *Stamens* inserted at ± middle of corolla tube and included; *filaments* ± 3–4 mm long, thin, and light greenish yellow; *anthers* ± 0.75–1.00 mm long, yellowish brown, and included. *Pistil* consisting of 4 carpels; *carpels* 2.0–2.5 mm long and light shiny green; *styles* 2.5–3.0 mm long; *stigmas* very slightly capitate, brownish, and included; *scales* linear, obscurely bifid, free, and ± 2 mm long. *Follicles* not seen. *Seeds* not seen. *Chromosome number*: $n = 17$ (Raadts, 1989).

Flowering time:

June–November, peaking in August (southern hemisphere).

Illustrations:

Raymond-Hamet & Marnier-Lapostolle (1964: Plate XXXV, figure 122, Plate XXXVI, figure 123 as '*Kalanchoe figuereidoi*'); Figueiredo et al. (2017: 192).

Common names:

Afrikaans: sebraplakkie.

English: zebra kalanchoe, in reference to the often confluent horizontal bands of coloured mottling on the leaves.

Geographical distribution:

Kalanchoe humilis occurs in eastern and south-tropical Africa, where it is, as far as is known, restricted to northern Mozambique, southern Tanzania (T8 sensu Flora of Tropical East Africa; Raadts, 1977: 150; Wickens, 1987: 56, page facing p. 66), and Malawi. The species was omitted from a preliminary checklist of Mozambican plants (Silva et al., 2004), even though its type locality falls in the country.

Distribution by country. Malawi, Mozambique, and Tanzania.

Habitat:

The preferred habitat of *Kalanchoe humilis* is not well-known. Wickens (1987: 56) records it as between rocks at an altitude of 1200–1410 m.

Conservation status:

Apparently not assessed.

Additional notes and discussion:

Taxonomic history and nomenclature. Of the 18 species of *Kalanchoe* that Britten (1871) treated in volume 2 of the *Flora of Tropical Africa*, he described almost half as new. *Kalanchoe humilis* is one such species. At least one other species, *K. brachyloba*, a name that Welwitsch suggested and that Britten (1871: 392) adopted, extends to southern Africa from tropical Africa (Smith & Figueiredo, 2017a).

As far as we are aware, the first mention of *Kalanchoe humilis* following its description was by Hamet (1907, 1908b) in his two-part monograph of *Kalanchoe*. In this work, he treated *K. humilis* as a "SPECIES NON SATIS NOTAE" (Hamet, 1908b: 39), rather than including it in his Group 13, where Berger (1930: 405) later placed it. Berger included *K. humilis* in his 'Section 2 *Crenatae*', which equates to Group 13 of Hamet (1907: 879, 1908b). Also, rather than including *K. prasina*, a species described by N.E. Brown in 1904, in the synonymy of *K. humilis*, Hamet (1907: 895) regarded it as being a synonym of *K. baumii*, which in turn is today treated as a synonym of *K. brachyloba* (Fernandes, 1983: 63; Smith & Figueiredo, 2017a).

Jacobsen (1970: 284, 1986: 613) also did not make a taxonomic connection between *Kalanchoe humilis* and *K. figueiredoi*, as he treated both as 'good' species and, clearly following Hamet (1907, 1908b), included *K. prasina* in the synonymy of *K. baumii*.

Fernandes (1980: 366, 1983: 56) clarified the identity and synonymy of *Kalanchoe humilis* and included both *K. prasina* and *K. figueiredoi* in the synonymy of the former.

Descoings (2006) included *Kalanchoe humilis* in his *K.* subg. *Calophygia*, together with other species that show intermediate characters between the other two subgenera.

In the protologue of the name *Kalanchoe figueiredoi* (Croizat, 1937), the specific epithet was spelled 'figuereidoi', a misspelling of the [sur-]name 'Figueiredo'. Croizat (1937: 363) stated "Je dois à l'empressement amical de M. le Prof. A.[-ntónio de] Figuereido [sic] de Gomes e Sousa, chargé par le Gouvernement portugais de l'exploration botanique du Mozambique, d'avoir reçu vers la fin de 1934 des rhizomes frais d'une Crassulacée récoltée, malheureusement sans notes à l'appui, aux environs de Metonia, Nyassa Portugais", which translates to English as "I am indebted to the friendly eagerness of Professor A.[-ntónio de] Figuereido [sic] de Gomes e Sousa, commissioned by the Portuguese Government for the botanical exploration of Mozambique, to have received, at the end of 1934, freshly harvested rhizomes of a Crassulaceae, unfortunately without supporting notes, from the vicinity of Metonia, Portuguese Nyassa [Niassa]"]. The territory of [the] Niassa [Company] covered the northern part of Mozambique, north of the Lurio River. In 1904, the Niassa Company established and headquartered itself in the town of Porto Amélia, which is now known as Pemba. In the running text explaining the eponymy of the epithet 'figuereidoi', Croizat therefore already misspelled the name of Gomes e Sousa, an error perpetuated when he published the botanical name, as *Kalanchoe figuereidoi*, three pages later.

The *ICN* is clear in that the original spelling of an epithet is to be retained, except for the correction of typographical or orthographical errors (McNeill et al., 2012: 127; Article 60.1). The correction from *'figuereidoi'* to *'figueiredoi'* does not affect the first syllable of the specific epithet; Jacobsen (1970: 284) therefore made the change.

In this regard, the spelling of the genus name *Huernia* R.Br. is a corollary. *Huernia* commemorates Justi[n]s Heurnius [van Herne] (1587–1652), an early Dutch collector of South African plants at the Cape of Good Hope. His name was misspelled by Brown when the name was published, but the transposition of the 'u' and 'e' is to be retained in the genus name and should not be corrected to 'Heurnia', as the change will affect the first syllable of the name (McNeill et al., 2012: 127; Article 60.1).

Diagnostic characters and close allies. Although named for its low-growing habit, the epithet 'humilis' also applies to the modest size of the flowers of the species, especially in comparison with the size of the flowers of most other *Kalanchoe* species from south-tropical and eastern Africa.

In his Key, Berger (1930: 405) applied the following sequence of characters to *K. humilis*: "**B.** Blätter flach. **Bb.** Röhre der Blumenkrone kürzer (als 30 mm; sehen **B.**). **Bbα.** Blätter sitzend oder kaum gestielt, kahl. **BbαI.** Blütenstand kahl. **BbαI2*.** Röhre der Blumenkrone 4,5 mm lang, Zipfel 2,5 mm lang. Blätterspatelig, 2 cm lang, 8 mm breit, oben gezähnt". [English: **B.** Leaves flat. **Bb.** Corolla tube shorter (than 30 mm; see **B.**). **Bbα.** Leaves sessile or barely pedicellate, glabrous. **BbαI.** Inflorescence glabrous. **BbαI2*.** Corolla tube 4,5 mm long, lobes 2,5 mm long. Leaf blade spathulate, 2 cm long, 8 mm wide, serrated apically.] The characters given under "BbαI2*." were used to separate *K. humilis* from *K. baumii* Engl. & Gilg. (= *K. brachyloba*; southern and south-tropical Africa) and *K. floribunda* Tul. (= *K. adelae* Raym.-Hamet; Comoro Islands). *Kalanchoe humilis* was therefore treated as having "Corolla tube 4.5 mm long, lobes 2.5 mm long. Leaf blade spathulate, 2 cm long, 8 mm wide, serrated apically". As understood today, both *K. brachyloba* and *K. adelae* are rather remotely related to *K. humilis*, and these three species cannot be confused. Following Descoings' classification (2006), *K. brachyloba* is placed in *K.* subg. *Kalanchoe*, while *K. adelae* and *K. humilis* are in *K.* subg. *Calophygia*.

Kalanchoe figueiredoi, a synonym of *K. humilis*, was described by Croizat (1937) based on plants grown in a greenhouse. The plants were grown from rhizomes that had been sent in 1934 by A.F. Gomes e Sousa from Mozambique, without collecting data. Croizat placed his species in Berger's group in 'Section 10 *Pumilae*', together with species from Madagascar, considering it quite distinct from the other African species. It was only much later that Fernandes (1980: 367) observed that the name was a synonym of *K. humilis* as the differences that Croizat saw as characteristic of a new species were only due to the material being in cultivation.

Horticulture and cultivation. *Kalanchoe humilis* is easy to propagate from stem cuttings and also sprouts roots and rosettes from leaves placed in a sandy, well-drained medium. The leaves of *Kalanchoe humilis* are highly decorative, being variously purplish brown mottled against a light-sea-green background. A few-leaved rosette in fact looks uncannily like a species of the genus *Adromischus* (Crassulaceae). However, if grown in shady positions, the leaves will eventually lose the mottling and turn a uniform sea-green colour.

The flowers of *Kalanchoe humilis* are small and insignificant, and being greenish purple, even the colour is not very striking. Part of the allure and curiosity value of the species is indeed in the tiny flowers that upon inspection with a ×10 hand lens look uncannily like Chinese lanterns.

Plants are especially well-suited to being grown in containers. Initially, plants will form small leafy rosettes, but in time, the elongating stems and branches will dangle over the edge of a pot.

27. KALANCHOE 'MARGRIT'S MAGIC'

Red chandelier plant (Figs. 12.27.1 to 12.27.7)

FIG. 12.27.1 Material of *Kalanchoe* 'Margrit's Magic' flowering profusely in South Africa's Gauteng province. Photograph: Gideon F. Smith.

***Kalanchoe* 'Margrit's Magic' Gideon F.Sm. & Figueiredo** [in Smith et al.] in *Bradleya* 36: 229 (2018b). **Representative specimen**: SOUTH AFRICA, GAUTENG PROVINCE.—2528 (Pretoria): suburb Weavind Park in Pretoria, (–CB), 25°44′01.04″S 28°16′09.74″E, 30 September 2017, *G.F. Smith & E. Figueiredo 53* (PRU).

Derivation of the cultivar name:

Named for Ms Margrit Bischofberger ([Winterthur, Switzerland] 1942–). Ms Bischofberger has an interest in a wide range of succulents but has devoted considerable attention and resources to studying *Echeveria* DC. (Crassulaceae). She was also an early supporter of, and regularly contributes to, the International Crassulaceae Network (ICN) [see: http://www.crassulaceae.ch/de/home], Sedum Society, and numerous other succulent plant societies.

Description:

Perennial, few- to many-leaved, multibranched, glabrous or finely pubescent, tuft-forming succulent, and to 60 cm tall. *Stems* brown to reddish brown, older internodes with longitudinal light brownish or greenish stripes, woody, somewhat brittle, few, unbranched or sparsely branched, erect to leaning, sometimes creeping, rooting along the way, leaning branches developing short, near-woody stilt-like roots, nodes thickened, round, sterile, and reproductive stems smooth to finely pubescent. *Leaves* opposite-decussate, subsessile to distinctly petiolate, green to variously infused with red, succulent, lower older ones spreading to horizontal to decurved, upper younger ones ± vertical, and papery on drying; *petiole* to 20 mm long, channelled above, and not clasping the stem; *blade* 1–3 × 1–2 cm, obovate to

FIG. 12.27.2 Once plants have flowered, during and often towards the end of the dry season, leaves on desiccating stems turn a deep red colour. Photograph: Gideon F. Smith.

FIG. 12.27.4 The flowers of *Kalanchoe* 'Margrit's Magic' are variously disposed on the sparsely to densely flowered inflorescences. The flowers are a crimson red colour and, in contrast to those of 'typical' *K. manginii*, one of the postulated parents of the cultivar, almost perfectly cylindrical. Photograph: Gideon F. Smith.

FIG. 12.27.3 The finely pubescent, red-margined leaves of *Kalanchoe* 'Margrit's Magic' are orbicular to cordate. Leaf blades are virtually glabrous. The crimson red flowers of *Kalanchoe* 'Margrit's Magic' are, like the leaf margins, finely pubescent. Photograph: Gideon F. Smith.

FIG. 12.27.5 The flowers of *Kalanchoe* 'Margrit's Magic' dry to a purple-red colour. Photograph: Gideon F. Smith.

orbicular to cordate or somewhat oblong, flat, and curved upwards towards margins; *apex* rounded-obtuse; *base* cuneate; *margins* entire to weakly crenate especially in upper ⅔, and pubescent. *Inflorescence* 18–20 cm tall; a terminal, branched, erect, apically sparse to dense, few- to many-flowered, flat-topped cyme with several dichasia; rounded when viewed from above; branches opposite; erect; subtended by very small leaflike bracts; and without leafy branchlets in axils; *peduncle* bright red and minutely white-hairy; *pedicels* slender, 8–10 mm long. *Flowers* erect to pendent, usually spreading, bright crimson red (tube and lobes), light green at level of calyx, and cylindrical to campanulate; *calyx* light reddish green, strongly infused with small red spots especially towards

sepal margins; *sepals* 4, ± separate, basally fused for ± 1 mm, ± 4–6 × 3–4 mm, triangular-lanceolate, acute-tipped, hardly contrasting against light green basal part of corolla tube, and minutely white-hairy; *corolla* 18–20 mm long, slightly enlarged above the middle, not twisted apically after anthesis, bright crimson red, light green lower down, minutely white-hairy, and drying purple red; *corolla tube* 16–18 mm long, cylindrical, distinctly 4-angled, box-shaped-square when viewed from below, bright crimson red, light green lower down, and minutely white-hairy; *lobes* 4.5–5.0 × 4.5–5.0 mm, ovate to suborbicular, rounded at apex, apiculate, and bright crimson red. *Stamens* inserted at about the middle of

FIG. 12.27.6 A demobilised cocopan (small cart running on narrow-gauge railway lines) planted with *Kalanchoe* 'Margrit's Magic'. The first inflorescences are about to develop. The cocopan is flanked by specimens of *Euphorbia tirucalli* L. Photograph: Gideon F. Smith.

the corolla tube and included; *filaments* 6–7 mm long, thin, and yellow; *anthers* 0.5 mm long and purplish brown. *Pistil* consisting of 4 carpels; *carpels* 6–7 mm long and light green; *styles* 8–9 mm long; *stigmas* capitate and whitish yellow; *scales* ± 2 mm long, narrowly columnar to slightly linear, and light yellowish green. *Follicles* brittle, grass spikelet-like, enveloped in dry purplish remains of corolla, dull whitish green, and 6–7 mm long. *Seeds* 0.50–0.75 mm long and light brown. *Chromosome number*: unknown.

Flowering time:
July–October, midwinter to spring, peaking in August (southern hemisphere).

Illustrations:
Smith et al. (2018b: figures 1, 4–6, and 8).

Common names:
Afrikaans: rooikandelaar.
English: red chandelier plant.

Geographical distribution:
A cultivar recorded from cultivation in South Africa, of which the parents are very likely Madagascan *Kalanchoe* species.

Habitat:
Not applicable.

Conservation status:
Not applicable.

Additional notes and discussion:
Taxonomic history and nomenclature. Many species of *Kalanchoe* from southern Africa and Madagascar are excellent garden subjects in their own right. However, as part of horticultural efforts to make even more attractive material available to a trade that continuously demands the introduction of new, marketable novelties, literally, hundreds of *Kalanchoe* cultivars have been selected and in many instances mass-produced; many of these are internationally distributed and sold. Some of these cultivars are derived from superior forms of the pure species, while others were selected from interspecific hybrids that were deliberately created, especially using Madagascan material (Graf, 1980: 682–689, 1637–1638; Brickell, 1998: 577–578; Bryant et al., 2005: 740–741).

As far as is known, *Kalanchoe* 'Margrit's Magic' is a cultivar peculiar to South Africa. It was for the first time recorded and described by Gideon F. Smith and Estrela Figueiredo (Smith et al., 2018b), following the *International code of nomenclature for cultivated plants (ICNCP or Cultivated Plant Code)* (Brickell et al., 2016).

Possible parentage: Based on its gross vegetative and reproductive morphology, the parents of *Kalanchoe* 'Margrit's Magic' are most certainly two or more *Kalanchoe* species or cultivars. It was postulated by Smith et al. (2018b) that the parents are the Madagascan *K. manginii*

FIG. 12.27.7   *Kalanchoe* 'Margrit's Magic' planted in a container.

Raym.-Hamet & H.Perrier (Hamet & Perrier de la Bâthie, 1912) and *K. pubescens* Baker (Baker, 1887: 470).

In some respects, the appearance of *Kalanchoe* 'Margrit's Magic' is reminiscent of material widely known as *Kalanchoe* 'Tessa'; the latter is a cultivar name that is deeply entrenched in the horticulture of *Kalanchoe* (Purveur & Harbour, 1996). However, *K.* 'Tessa' is a different hybrid of which the parents are *K. manginii* and *K. gracilipes* (Baker) Baill.

Kalanchoe manginii is a variable species, and numerous selections, some of which are very similar in appearance, have been given formal cultivar names or are sold in plant nurseries and florist shops under informal names. Two other cultivars based on *K. manginii* also to some degree resemble *K.* 'Margrit's Magic', namely, *K. manginii* 'Mirabella' and *K. manginii* 'Red Bells'.

Diagnostic characters. Plants treated here as the formally described *Kalanchoe* cultivar, *Kalanchoe* 'Margrit's Magic', not only clearly have characters of 'typical' *Kalanchoe manginii* but also differ from it in several respects: the flowers are not perfectly pendent, some being borne horizontally or even vertically; the flowers are intensely crimson red (rather than reddish pink), more or less tubular, and not as campanulate as in *K. manginii*, and all the

plant parts are minutely pubescent (hardly observable without a ×10 magnifying glass) and not predominantly glabrous. *Kalanchoe manginii* is virtually entirely glabrous, while *K. pubescens* is entirely pubescent. The pubescence inherited from *K. pubescens* is especially evident on the leaf margins of *Kalanchoe* 'Margrit's Magic', while the leaf blades are virtually glabrous, as in *K. manginii*. The leaf margins of *K. manginii* are most commonly entire, while those of *K. pubescens* are crenate-dentate; those of *K.* 'Margrit's Magic' are intermediate, being weakly crenate-dentate. In inflorescence architecture, *K.* 'Margrit's Magic' tends to be closer to *K. pubescens*, also in that the inflorescence and flowers of this cultivar described from South Africa are completely minutely pubescent.

Horticulture and cultivation. *Kalanchoe* 'Margrit's Magic' is very easy in cultivation. It grows well when planted directly in the soil or in small or large containers and in hanging baskets, especially when these are kept in dappled shade but bright light. Material thrives outdoors in mild climates and also grows well indoors under comparatively low-light conditions. Horticulturally, it performs well in subtropical climates and also under near continental-type climatic conditions where temperatures drop slightly below 0°C.

28. *KALANCHOE PORPHYROCALYX* (BAKER) BAILL.

Red-cup kalanchoe (Figs. 12.28.1 to 12.28.4)

FIG. 12.28.1   *Kalanchoe porphyrocalyx* remains low-growing and thrives terrestrially or as an epiphyte. Photograph: Gideon F. Smith.

Kalanchoe porphyrocalyx **(Baker) Baill.** in *Bull. Mens. Soc. Linn. Paris* 1(59): 469 [not p. 449 as claimed by Descoings, 2003: 170] (1885). Hamet (1908a, b: 41 (57)); Hamet (1910c: 49); Raymond-Hamet & Marnier-Lapostolle (1964: 95; Plate III, Figure N; Plate XXXIV, figures 117–118); Jacobsen (1970: 288); Jacobsen (1986: 624); Rauh (1995: 116, 155); Boiteau & Allorge-Boiteau (1995: 78); Descoings (2003: 170). **Type**: Central Madagascar, October 1882, *Baron 1708* (K K000232804, holo-).

Synonyms:

Kitchingia porphyrocalyx Baker in *J. Linn. Soc., Bot.* 20: 142 (1883). **Type**: as above.

Kalanchoe sulphurea Baker in *J. Linn. Soc., Bot.* 22: 471 (1887). **Type**: Central Madagascar, received November 1885, *Baron 4180* (K K000232793, holo-).

Kitchingia sulphurea Baker in *J. Linn. Soc., Bot.* 22: 471 (1887), nom. illeg. **Type**: as for *Kalanchoe sulphurea*.

Bryophyllum porphyrocalyx (Baker) A.Berger in Engler & Prantl, *Nat. Pflanzenfam.*, ed. 2 18a: 412 (1930). **Type**: as for *Kalanchoe porphyrocalyx*.

Bryophyllum sulphureum (Baker) A.Berger in Engler & Prantl, *Nat. Pflanzenfam.*, ed. 2 18a: 412 (1930). **Type**: as for *Kalanchoe sulphurea*.

Kalanchoe porphyrocalyx var. *sulphurea* (Baker) Boiteau & Mannoni in *Cactus (Paris)* 17–18: 58 (1948). Jacobsen (1970: 288); Rauh (1995: 155). **Type**: as for *Kalanchoe sulphurea*.

Kalanchoe porphyrocalyx var. *sambiranensis* Humbert ex Boiteau & Mannoni in *Cactus (Paris)* 17–18: 58 (1948). Jacobsen (1970: 288); Jacobsen (1986: 624). **Type**: Madagascar, Bassin supérieur du Sambirano, crête près des sources du Sambirano, 1800 m alt., *Humbert 18607* (P P00374213 & P00374214).

Kalanchoe porphyrocalyx var. *typica* Boiteau & Mannoni in *Cactus (Paris)* 17–18: 58 (1948), not validly published (Art. 24.3).

Derivation of scientific name:

From the Greek *porphyreos*, purplish red, and *kalyx*, calyx, referring to the red to deep purple calyx.

Description:

Perennial, tuft-like, smooth, sparsely to much-branched shrublets, with small, untidily rounded

FIG. 12.28.2 The leaves of *Kalanchoe porphyrocalyx* are a pleasant light to midgreen colour, and the leaf margins are crenate. Photograph: Gideon F. Smith.

FIG. 12.28.3 The inflorescences of *Kalanchoe porphyrocalyx* are comparatively few-flowered. Photograph: Gideon F. Smith.

canopies, succulent, and to 40 cm tall. *Stems* thin, light green to light brownish green ± erect, sturdy, and leaf scars obvious when young. *Leaves* many, variable in shape, sparsely to densely arranged towards upper ends of branches, erect to slanted away from branches at a 45

degree angle, succulent, glabrous, petiolate to subsessile, lengthwise-folded along midrib, light green, and not waxy; *petiole* same colour as leaf blade, succulent, and 1–10 mm long; *blade* 20–50 × 5–20 mm, exceedingly variable from near-linear to near-orbicular, mostly oblong-ovate, and succulent; *base* narrow and attenuate; *apex* rounded-obtuse; *margins* irregularly toothed with coarse crenations and concolourous or faintly red-infused. *Inflorescence* a terminal, apically branched, often one- to few-flowered head-shaped corymb, 2–10 cm tall; erect; peduncle ± straight; and deep red purple; *pedicels* 5–15 mm long and glabrous. *Flowers* 10–25 mm long, pendent, glabrous or obsolescently pubescent, subtended by small succulent bracts, red to deep purple red, lobes or margins only of corolla lobes light yellowish green and waxy bloom absent; *calyx* basally fused, red to deep purple red, and occasionally light greenish purple; *sepals* 5–6 mm long, free portion lanceolate- to deltoid-triangular, fused for ± 2–3 mm, acute, curved away from the corolla tube at a 90 degree angle, and glabrous or rarely very finely pubescent; *corolla* red to deep purple red (tube), lobes or lobe margins only light yellowish green, and waxy bloom absent; *tube* 18–21 mm long, more or less campanulate to cylindrical-urceolate, and flared at the mouth; *lobes* ± 5–6 × 6–7 mm, ovate, apically obtuse-flattened with a slight indentation, and yellowish green inside. *Stamens* inserted very low-down in corolla tube at ± lower end of carpels and included; *filaments* ± 20 mm long, thin, and light yellowish green; *anthers* ± 1 mm long, black, and included. *Pistil* consisting of 4 carpels; *carpels* 7–8 mm long and light shiny green; *styles* 15–17 mm long; *stigmas* very slightly capitate, brownish, and included; *scales* rectangular, free, ± 2 × 1 mm, and distinctly indented above. *Follicles* not seen. *Seeds* not seen. *Chromosome number*: unknown.

Flowering time:
July–September (southern hemisphere).

Illustrations:
Raymond-Hamet & Marnier-Lapostolle (1964: Plate III, Figure N; Plate XXXIV, figures 117–118); Rauh (1995: 156, figure 406); Rauh (1998: 344, bottom right).

Common names:
Afrikaans: rooikoppie.
English: red-cup kalanchoe.

Geographical distribution:
Central and northern Madagascar (see Boiteau & Allorge-Boiteau, 1995: 199, Map 7).
Distribution by country. Madagascar.

Habitat:
It is treated as a montane (cloud) forest species by Rauh (1995: 147).

Conservation status:
Not known.

Additional notes and discussion:
Taxonomic history and nomenclature. In his monograph of *Kalanchoe*, Hamet (1908a, b: 41 (57)) treated *Kalanchoe*

FIG. 12.28.4   The flowers are pendent and more or less glabrous. The colour of the corolla tube varies slightly from red to a deep purple red, while the corolla lobes are usually light yellowish green. Photograph: Gideon F. Smith.

porphyrocalyx as one of the "species non satis notæ". However, two years later (Hamet, 1910c: 49), he accepted the species and placed it in his Group 9. The variation found in the species resulted in the description of another species (*K. sulphurea*) and an infraspecific taxon (*Kalanchoe porphyrocalyx* var. *sambiranensis*). However, these were eventually considered to be synonyms of *K. porphyrocalyx*.

Identity and close allies. The calyx and corolla of *Kalanchoe porphyrocalyx* that are often the same colour are distinctive features for the species. However, like virtually all species included in the genus, *K. porphyrocalyx* shows considerable variation in vegetative and reproductive morphology.

The Madagascan *Kalanchoe gracilipes* (Baker) Baill. has a very similar growth habit to that of *K. porphyrocalyx* (Rauh, 1995: 151), but its calyx and corolla are different colours.

Along with *K. uniflora*, *K. porphyrocalyx* was included in Groupe IV Epidendreae by Boiteau & Allorge-Boiteau (1995: 16).

Cultivation. This species has been recorded as sometimes being epiphytic, but it also grows terrestrially. It responds well to being grown in hanging baskets.

29. *KALANCHOE PUMILA* BAKER

Flower dust kalanchoe (Figs. 12.29.1 to 12.29.3)

FIG. 12.29.1 *Kalanchoe pumila* grown as a groundcover outside the Royal Exhibition Building and Carlton Gardens, Melbourne, Australia. Plants remain low-growing and thrive in full sun or dappled shade. Photograph: Gideon F. Smith.

Kalanchoe pumila **Baker** in *J. Linn. Soc., Bot.* 20: 139 (1883). Hamet (1908b: 25); Raymond-Hamet & Marnier-Lapostolle (1964: 21; Plate VI, figures 11–12); Jacobsen (1970: 289); Jacobsen (1986: 626); Boiteau & Allorge-Boiteau (1995: 174); Rauh (1995: 116, 157); Brickell (1998: 578); Brickell (2003: 594); Descoings (2003: 171). **Type**: Madagascar, *Baron 2117* (K K000232869 holo-; P P00431387 iso- [fragment]).

Synonyms:

Kalanchoe multiceps Baill. in *Bull. Mens. Soc. Linn. Paris* 1: 469 (1885). **Type**: Madagascar, Nord Betsileo, Sirabé, *Hildebrandt 3576* (P P00431388, holo-; P P00431389, BM BM000645878, K K000232870, iso-).

Kalanchoe brevicaulis Baker in *J. Linn. Soc., Bot.* 22: 470 (1887). **Type**: Madagascar, *Baron 3542* (BM000645886, holo-; P00431386, iso-).

Kalanchoe pumila forma *venustior* Boiteau ex L.Allorge in Boiteau & Allorge-Boiteau, Kalanchoe *Madagascar:*

175 (1995). **Type**: Antsirabe, *Herb. Jard. Boit. 2058* (P, holo-not seen, not listed in P database online).

Derivation of the scientific name:

From the Latin 'pumilus', 'small', for the low-growing stature of plants.

Description:

Perennial, many-leaved, branched, glabrous, small, epiphytic, lithophytic or terrestrial, succulent, and to 30 cm tall. *Stems* branched, erect to arched to leaning, ± round in cross-section, white-pruinose, and raised scars of abscised leaves conspicuous. *Leaves* opposite-decussate, sessile or very shortly petiolate, purplish pink, covered with a very fine dusty white waxy layer, succulent, and erect; *petiole* absent or very short and leaves not clasping the stem; *blade* 2.5–4.5 × 1.5–2.5 cm, obovate, and flat; *base* cuneate; *apex* rounded-obtuse; *margins* from midleaf coarsely crenate or undulate-crenate into rounded harmless teeth and purplish red. *Inflorescence* short, erect, few-

FIG. 12.29.2 The leaves of *Kalanchoe pumila* are sessile or very shortly petiolate and purplish pink but covered with a very fine, dusty, white-waxy layer. Photograph: Gideon F. Smith.

flowered, apically dense, corymbose, and greyish white waxy; *pedicels* 6–10 mm long and slender. *Flowers* erect, violet-pink, and purple veined; *calyx* green to strongly reddish purple-infused; *sepals* 3–5 × 1.0–2.5 mm, abruptly triangular to lanceolate, ± separate, basally fused for ± 1 mm, acute, and white-pruinose; *corolla* 6–11 mm long, campanulate, violet-pink, and purple veined; *tube* 5–10 mm long, suburceolate, violet-pink, and purple veined; *lobes* 5–10 × 3–4 mm, prominent, spreading to strongly recurved, partly obscuring the corolla tube, ovate to oblong-ovate, and mucronate. *Stamens* inserted well above the middle of the corolla tube and included to slightly exserted; *filaments* short; *anthers* 0.40–0.75 mm long, reniform, and conspicuously yellow. *Pistil* consisting of 4 carpels; *carpels* 8–11 mm long; *styles* short; *stigmas* capitate; *scales* oblong to rectangular. *Follicles* not seen. *Seeds* not seen. *Chromosome number*: 2*n* = 40 (Van Voorst & Arends, 1982: 580); 2*n* = 36 (Friedmann, 1971).

Flowering time:
August–October (southern hemisphere).

Illustrations:
Raymond-Hamet & Marnier-Lapostolle (1964: 21; Plate VI, figures 11–12); Jacobsen (1986: 626, figure 882); Rauh (1995: 92, figure 222, 157–158,

figures 413–416); Boiteau & Allorge-Boiteau (1995: 173); Bryant et al. (2005: 741, top).

Common names:
Afrikaans: meelblomplakkie and pienkkalanchoe.
English: flower dust kalanchoe.

Geographical distribution:
The species occurs in mountainous parts of the central plateau in the central-east Madagascar (see Map 18 in Boiteau & Allorge-Boiteau, 1995; Rauh, 1995). It is widely cultivated as a container plant, but has not become naturalised.

Distribution by country. Madagascar.

Habitat:
Kalanchoe pumila grows as a lithophyte and epiphyte or terrestrially in its mountainous habitat in central-east Madagascar.

Conservation status:
Not applicable.

Additional notes and discussion:
Taxonomic history and nomenclature. Kalanchoe pumila was described in 1883 by Baker, based on a specimen (*Baron 2117*), the only one cited, collected in Central Madagascar by Reverend Richard Baron (1847–1907), an English botanist and missionary. Interestingly, in the

FIG. 12.29.3   *Kalanchoe pumila* sold as a hanging basket subject in the Garden Shop of the Missouri Botanical Garden, St. Louis, USA. Photograph: Gideon F. Smith.

description, no mention is made of the white, floury wax that so prominently covers virtually all parts of the plant.

Four years later, in 1887, Baker published *Kalanchoe brevicaulis*, one of the synonyms of *K. pumila*. He also noted that *K. multiceps*, which was described by Baillon in 1885, two years earlier therefore, belongs in the synonymy of *K. pumila* and that his newly described *K. brevicaulis* is "…near…" *K. pumila*. *Kalanchoe brevicaulis* and *K. multiceps* were soon placed in the synonymy of *K. pumila* (Hamet, 1908b: 25). This interpretation has been followed since.

A forma, *Kalanchoe pumila* forma *venustior*, was described in Boiteau & Allorge-Boiteau (1995: 173), but it was based on plants that existed only in cultivation, and the authors admitted that it could be a result of a mutation that persisted through vegetative propagation.

Identity and close allies. *Kalanchoe pumila* is easy to identify given its low-growing stature, purplish pink leaves covered in a white, floury coating and clusters of large, erectly borne to spreading, purplish pink flowers.

Boiteau & Allorge-Boiteau (1995: 16) included *Kalanchoe pumila* in Group XI Alpestres. The other Madagascan species, *K. rhombopilosa* Mannoni & Boiteau, included in the Group is not as well-known in cultivation in South Africa.

Cultivation. The species is very easy in cultivation. It works well as a groundcover in mild, subtropical, or Mediterranean climates. It also grows well in hanging baskets. Plants sometimes grow as lithophytes and prefer well-drained substrates (Rauh, 1995: 81). When watering the plant care should be taken not to wash off the white, waxy coating, which much adds to the charm of the species in cultivation.

Kalanchoe pumila has been used as one parent in producing some of the strikingly beautiful hybrids with *K. blossfeldiana* that are sold in florist shops and plant nurseries.

In cultivation, *Kalanchoe pumila* can be shy to flower. However, with the white, floury coating on its leaves, the species is very striking vegetatively, and it is mostly not grown for its flowers.

30. KALANCHOE TOMENTOSA BAKER

Panda plant (Figs. 12.30.1 to 12.30.10)

FIG. 12.30.1 *Kalanchoe tomentosa* is easy to identify given the conspicuous hairs that cover all the plant parts. Photograph: Gideon F. Smith.

Kalanchoe tomentosa **Baker** in *J. Bot. (British & Foreign)* 11: 110 (1882). Hamet (1908b: 31); Raymond-Hamet & Marnier-Lapostolle (1964: 37; Plate XI, figures 29–30, Plate XII, figure 31); Jacobsen (1970: 290); Jacobsen (1986: 631); Boiteau & Allorge-Boiteau (1995: 166); Rauh (1995: 92, 116, 219–220); Brickell (1998: 578); Brickell (2003: 594); Bryant et al. (2005: 741); Descoings (2003: 178). **Type**: Madagascar, Central Madagascar, chiefly in Betsileo-land, recd. July 1880, *Baron 247* (K without barcode, holo-; P P00374199, iso-).

Nomenclature note. Hamet (1908b: 31) stated "[R. Baron, no 3560.—Echantillon authentique!]" and also cited two further specimens, "R. Baron, n° 247]" and "[D. Cowan!]". However, *R. Baron 247* is the only collection cited by Baker (1882: 110) in the protologue of the name. The specimen *Baron 3560* (K K000232863) from Central Madagascar, received in November 1885, is data-based and labelled as type at K, but it is not part of the original material as it was received on a later date. It cannot take precedence over *R. Baron 247*, which is the type collection.

Derivation of the scientific name:

From the Latin 'tomentosus', 'hairy', for the hairy leaves and generally hairy or felted appearance of the plants.

Description:

Perennial, many-leaved, sparsely branched, entirely hairy, medium-sized, terrestrial, succulent, and to 1 m tall. *Stem* branched, basally somewhat woody and covered in longitudinally flaking, yellowish bark, erect to

FIG. 12.30.2   *Kalanchoe tomentosa* grown as a potplant near Serra de Santo António in central Portugal. Photograph: Gideon F. Smith.

FIG. 12.30.3   Old stems of *Kalanchoe tomentosa* are covered in a flaky, yellowish bark. Note the sprouting leaves that accumulated on the ground around the stem. Photograph: Gideon F. Smith.

FIG. 12.30.4   This form of *Kalanchoe tomentosa* has more or less oval leaves. Photograph: Gideon F. Smith.

arched to leaning, ± round in cross-section, and scars of abscised leaves obscure. *Leaves* alternate, often rosulate, sessile, dull light green to dull bluish green, distinctly succulent, and erect; *petiole* absent and leaves not clasping the stem; *blade* 2.5–8.0 × 1.0–2.5 cm, obovate, ovate, oblong, or subcylindrical, concave above, and convex below; *base* cuneate; *apex* rounded-obtuse; *margins* often somewhat upcurved, entire, from ± midleaf coarsely crenate with rounded, and harmless light- to dark brown teeth; all plant parts, especially the leaves, covered with hairs; *hairs* fine to coarse; erectly spreading; and white,

FIG. 12.30.5   A form of *Kalanchoe tomentosa* with white silvery hairs and somewhat more elongated leaves. Photograph: Gideon F. Smith.

FIG. 12.30.7   A form of *Kalanchoe tomentosa* of which all the plant parts are covered by very long, prominently visible hairs. It is horticulturally known as *K. tomentosa* 'Long Hairs' or as *K. tomentosa* 'Super Fuzzy'. Photograph: Gideon F. Smith.

FIG. 12.30.6   This form of *Kalanchoe tomentosa* has somewhat darker brown leaves and similarly darker notches on the leaf margins and is sometimes seen for sale as *K. tomentosa* 'Chocolate Lips'. Photograph: Gideon F. Smith.

FIG. 12.30.8   Detached leaves of *Kalanchoe tomentosa* readily sprout roots and small plantlets. Photograph: Estrela Figueiredo.

brown, or reddish brown. *Inflorescence* ± 70 cm tall, erect to leaning, few-flowered, apically dense, and corymbose; *pedicels* 5–10 mm long and slender. *Flowers* erect to spreading, yellowish brown to greenish, often somewhat purplish, and hairy; *calyx* yellowish brown to greenish, often somewhat purplish, and hairy; *sepals* 3–5 × 1.0–2.5 mm, ± deltoid, ± separate, basally fused for ± 1 mm, and obtuse; *corolla* 8–13 mm long, campanulate to very slightly urceolate, yellowish brown to greenish, and sometimes somewhat purplish especially on the inside of the flower mouth; *tube* 9–12 mm long and campanulate to very slightly urceolate; *lobes* 3–4 × 4–6 mm, prominent, deltoid to round, erectly spreading, and mucronate. *Stamens* inserted at ± the middle of the corolla tube and included; *filaments* short and filiform;

anthers ± 1 mm long, yellow, ovate, and visible inside the mouth. *Pistil* consisting of 4 carpels; *carpels* 7–11 mm long; *styles* short; *stigmas* capitate; *scales* rectangular. *Follicles* not seen. *Seeds* not seen. *Chromosome number*: $2n = 34$ (Baldwin, 1938: 576).

Flowering time:

Variable, including in winter in the southern hemisphere.

Illustrations:

Raymond-Hamet & Marnier-Lapostolle (1964: 37; Plate XI, figures 29–30, Plate XII, figure 31); Jacobsen (1970: Plate 106,1); Jacobsen (1986: 361, figure 414; 631, figure 894); Sajeva & Costanzo (1994: 154); Rauh (1995: 92, figures 223, 219–220, figures 609–617); Boiteau & Allorge-Boiteau (1995: 165); Bryant et al. (2005: 720–721).

FIG. 12.30.9   *Kalanchoe tomentosa* (left) and *K. luciae* (large, red-infused leaves) grown on a windowsill in Santa Luzia, near Tavira, Algarve, southern Portugal.   Photograph: Gideon F. Smith.

FIG. 12.30.10   This form of *Kalanchoe tomentosa* is reminiscent of the cultivar *K. tomentosa* 'Caramel'. It is here sold in the Garden Shop of the Missouri Botanical Garden, St Louis, USA. Photograph: Gideon F. Smith.

Common names:

Afrikaans: harige-plakkie and harige-kalanchoe.

English: hairy kalanchoe, panda plant, and Widow's thrill.

Geographical distribution:

The species occurs in mountainous parts in east-central Madagascar (see Map 19 in Boiteau & Allorge-Boiteau,

1995; Rauh, 1995). Given its heavily felted leaves, it is popular and widely cultivated in southern Africa and beyond as a novelty container plant, but has not become naturalised.

Distribution by country. Madagascar.

Habitat:

Kalanchoe tomentosa is "…found on granite rocks, above all on the inselbergs" in east-central Madagascar.

Conservation status:

Not applicable.

Additional notes and discussion:

Taxonomic history and nomenclature. Baker (1882) described *Kalanchoe tomentosa* based on a collection (*Baron 247*), the only one cited, collected in central Madagascar by Reverend Richard Baron (1847–1907), an English botanist and missionary. In the description, Baker (1882) makes prominent mention of the "short brown spreading hairs" with which the plants are "densely coated throughout", so unambiguously establishing this distinctive character of the species.

Unlike most species of *Kalanchoe*, including such distinctive ones such as *K. pumila*, not a single synonym applies to *K. tomentosa*.

Identity and close allies. *Kalanchoe tomentosa* is easy to identify given the hairs that cover all the plant parts. The leaves of *K. tomentosa* are arranged alternately, rather than opposite-decussate as in the majority of *Kalanchoe* species.

Boiteau & Allorge-Boiteau (1995: 16) placed *Kalanchoe tomentosa* in their Group X Lanigerae, the group in which *K. beharensis* was also included.

Cultivation. The species is very easy in cultivation. It does best in sunny, dry positions; if grown in places with high humidity, the branches tend to form clusters of adventitious roots, which some cultivators find unsightly. Plants will tolerate considerable overwatering, as long as the growth medium is well-drained.

Plants grow well from stem and branch cuttings, and where they fall, detached leaves will easily strike root and develop quick-growing plantlets.

Numerous forms of this polymorphic species are available in horticulture, the most desirable of which are either completely covered in very long hairs or have pronounced brown 'teeth' on the margins of the upper parts of the leaves. Cultivated plants are sometimes seen under the designation 'Kalanchoe pilosa Hort.' (Jacobsen, 1986: 631).

Kalanchoe tomentosa does very well as a subject for container gardening; containers should be filled with a well-drained soil mixture.

In cultivation, the species can be shy to flower. However, the species is very striking, even in the vegetative phase and not primarily grown for its flowers.

Acknowledgements

In southern Africa, many species of *Kalanchoe* remain well camouflaged in their natural habitats, such as here near Thabazimbi in South Africa's North-West province where *Kalanchoe paniculata* only really becomes noticeable when its 1 m-tall inflorescence that carries hundreds of small, yellow flowers rise above the grass layer in the bushveld. Photograph: Gideon F. Smith.

We are grateful to Margrit Bischofberger (Switzerland), Christo Botha (South Africa), Kate Braun (Swaziland), Adam Braziel (USA), Dr António Coutinho (Portugal), Dr Tanza Crouch (South Africa), Callie de Wet (South Africa), Tony Dold (South Africa), Adri du Preez (South Africa), Anne-Lise Fourie (South Africa), Arnold Frisby (South Africa), Dr Lorenzo Gallo (Italy), Prof. Norbert Hahn (South Africa), Martin Heigan (South Africa), Dr Mark Mort (USA), Roy Mottram (UK), Alec Naidoo (South Africa), Solomon Nkoana (South Africa), Delia Oosthuizen (South Africa), Jason Sampson (South Africa), Ray Stephenson (UK), Betsie Steyn (South Africa), and Steve Woodhall (South Africa) for useful discussions on kalanchoes and the family Crassulaceae in general and/or for providing us with photographs. Particular thanks are extended to Prof. Neil R. Crouch (South Africa) for the use of his photographs and freely sharing material of southern African kalanchoes, Dr Hester Steyn (South Africa) for kindly preparing the maps, and Jaco van Wyk (South Africa) for graphics support. Dr Colin C. Walker is thanked for kindly writing the Foreword to the book.

Bibliography

Adams, J., 1976. *Wild flowers of the Northern Cape.* The Department of Nature and Environmental Conservation of the Provincial Administration of the Cape of Good Hope, Cape Town.

Adanson, M., 1763. *Familles des Plantes.* Seconde Partie. Etat actuel de la Botanike. Chez Vincent, Imprimeur-Libraire de M^gt le Comte de Provence, rue S. Severin, Paris. [see: https://www.biodiversitylibrary.org/item/6958#page/794/mode/1up].

Adendorff, R., n.d. *Flower terms and common flower names/Blomterme en algemene blomname.* Self-published, Ladysmith.

Agrawal, A.A., Petschenka, G., Bingham, R.A., Weber, M.G. & Rasmann, S., 2012. Toxic cardenolides: chemical ecology and coevolution of specialized plant–herbivore interactions. *New Phytologist* 194, 28–45.

Aida, R. & Shibata, M., 2002. High frequency of polyploidization in regenerated plants of *Kalanchoe blossfeldiana* cultivar 'Tetra Vulcan'. *Plant Biotechnology* 19, 329–334.

Akinsulire, O.R., Aibin, I.E., Adenipekun, T., Adelowotan, T. & Odugbemi, T., 2007. In vitro antimicrobial activity of crude extracts from plants *Bryophyllum pinnatum* and *Kalanchoe crenata. African Journal of Traditional, Complementary and Alternative Medicines* 4, 338–344.

Akulova-Barlow, Z., 2009. Kalanchoe: beginner's delight, collector's dream. *Cactus and Succulent Journal (US)* 81, 268–276.

Albuquerque, S., Brummitt, R.K. & Figueiredo, E., 2009. Typification of names based on the Angolan collections of Friedrich Welwitsch. *Taxon* 58, 641–646.

Allorge-Boiteau, L., 1996. Madagascar centre de spéciation et d'origine du genre *Kalanchoe* (Crassulaceae). In: W.R. Lourenço (Ed.), *Biogéographie de Madagascar, Actes du Colloque International Biogéographie de Madagascar.* Société de Biogéographie – Muséum – Orstom. Paris (France) du 26 au 28 septembre 1995. Orstom Éditions, Paris, pp. 137–146.

Anderson, L.A., Schultz, R.A., Joubert, J.P., Prozesky, L., Kellerman, T.S., Erasmus, G.L. & Procos, J., 1983. Krimpsiekte and acute cardiac glycoside poisoning in sheep caused by bufadienolides from the plant *Kalanchoe lanceolata* [Forssk.] Pers. *The Onderstepoort Journal of Veterinary Research* 50, 295–300.

Anderson, L.A., Steyn, P.S. & Van Heerden, F.R., 1984. The characterization of two novel bufadienolides, lanceotoxins A and B from *Kalanchoe lanceolata* [Forssk.] Pers. *Journal of the Chemical Society, Perkin Transactions* 1, 1573–1575.

Andrews, H.[C.], 1798. *The botanist's repository, for new, and rare plants. Containing coloured figures of such plants, as have not hitherto appeared in any similar publication; with all their essential characters, botanically arranged, after the sexual system of the celebrated Linnaeus; in English and Latin. To each description is added, a short history of the plant, as to its time of flowering, culture, native place of growth, when introduced, and by whom. The whole executed by Henry Andrews, author of the coloured engravings of heaths, in folio* Printed by T. Bensley, and published by the author, No. 5, Knightsbridge. To be had of J. White, Fleet-street, and all the Booksellers, London. [see: https://www.biodiversitylibrary.org/item/111072#page/94/mode/1up].

Anonymous, 1948. One of the old school. [Biographical note on Mr Pieter Koch]. *Farming in South Africa* 23 (1), 2.

Archetti, M., Döring, T.F., Hagen, S.B., Hughes, N.M., Leather, S.R., Lee, D.W., Lev-Yadun, S., Manetas, Y., Ougham, H.J., Schaberg, P.G. & Thomas, H., 2009. Unravelling the evolution of

autumn colours: an interdisciplinary approach. *Trends in Ecology & Evolution* 24, 166–173.

Arnold, M. (Ed.), 2001. *South African botanical art.* Fernwood Press, Vlaeberg, in association with Art Link (Pty) Ltd, Saxonwold.

Astle, W.L., Phri, P.S.M. & Prince, S.D., 1997a. Annotated checklist of the flowering plants and ferns of the South Luangwa National Park, Zambia. *Kirkia* 16 (2), 109–160.

Astle, W.L., Phiri, P.S.M. & Prince, S.D., 1997b. A dictionary of the flowering plants and ferns of the South Luangwa National Park, Zambia. *Kirkia* 16 (2), 161–203.

Baillon, M.H., 1885 [séance du 18 Février, 1885]. Liste des plantes de Madagascar (suite de la page 464). [*Kalanchoe* § *Kitchingia*]. *Bulletin Mensuel de la Société Linnéenne de Paris (Paris)* 1 (59), 465–472. [see: https://www.biodiversitylibrary.org/item/41447#page/60/mode/1up].

Baker, E.G., 1899. Rhodesian Polypetalae. *The Journal of Botany (British and Foreign)* 37 (442), 422–438.

Baker, H.G. & Baker, I., 1983. Floral nectar sugar constituents in relation to pollinator type. In: C.E. Jones & R.J. Little (Eds), *Handbook of experimental pollination biology.* Van Nostrand Reinhold, New York, pp. 117–141.

Baker, J.G., 1881 [21 February 1881]. Notes on a collection of flowering plants made by L. Kitching, Esq, in Madagascar in 1879 (Plates VII. & VIII.). *Journal of the Linnean Society, Botany* 18, 264–281.

Baker, J.G., 1882. Contributions to the flora of central Madagascar. (continued from p. 70.) *The Journal of Botany (British and Foreign)* 11 (New Series; volume 20 of entire series), 109–114 (to be continued). [see: https://www.biodiversitylibrary.org/item/35873#page/126/mode/1up].

Baker, J.G., 1883 [1884; published 24 March 1883]. Contributions to the Flora of Madagascar.—Part I. Polypetalæ [Plates XXII & XXIII]. *Journal of the Linnean Society, Botany (London)* 20, 87–158. [see: https://www.biodiversitylibrary.org/item/8375#page/94/mode/1up].

Baker, J.G., 1887. Further contributions to the Flora of Madagascar. *Journal of the Linnean Society, Botany (London)* 22, 441–537. [see: https://www.biodiversitylibrary.org/item/8378#page/448/mode/1up].

Baker, J.G., 1895. Diagnoses africanae IX. *Bulletin of Miscellaneous Information (Royal Botanic Gardens, Kew)* 1895 (107), 288–293.

Baldwin, J.T. Jr., 1938. *Kalanchoe*: the genus and its chromosomes. *American Journal of Botany* 25, 572–579. [see: http://www.jstor.org/stable/2436516].

Balfour, I.B., 1888. Botany of Socotra. *Transactions of the Royal Society of Edinburgh* 31, 1–446.

Baluška, F., Mancuso, S. & Volkman, D. (Eds), 2006. *Communication in plants: neuronal aspects of plant life.* Springer, Berlin.

Bandeira, S., Bolnick, D. & Barbosa, F., 2007. *Flores nativas do Sul de Moçambique/Wild flowers of southern Mozambique.* Universidade do Eduardo Mondlane, Departamento de Ciências Biológicas, Maputo, Mozambique.

Bargel, H., Koch, K., Cerman, Z. & Neinhuis, C., 2006. Evans Review No. 3. Structure-function relationships of the plant cuticle and cuticular waxes—a smart material? *Functional Plant Biology* 33, 893–910.

Barkhuizen, B.P., 1978. *Succulents of southern Africa with specific reference to the succulent families found in the Republic of South Africa and South West Africa.* Purnell & Sons Publishers, Cape Town.

Barnes, J.E. & Turton, L.M. (Comps) [with additions and corrections to April 1990 made by L.M. Turton and E. Kalake.], 1994. *A list of the flowering plants of Botswana in the herbaria at the National Museum, Sebele and University of Botswana 1986.* Revised edition. The Botswana Society, Gaborone, and The National Museum, Monuments and Art Gallery, Gaborone.

Barthlott, W. & Neinhuis, C., 1997. Purity of the sacred lotus or escape from contamination in biological surfaces. *Planta* 202, 1–7.

Barthlott, W., Neinhuis, C., Cutler, D., Ditsch, F., Meusel, I., Theisen, I. & Wilhelmi, H., 1998. Classification and terminology of plant epicuticular waxes. *Botanical Journal of the Linnean Society* 126, 237–260.

Batten, A. & Bokelmann, H., 1966. *Wild flowers of the Eastern Cape Province.* Books of Africa (Pty) Ltd, Cape Town.

Bawa, K.S., 1983. Patterns of flowering in tropical plants. In: C.E. Jones & R.J. Little (Eds), *Handbook of experimental pollination biology.* Van Nostrand Reinhold, New York, pp. 394–410.

Beentje, H., 2010. *The Kew plant glossary. An illustrated dictionary of plant terms.* Kew Publishing, Royal Botanic Gardens, Kew.

Bell, A.D., 2008. *Plant form: an illustrated guide to flowering plant morphology.* Timber Press, Portland.

Berger, A., 1930. Crassulaceae. Unterfam. II. Kalanchoideae. 6. *Kalanchoe* In: A. Engler & K. Prantl (Eds), *Die Naturlichen Pflanzenfamilien.* 2nd Edition, 18a, 402–412. Verlag von Wilhelm Engelmann, Leipzig.

Bews, J.W., 1921. *An introduction to the flora of Natal and Zululand.* City Printing Works, Pietermaritzburg.

Bischofberger, M., 2015. *Kalanchoe* 'Vivien'. *Sedum Society Newsletter* 112, 40–43.

Black, C.C. & Osmond, C.B., 2003. Crassulacean acid metabolism photosynthesis: 'working the night shift'. *Photosynthesis Research* 76, 329–341.

Blackmore, S. & Tootill, E. (Eds), 1984. *The Penguin dictionary of botany.* Penguin Books, London.

Blanco, F.M., 1845. *Flora de Filipinas: segun el sistema sexual de Linneo.* Edition: 2. impresion, corr. y aum. D. Miguel Sanchez, Manila. [see: https://www.biodiversitylibrary.org/item/102888#page/281/mode/1up].

Blossfeld, R., 1934. *Kalanchoe Blossfeldiana*, V. Poellnitz Sp. N. *The Cactus Journal (Great Britain)* 3 (2), 36–37.

Boiteau, P., 1947. Les plantes grasses de Madagascar. *Cactus* 12, 5–10.

Boiteau, P. & Allorge-Boiteau, L., 1995. Kalanchoe *(Crassulacées) de Madagascar. Systématique, écophysiologie et phytochimie.* ICSN, CNRS, 91198 Gif-sur-Yvette, Éditions Karthala, Paris.

Boiteau, P. & Mannoni, O., 1948. Les *Kalanchoe* (suite). *Cactus (Paris)* 13, 7–10. figures 1–6; 14, 23–28, 6 figures; 15–16: 37–42, 5 figures; 17–18: 57–58, 2 figures.

Boiteau, P. & Mannoni, O., 1949. Les *Kalanchoe* (suite). *Cactus (Paris)* 19, 9–14. 7 figures; 20, 43–46, 5 figures; 21: 69–76, 6 figures; 22: 113–114, 2 figures.

Bolnick, D., 1995. *A guide to the common wild flowers of Zambia and neighbouring regions.* MacMillan Education Ltd, London.

Bonner, W. & Bonner, J., 1948. The role of carbon dioxide in acid formation by succulent plants. *American Journal of Botany* 35, 113–117.

Borges, P.A.V., Cunha, R., Gabriel, R., Martins, A.F., Silva, L. & Vieira, V., 2005. *Listagem da fauna e flora (Mollusca e Arthropoda) (Bryophyta, Pteridophyta e Spermatophyta) terrestres dos Açores Horta, Angra do Heroismo & Ponta Delgada.* Intermezzo, Lisboa.

Botha, C.[J.], 2016. Potential health risks posed by plant-derived cumulative neurotoxic bufadienolides in South Africa. *Molecules* 21 (3), 348. [see: https://doi.org/10.3390/molecules21030348].

Botha, C.J., 2013. Krimpsiekte in South Africa: historical perspectives. *Journal of the South African Veterinary Association* 84, 1–5.

Botha, C.J. & Penrith, M.L., 2009. Potential plant poisonings in dogs and cats in southern Africa. *Journal of the South African Veterinary Association* 80, 63–74.

Bourgaud, F., Gravot, A., Milesi, S. & Gontier, E., 2001. Production of plant secondary metabolites: a historical perspective. *Plant Science* 161, 839–851.

Braun, K.P., Dlamini, S.D., Mdlala, D.R., Methule, N.P., Dlamini, P.W. & Dlamini, M.S. (Comp.), 2004. *Swaziland flora checklist. Swaziland National Herbarium & National Botanical Institute, South Africa.* Southern African Botanical Diversity Network Report No. 27, i–x, 1–113. Southern African Botanical Diversity Network (SABONET), Pretoria.

Breedlove, D.E., 1986. Flora de Chiapas. *Listados Florísticos México* 4 (i–v), 1–246.

Brickell, C. (Ed.), 1990. *The Royal Horticultural Society gardeners' encyclopedia of plants and flowers.* Dorling Kindersley, London.

Brickell, C. (Editor-in-Chief), 1998. *The Royal Horticultural Society A–Z Encyclopedia of garden plants.* Corrected reprint. Dorling Kindersley, London.

Brickell, C. (Editor-in-Chief), 2003. *The Royal Horticultural Society A–Z Encyclopedia of garden plants.* Volume 2 K–Z. Revised edition. Dorling Kindersley, London.

Brickell C.D., Alexander, C., Cubey, J.J., David, J.C., Hoffman, M.H.A., Leslie, A.C., Malécot, V. & Xiaobai, Jin, 2016 [June]. *International code of nomenclature for cultivated plants (ICNCP or Cultivated Plant Code), incorporating the Rules and Recommendations for naming plants in cultivation adopted by the International Union of Biological Sciences and the International Commission for the Nomenclature of Cultivated Plants.* 9th Edition. International Society for Horticultural Science (ISHS), Leuven, Belgium. *Scripta Horticulturae* Number 18.

Britten, J., 1871. Order L. Crassulaceae. In: D. Oliver (Ed.), *Flora of Tropical Africa (Leguminosae to Ficoideae)* 2, 385–401. L. Reeve & Co., London. [see: https://www.biodiversitylibrary.org/item/60248#page/397/mode/1up].

Brown, N.E., 1902. *Kalanchoe diversa* N.E. Brown (*n. sp.*). *The Gardeners' chronicle* 32, 210.

Brown, N.E., 1904. New or noteworthy plants. [*Kalanchoe prasina*, N.E. Brown (n. sp.)†]. *The Gardeners' Chronicle* Series 3 [No. 901; Saturday, April 02, 1904] 35, 211.

Brown, N.E., 1909. XIV The flora of Ngamiland. *Bulletin of Miscellaneous Information Royal Botanic Gardens, Kew* 1909, 81–146.

Brown, N.E., 1912. 1393. *Kalanchoe ellacombei* N.E. Br. In: Diagnoses Africanae L. *Bulletin of Miscellaneous Information (Royal Botanic Gardens, Kew)* 1912 (7), 329.

Brown, N.E., 1913. 1436. *Kalanchoe sexangularis. Bulletin of Miscellaneous Information, Kew* 1913 (3), 120–121.

Bruce, E.A., 1948. *Kalanchoe marmorata.* Eritrea, Abyssinia, Somalia. Crassulaceae. *The Flowering Plants of Africa* 27, plate 1049.

Bruce, E.A., 1949. *Kalanchoe brachycalyx.* Abyssinia. Crassulaceae. *The Flowering Plants of Africa* 27, plate 1052.

Bruce, E.A., 1950. *Kalanchoe densiflora.* Sudan, Uganda, Kenya. *The Flowering Plants of Africa* 28, plate 1089.

Brummitt, R.K. & Powell, C.E., 1992. *Authors of plant names. A list of authors of scientific names of plants, with recommended standard forms of their names, including abbreviations.* Royal Botanic Gardens, Kew.

Bryant, G., Bryant, K., Rutherford, B. & Vogan, R., 2005. Cacti & Succulents. *Kalanchoe.* In: L. Barnard, A. Edwards, K. Etherington, D. Imwold, E. King, M. Malone & J. Parker (Eds), *The ultimate plant book.* New Holland, London.

Bullock, A.A., 1952. South African poisonous plants. *Kew Bulletin* 7 (1), 117–129.

Burgoyne, P.M., 2003. *Kalanchoe.* In: G. Germishuisen & N.L. Meyer (Eds), *Plants of southern Africa: an annotated checklist.* Strelitzia 14, 407–408. National Botanical Institute, Pretoria.

Burgoyne, P.M., 2006. Crassulaceae. In: G. Germishuisen, N.L. Meyer, Y. Steenkamp & M. Keith (Eds), *A checklist of South African plants.* SABONET Report No. 41, 336–369. SABONET, Pretoria.

Burkill, H.M., 1985. *The useful plants of West Tropical Africa.* 2nd edition. Volume 1. Families A–D. Royal Botanic Gardens, Kew, Richmond.

Burtt Davy, J., 1926. *A manual of the flowering plants and ferns of the Transvaal with Swaziland.* Longmans, Green & Co, London.

Burtt-Davy, J. & Pott-Leendertz, R., 1912. A first check-list of the flowering plants and ferns of the Transvaal and Swaziland. Reprinted from the *Annals of the Transvaal Museum* May 1912, 119–182.

Byalt, V.V., 2000. *Bryophyllum calcicola* (H.Perrier) V.V. Byalt. *Novosti Sistematiki Vysshikh Rastenii* 32, 51.

Byalt, V.V., 2008. New combinations in the genera *Bryophyllum* and *Kalanchoë* (Crassulaceae). *Botanicheskii Zhurnal* 93 (3), 461–465. (in Russian).

Čapek, K., 1931. *The gardener's year* (with illustrations by J. Čapek). George Allen & Unwin, London [Original Czech edition published in 1929. Republished in English in 2002 by Modern Library, New York, in the Modern Library Gardening Edition series, with an Introduction to the series by M. Pollan, and an Introduction to the book by V. Klinkenberg].

Castaneda-Ovando, A., Lourdes Pacheco-Hernández, M., de Páez-Hernández, M.E., Rodríguez, J.A. & Galán-Vidal, C.A., 2009. Chemical studies of anthocyanins: a review. *Food chemistry* 113, 859–871.

Chalker-Scott, L., 1999. Environmental significance of anthocyanins in plant stress responses. *Photochemistry and Photobiology* 70, 1–9.

Chernetskyy, M.A., 2012. The role of morpho-anatomical traits of the leaves in the taxonomy of Kalanchoideae Berg. subfamily (Crassulaceae DC.). *Modern Phytomorphology* 1, 15–18.

Chiovenda, E., 1916. *Resultati Scientifici della missione Stefanini-Paoli nella Somalia Italiana.* Volume I. Le collezioni botaniche. Galletti e Cocci, Firenze.

Christoffels, J., 2016. *Kalanchoe sexangularis* N.E.Br. PlantZAfrica: Plants of the Week, South African National Biodiversity Institute. [see: http://pza.sanbi.org/kalanchoe-sexangularis Accessed 30 May 2018].

Clarke, C.B., 1878 [July 1878]. Order LIII. Crassulaceae. 4. *Kalanchoe,* Adans. In: J.D. Hooker, *Flora of British India* [Sabiaceae to Cornaceae]. 2 (5), 414–416. L. Reeve & Co., 6 Henrietta Street, Covent Garden, London. [see: https://www.biodiversitylibrary.org/item/13815#page/419/mode/1up].

Clarke, H. & Charters, M., 2016. *The illustrated dictionary of southern African plant names.* Jacana Media (Pty) Ltd, Auckland Park, Johannesburg, South Africa.

Cole, D.T., 1995. *Setswana—animals and plants (Setswana—Ditshedi le ditlhare).* The Botswana Society, Gaborone, Botswana.

Collenette, S., 1985. *An illustrated guide to the flowers of Saudi Arabia.* Meteorology and Environmental Protection Administration, Kingdom of Saudi Arabia. Flora Publication No. 1. Scorpion Publishing Ltd, Buckhurst Hill, Essex.

Compton, R.H., 1966. An annotated checklist of the flora of Swaziland. *Journal of South African Botany Supplementary Volume* No. VI, 1–191.

Compton, R.H., 1967. Plantae novae Africanae. Series XXXII. *Journal of South African Botany* 33, 293–304.

Compton, R.H., 1975. Plantae novae Africanae. "Ex Africa semper aliquid novi"—Pliny. Series XXXIII. *Journal of South African Botany* 41, 47–50. [see: https://archive.org/stream/journalofsouthaf41unse#page/46/mode/2up].

Compton, R.H., 1976. The Flora of Swaziland. *Journal of South African Botany Supplementary Volume* No. XI, 1–684.

Condy, G. & Rourke, J.[P.], 2001 Concise dictionary of South African botanical artists. In: M. Arnold (Ed.), *South African botanical art.* Fernwood Press, Vlaeberg, in association with Art Link (Pty) Ltd, Saxonwold, pp. 185–207.

Conte, C.A., 2004. *Highland sanctuary: environmental history in Tanzania's Usambara Mountains.* Ohio University Press, Athens, USA.

Constantin, M.M. & Poisson, H., 1908. *Katafa, Geaya et Macrocalyx,* trois plantes nouvelles se Madagascar. *Comptes rendus hebdomadaires des séances de l'Académie des sciences* 147, 635–637.

Court, D., 2010. *Succulent flora of southern Africa. Revised edition.* Struik Nature, an imprint of Random House Struik (Pty) Ltd, Cape Town.

Coutinho, A. Pereira, 2007. Rosette Batarda Fernandes. Published online at: https://www.uc.pt/herbario_digital/history/rosette.

Craven, P., Maggs-Kölling, G., Mannheimer, C., Austaller, S., Bartsch, S., Klaassen, E., Uiras, M. & Kolberg, H. (Comps & Eds), 1999. *A checklist of Namibian plant species.* Southern African Botanical Diversity Network Report No. 7. Southern African Botanical Diversity Network [SABONET]. Windhoek, Namibia.

Croizat, L., 1937. Une nouvelle espèce *Kalanchoe* du Mozambique. *Bulletin du Jardin botanique de l'État a Bruxelles* 14 (4), 363–366.

Crouch, N.R., Figueiredo, E. & Smith, G.F., 2016a. Occurrence of the little-known *Kalanchoe leblanciae* Raym.-Hamet (Crassulaceae) confirmed in South Africa. *Bradleya* 34, 70–76.

Crouch, N.R. & Smith, G.F., 2007. Crassulaceae. *Bryophyllum proliferum* naturalised in KwaZulu-Natal, South Africa. *Bothalia* 37, 206–208.

Crouch, N.R. & Smith, G.F., 2009. *Kalanchoe crenata* subsp. *crenatas.* Crassulaceae. Southern, eastern and tropical Africa. *Flowering Plants of Africa* 61, 62–68.

Crouch, N.R., Smith, G.F., Klopper, R.R., Figueiredo, E., McMurtry, D. & Burns, S., 2015. Winter-flowering maculate aloes from the Lowveld of southeastern Africa: notes on *Aloe monteiroae* Baker (Asphodelaceae: Alooideae), the earliest name for *Aloe parvibracteata* Schönland. *Bradleya* 33, 147–155.

Crouch, N.R., Smith, G.F., Walters, M. & Figueiredo, E., 2016b. *Kalanchoe winteri* Gideon F.Sm., N.R.Crouch & Mich.Walters (Crassulaceae), a new species from the Wolkberg Centre of Endemism, South Africa. *Bradleya* 34, 217–224.

Cufodontis, G., 1957. Erster Versuch Einer entwirrung des Komplexes "*Kalanchoë laciniata* (L.) DC." *Bulletin du Jardin botanique de l'État a Bruxelles* 27 (4), 709–718.

Cufodontis, G., 1958. Systematische Bearbeitung der in Süd-Äthiopien gesammelten Pflanzen. *Senckenbergiana Biologica* 39 (1/2), 103–126.

Cufodontis, G., 1965. The species of *Kalanchoe* occurring in Ethiopia and Somalia Republic. *Webbia* 19 (2), 711–744.

Cufodontis, G., 1969. Über *Kalanchoë integra* (Med.) O. Kuntze und ihre Beziehung zu *K. crenata* (Andr.) Haworth. *Österreichische Botanische Zeitschrift* 116, 312–320.

Currey, C. & Erwin, J., 2010. Variation among *Kalanchoe* species in their flowering responses to photoperiod and short-day cycle number. *The Journal of Horticultural Science and Biotechnology* 85, 350–354.

Currey, C.J. & Erwin, J.E., 2011. Photoperiodic flower induction of several *Kalanchoe* species and ornamental characteristics of the flowering species. *HortScience* 46, 35–39.

Curson, H.H., 1928. Some little known South African poisonous plants and their effects on stock. *Report on Veterinary Research, Union of South Africa* 13 & 14, 205–229.

Dalzell, N.A., 1852. Contributions to the botany of Western India. *Hooker's journal of botany and Kew Garden miscellany* 4, 341–347.

De Candolle, A.P., 1802 [June–July 1802]. *Plantarum historia succulentarum. Histoire des plantes grasses.* Volume 2. Part 17. Paris. [*Plantes grasses* de P.J. Redouté peintre du Muséum National D'histoire naturelle, Décrites par A.P. Decandolle, Membre de la Société des Sciences Naturelles Genève, etc. Livraison. Prix, 3o francs la Livraison.—Il en paraîtra une aque mois. A. Paris, Chez Garnery, Libraire, rue de Seine, ancien Hôtel Mirabeau, Ant. Aug. Renouard, Libraire, rue S. André-des-Arcs, n°. 42. A Paris et à Strasbourg, chez les frères Levrault, Libraires. AN X.] [see: https://www.biodiversitylibrary.org/item/9909#page/220/mode/1up].

De Candolle, A.P., 1828 [mid-March 1828]. *Prodromus systematis naturalis regni vegetabilis, sive enumeratio contracta ordinum, generum, specierumque plantarum huc usque cognitarum, juxta methodi naturalis normas digesta; Auctore Aug. Pyramo de Candolle. Pars Tertia sistens Calyciflorarum Ordines XXVI.* [X. *Kalanchoe* Adans, pp. 394–395; XI. *Bryophyllum* Salisb. 395–396]. M. DCCC. XXVIII. Parisiis, Sumptibus Sociorum Treuttel et Würtz, rue de Bourbon, n° 17. Venitque in

Eorumdem bibliopoliis Argentorati et Londini. [see: https://www.biodiversitylibrary.org/item/7152#page/406/mode/1up].

De Jong, T.J., Klinkhamer, P.G.L. & Van Staalduinen, M.J., 1992. The consequences of pollination biology for selection of mass or extended blooming. *Functional Ecology* 6, 606–615.

De Wildeman, E., 1913. Decades novarum specierum florae katangensis, XII–XIV. *Repertorium specierum novarum regni vegetabilis* 12, 289–298.

De Wildeman, E., 1921. Notes sur la flore du katanga. *Annales de la Société scientifique de Bruxelles* 40, 69–128.

Descoings, B., 2003. Kalanchoe. In: U. Eggli (Ed.), *Illustrated handbook of succulent plants: Crassulaceae*. Springer Verlag, Berlin, pp. 143–181.

Descoings, B., 2006. Le genre *Kalanchoe* structure et définition. *Journal de Botanique de la Société Botanique de France* 33, 3–28.

Desmond, R., 1994. *Dictionary of British and Irish Botanists and horticulturists*. CRC Press, Florida.

Dietrich, D.N.F., 1840. *Synopsis plantarum seu enumeratio systematica plantarum plerumque adhuc cognitarum cum differentiis specificis et synonymis selectis ad modum Persoonii elaborata, auctore Dr. David Dietrich. Sectio Secunda. Classis V–X.* Sumtibus et typis Bernh. Frieder. Voigtii, Vimariae.

Dimo, T., Fotio, A.L., Nguelefack, T.B., Asongalem, E.A. & Kamtchouing, P., 2006. Antiinflammatory activity of leaf extracts of *Kalanchoe crenata* Andr. *Indian Journal of Pharmacology* 38, 115–119.

Dinter, K., 1909. *Deutsch-Südwest-Afrika; Flora, Forst- und landwirtschaftliche Fragmente.* Theodor Oswald Weigel, Leipzig.

Dinter, [M.]K., 1922. LXIX. Index der aus Deutsch-Südwestafrika bis zum Jahre 1917 bekannt gewordenen Pflanzenarten. XII. *Repertorium specierum novarum regni vegetabilis. Centralblatt für Sammlung und Veröffentlichung von Einzeldiagnosen neuer Pflanzen,* 18, 423–444.

Dinter, [M.]K., 1923a. XXVII. Beiträge zur Flora von Südwestafrika. I. *Repertorium specierum novarum regni vegetabilis* 19, 122–160.

Dinter, K., 1923b. Sukkulentenforschung in Südwestafrika. Erlebnisse und Ergebnisse meiner Reise im Jahre 1922. *Repertorium specierum novarum regni vegetabilis Beihefte* 23, 1–80.

Dold, A.P. & Cocks, M.L., 1999. Preliminary list of Xhosa plant names from Eastern Cape, South Africa. *Bothalia* 29 (2), 267–292.

Drake del Castillo, E., 1903. Note sur les plantes recueillies par M. Guillaume Grandidier dans le sud de Madagascar, en 1898 et 1901. *Bulletin du Muséum d'Histoire Naturelle* 9, 35–46.

Dreyer, L.L. (Comp.), 1997. Crassulaceae DC. (Stonecrop or "Plakkie" family). In: G.F. Smith, E.J. van Jaarsveld, T.H. Arnold, F.E. Steffens, R.D. Dixon & J.A. Retief (Eds), *List of southern African succulent plants*. Umdaus Press, Pretoria, pp. 56–63.

Dyer, R.A., 1942. Hoofstuk IV. Die aankweek van vetplante. In: I.C. Verdoorn (Ed.), *'n Inleiding tot plantkunde en tot enige Transvaalse veldblomme*. J.L. van Schaik, Bpk, Pretoria, pp. 131–140.

Dyer, R.A., 1947. *Kalanchoe paniculata*. Rhodesia, Transvaal, Orange Free State. *The Flowering Plants of Africa* 26, plate 1007.

Dyer, R.A., 1979. *Kalanchoe robusta*. Socotra. Crassulaceae. *The Flowering Plants of Africa* 45, plate 1783.

Ecklon, C.F. & Zeyher, C.L.P., 1837. *Enumeratio plantarum Africae australis extratropicae quae collectae, determinatae et expositae a Christiano Friederico Ecklon & Carolo Zeyher. Pars III.* Sumtibus auctorum. Prostat apud Perthes & Besser, Hamburgi. [see: https://www.biodiversitylibrary.org/item/152#page/164/mode/1up].

Edmonds, H. & Marloth, R. [revised by Bretland Farmer, J.], 1909. *Elementary botany for South Africa. Theoretical and practical.* New edition (1903), new impression. Longmans, Green and Co., London, [revised by Bretland Farmer, J.].

Edwards, D., 1967. *A plant ecological survey of the Tugela River basin.* Natal Town and Regional Planning Reports 10 / Botanical Survey of South Africa Memoir No. 36. Town and Regional Planning Commission, Natal, city where published not stated. [Published in at least two different versions. One other version was published by the Department of Agricultural Technical Services, Botanical Research Institute, city where published not stated.]

Edwards, S., 1976. *Some wild flowering plants of Ethiopia: an introduction.* Addis Ababa Press, place of publication not stated.

Eggli, U., 1987. Neue Taxa der Gattung *Rosularia* (De Candolle) Stapf (Crassulaceae—Sedoideae). *Kakteen und andere Sukkulenten* 38, 134–138.

Eggli, U., 1988. A monographic study of the genus *Rosularia* (Crassulaceae). *Bradleya* 6 (Supplement), 1–119.

Eggli, U., 1992. Nomenclatural notes on two genera of Crassulaceae and a new combination. *Bradleya* 10, 83–84.

Eggli, U. (Comp.), 1993. *Glossary of botanical terms with special reference to succulent plants, including German equivalents,* British Cactus & Succulent Society, Richmond, Surrey, pp. 1–109.

Eggli, U. (Ed.), 2003. *Illustrated handbook of succulent plants: Crassulaceae.* Springer-Verlag, Berlin.

Eggli, U. & Newton, L.E., 2004. *Etymological dictionary of succulent plant names.* Springer-Verlag, Berlin.

Eggli, U., 't Hart, H. & Nyffeler, R., 1995. Toward a consensus classification of the *Crassulaceae*. In: H. 't Hart & U. Eggli (Eds), *Evolution and systematics of the Crassulaceae*. Backhuys Publishers, Leiden, pp. 173–192.

Ehle, M., Patel, C. & Giugliano, R.P., 2011. Digoxin: clinical highlights: a review of digoxin and its use in contemporary medicine. *Critical Pathways in Cardiology* 10, 93–98.

Engelmann, W., 1960. Endogene Rhythmik und photoperiodische Blühinduktion bei *Kalanchoe. Planta* 55, 496–511.

Engler, A., 1892. *Über die Hochgebirgsflora des tropischen Afrika.* Verlag der Königl. Akademie der Wissenschaften in Commission bei G. Reimer, Berlin.

Engler, A., 1895. *Die Pflanzenwelt Ost-Afrikas und der Nachbargebiete.* Theil C. D. Reimer, Berlin.

Engler, A., 1901. Berichte über die botanischen Ergebnisse der Nyassa-See- und Kinga-Gebirgs-Expedition der Harmann- und Elise- geb. Heckmann-Wentzel-Stiftung. IV. *Botanische Jahrbücher fur Systematik, Pflanzengeschichte und Pflanzengeographie* 30, 289–445.

Engler, A., 1902. Araceae, Liliaceae, Moraceae, Hydnoraceae etc. in Harar, territorio Galla et in Somalia a DD. Robecchi-Bricchetti et D. Riva lectae. *Annuario del Reale Istituto Botanico di Roma* 9, 243–256.

Engler, A., 1906. Beiträge zur Kenntniss der Pflanzenformationen von Transvaal und Rhodesia. *Sitzungsberichte der Königlich Preussischen Akademie der Wissenschaften* 1906, 866–906.

Engler, A. & Diels, L., 1907. Beiträge zur Flora von Afrika. XXX. Crassulaceae africanae. *Botanische Jahrbücher für Systematik Pflanzengeschichte und Pflanzengeographie* 39, 462–468.

Evert, R.F., 2006. *Esau's plant anatomy. Meristems, cells, and tissues of the plant body: their structure, function and development.* 3rd edition. John Wiley & Sons, Hoboken (NY).

Farmer, E.E., 2014. *Leaf defence.* Oxford University Press, Oxford.

Federle, W., Maschwitz, U., Fiala, B., Riederer, M. & Hölldobler, B., 1997. Slippery ant-plants and skilful climbers: selection and protection of specific ant partners by epicuticular wax blooms in *Macaranga* (Euphorbiaceae). *Oecologia* 112, 217–224.

Fernandes, R.B., 1978. Crassulaceae africanae novae vel minus cognitae. *Boletim da Sociedade Broteriana* 2ª série 52, 165–220.

Fernandes, R.B., 1980. Notes sur quelques espèces du genre *Kalanchoe* Adans. *Boletim da Sociedade Broteriana* 2ª série 53, 325–442.

Fernandes, R.B., 1982a. Fam. 70. Crassulaceae. In: R.B. Fernandes & E.J. Mendes (Eds), *Conspectus florae angolensis*. Instituto de Investigação Científica Tropical & Junta de Investigações Científicas do Ultramar, Lisboa, pp. 1–39.

Fernandes, R.B., 1982b. Crassulaceae africanae novae vel minus cognita—II. *Boletim da Sociedade Broteriana, 2ª série* 55, 95–116.

Fernandes, R.[B.], 1983. 67. Crassulaceae. In: E. Launert (Ed.), *Flora zambesiaca* 7 (1). Managing Committee on behalf of the contributors to Flora zambesiaca, London, pp. 3–71.

Figueiredo, E., Soares, M., Seibert, G., Smith, G.F. & Faden, R.B., 2009. The botany of the Cunene–Zambezi Expedition with notes on Hugo Baum (1867–1950). *Bothalia* 39, 185–211.

Figueiredo, E. & Smith, G.F., 2008. *Plants of Angola/Plantas de Angola.* Strelitzia 22. South African National Biodiversity Institute, Pretoria, pp. i–vi; 1–282.

Figueiredo, E. & Smith, G.F., 2010. What's in a name: epithets in *Aloe* L. (Asphodelaceae) and what to call the next new species. *Bradleya* 28, 79–102.

Figueiredo, E. & Smith, G.F., 2012. *Common names of Angolan plants.* Inhlaba Books, Pretoria.

Figueiredo, E. & Smith, G.F., 2015. Types to the rescue as technology taxes taxonomists, or the New Disappearance. *Taxon* 64, 1017–1020.

Figueiredo, E. & Smith, G.F., 2017a. *Common names of Angolan Plants.* 2nd edition. Protea Book House, Pretoria.

Figueiredo, E. & Smith, G.F., 2017b. Notes on the discovery and type of *Kalanchoe rotundifolia* (Haw.) Haw. (Crassulaceae). *Bradleya* 35, 106–112.

Figueiredo, E. & Smith, G.F., 2017c. (56) Request for a binding decision on the descriptive statement associated with *Kalanchoe delagoensis* (Crassulaceae). *Taxon* 66, 771.

Figueiredo, E. & Smith, G.F., 2018. Typification of the name *Bryophyllum prolifera* Bowie ex Hook., basionym of *Kalanchoe prolifera* (Bowie ex Hook.) Raym.-Hamet (Crassulaceae). *Bradleya* 36, 22–24.

Figueiredo, E., Smith, G.F. & Bingre do Amaral, P., 2017. Notes on António de Figueiredo Gomes e Sousa, a near-forgotten collector of succulent plants in Mozambique. *Bradleya* 35, 186–194.

Figueiredo, E., Smith, G.F. & Crouch, N.R., 2016. The taxonomy and type of *Kalanchoe sexangularis* N.E.Br. (Crassulaceae). *Bradleya* 34, 92–99.

Figueiredo, E., Smith, G.F. & Nyffeler, R., 2013. August Wulfhorst (1861–1936) and his overlooked contributions on the flora of Angola. *Candollea* 68, 123–131.

Fondjo, F.A., Kamgang, R., Oyono, J.L. & Yonkeu, J.N., 2012. Anti-dyslipidemic and antioxidant potentials of methanol extract of *Kalanchoe crenata* whole plant in streptozotocin-induced diabetic nephropathy in rats *Tropical Journal of Pharmaceutical Research* 11, 767–775.

Forbes, H., 1941. Newly described species and new combinations. *Bothalia* 4 (1), 37–39, 1 plate.

Forbes, V.S. (Ed.), 1986. *Carl Peter Thunberg: Travels at the Cape of Good Hope, 1772–1775.* VRS Second Series No. 17. Van Riebeeck Society, Cape Town.

Forsskål, P., 1775. *Flora Ægyptiaco-Arabica. Sive descriptiones plantarum, quas per Ægyptum inferiorem et Arabiam felicem detexit* […]. Ex officina Mölleri, aulæ Typographi. Prostat apud Heineck et Faber, Hauniæ. [see: https://www.biodiversitylibrary.org/item/29348#page/257/mode/1up].

Foxcroft, L.C., Richardson, D.M. & Wilson, J.R.U., 2008. Ornamental plants as invasive aliens: problems and solutions in Kruger National Park, South Africa. *Environmental Management* 41, 32–51. https://doi.org/10.1007/s00267-007-9027-9.

Foxcroft, L.C., Van Wilgen, N.J., Baard, J.A. & Cole, N.S., 2017. Biological invasions in South African National Parks. *Bothalia* 47 (2) https://doi.org/10.4102/abc.v47i2.2158. pp. 1–12.

Friedmann, F., 1971. Sur de nouveaux nombres chromosomiques dans le genre *Kalanchoë* (Crassulacees) à Madagascar. *Condollea* 26, 103–107.

Friedrich, H., 1968. Crassulaceae. In: H. Merxmüller (Ed.), *Prodromus einer Flora von Südwestafrika* 52. Verlag von J. Cramer, Lehre.

Fu, Kunjun (Fu, Kun-tsun) & Gilbert, M.G., 2001 [June]. 2. *Bryophyllum* Salisbury, Parad. Lond. t. 3. 1805. *Flora of China* 8. Print edition copyright Missouri Botanical Garden Press, St. Louis, USA, and Science Press, Beijing, China, the Flora of China. Published online by the Flora of China Project, pp. 1–204. [see: http://www.efloras.org/florataxon.aspx?flora_id=2&taxon_id=104753. Accessed 9 February 2018].

Fu, Kunjun (Fu, Kun-tsun), Gilbert, M.G. & Ohba, H., 2001 [June] 3. *Kalanchoe* Adanson, Fam. Pl. 2: 248. 1763. *Flora of China* 8. Print

edition copyright Missouri Botanical Garden Press, St. Louis, USA, and Science Press, Beijing, China, the Flora of China. Published online by the Flora of China Project, pp. 204–205. [see: http://www.efloras.org/florataxon.aspx?flora_id=2&taxon_id=116917. Accessed 9 February 2018].

Gehrig, H., Gaußmann, O., Marx, H., Schwarzott, D. & Kluge, M., 2001. Molecular phylogeny of the genus *Kalanchoe* (Crassulaceae) inferred from nucleotide sequences of the ITS-1 and ITS-2 regions. *Plant Science* 160, 827–835.

Germishuizen, G. (text) & Fabian, A. (paintings), 1982. *Transvaal wild flowers.* MacMillan South Africa (Pty) Ltd, Johannesburg.

Germishuizen, G. (text) & Fabian, A. (paintings), 1997. *Wild flowers of northern south Africa.* Fernwood Press, Vlaeberg.

Gibson, J.M., 1978. *Wild flowers of Natal (inland region).* The Trustees of the Natal Publishing Trust Fund, Durban.

Gill, K. & Engelbrecht, A., 2012. *Wild flowers of the Magaliesberg.* Published by Kevin Gill, Sandton, [Johannesburg].

Githens, T.S., 1949. *Drug plants of Africa.* African Handbooks. Volume 8. University of Pennsylvania Press, Philadelphia.

Given, D.R., 1984. Checklist of dicotyledons naturalized in New Zealand 17. Crassulaceae, Escalloniaceae, Philadelphaceae, Grossulariaceae, Limnanthaceae. *New Zealand Journal of Botany* 22, 191–193. [see: https://doi.org/10.1080/0028825x.1984.10425250].

Gledhill, E., 1981. *Veldblomme van Oos-Kaapland.* Die Departement van Natuur- en Omgewingsbewaring van die Kaapse Provinsiale Administrasie, Kaapstad.

Gontcharova, S.B. & Gontcharov, A.A., 2007. Molecular phylogenetics of Crassulaceae. *Genes, Genomes and Genomics* 1 (1), 40–46.

González de León, S., Herrera, I. & Guevara, R., 2016. Mating system, population growth, and management scenario for *Kalanchoe pinnata* in an invaded seasonally dry tropical forest. *Ecology and Evolution* 6, 4541–4550.

Graf, A.B., 1980 [January]. *Exotica series 3. Pictorial cyclopedia of exotic plants from tropical regions.* [12,000 illustrations, 204 plants in colour, guide to care of plants indoors, horticultural colour guide, plant geography]. 10th edition. Roehrs Company Inc., E. Rutherford, N.J. 07073, USA.

Gregory, M., 1998. Crassulaceae. In: D.F. Cutler & M. Gregory (Eds), *Anatomy of the Dicotyledons.* 2nd edition. Clarendon Press, Oxford, pp. 201–220. IV. *Saxifragales (sensu Armen Takhajan 1983).*

Guillot Ortiz, D., Laguna Lumbreras, E., López-Pujol, J., Sáez, L. & Puche, C., 2014. *Kalanchoe ×houghtonii* 'Garbí'. *Bouteloua* 19, 99–128.

Guillot Ortiz, D., Laguna Lumbreras, E. & Rosselló, J.A., 2009. La familia Crassulaceae en la flora alóctona valenciana. *Monografías de la revista Bouteloua* 4, 1–106.

Gunn, M. & Codd, L.E., 1981. *Botanical exploration of southern Africa. An illustrated history of early botanical literature on the Cape flora. Biographical accounts of the leading plant collectors and their activities in southern Africa from the days of the East India Company until modern times.* A.A. Balkema, Cape Town.

Guralnick, L.J. & Gladsky, K., 2017. Crassulacean acid metabolism as a continuous trait: variability in the contribution of crassulacean acid metabolism (CAM) in populations of *Portulacaria afra Heliyon*, 3 (4). e00293.

Hadrava, J. & Miklánek, M., 2007a. The Czech houseleeks. The genus *Sempervivum* L. *Kaktusy* XXXXIII. Special 1, 7–20, 35.

Hadrava, J. & Miklánek, M., 2007b. The Czech houseleeks. The genus *Jovibarba* Opiz. *Kaktusy* XXXXIII. Special 1, 21–35.

Hahn, N., 2002. *Endemic flora of the Soutpansberg.* Master of Science thesis, School of Botany and Zoology, Faculty of Science and Agriculture, University of Natal [now University of KwaZulu-Natal], Pietermaritzburg.

Hahn, N., 2017. Endemic flora of the Soutpansberg, Blouberg and Makgabeng. *South African Journal of Botany* 113, 324–336.

Hallé, F., Oldeman, R.A. & Tomlinson, P.B., 1978. *Tropical trees and forests: an architectural analysis.* Springer, Berlin.

Hamet, R., 1906. Note sur une nouvelle espèce de *Kalanchoe*. *Journal de Botanique* 20, 109–111.

Hamet, R., 1907. Monographie du genre *Kalanchoe. Bulletin de l'Herbier Boissier, série 2* 7, 869–900.

Hamet, R., 1908a. *Kalanchoe luciae* sp. nov. *Bulletin de l'Herbier Boissier, série 2* 8, 254–257.

Hamet, R., 1908b. Monographie du genre *Kalanchoe.* (Suite et fin). *Bulletin de l'Herbier Boissier, série 2* 8, 17–48.

Hamet, R., 1910a. Sur quelques *Kalanchoe* peu connus. *Bulletin de la Société Botanique de France, Quatrième series* 57, 18–24.

Hamet, R., 1910b. *Kalanchoe aliciae* sp. nova et *K. beharensis* Drake del Castillo. *Bulletin de la Société Botanique de France* 57, 191–194.

Hamet, R., 1910c. Sur quelques *Kalanchoe* peu connus (Suite). *Bulletin de la Société Botanique de France, Quatrième series* 57, 49–54.

Hamet, R., 1912a. Sur un nouveau *Kalanchoe* de la baie de Delagoa. *Repertorium specierum novarum regni vegetabilis* 11, 292–294.

Hamet, R., 1912b. Observations sur le *Kalanchoe tubiflora* nom. nov. *Beihefte zum botanischen Centralblatt* 29, 41–44.

Hamet, R., 1915. Sur un *Kalanchoe* Nouveau de l'Herbier de l'Albany Museum de Grahamstown [*Kalanchoe rogersii* Raym.-Hamet]. *Records of the Albany Museum* 3, 127–129.

Hamet, R. & Perrier de la Bâthie, J.M.H.A., 1912. Contribution à l'étude des Crassulacées malgaches. *Annales des sciences naturelles. Botanique, sér. 9*, 16, 361–377. [see: https://www.biodiversitylibrary.org/item/24564#page/770/mode/1up].

Hansen, A. & Sunding, P., 1993. Flora of Macaronesia. Checklist of vascular plants. 4th revised edition. *Sommerfeltia* 17, 1–297.

Harborne, J.B., 1994. *Introduction to ecological biochemistry.* 4th edition. Academic Press, London.

Harding, W., 1992. *Saxifrages. A gardener's guide to the genus.* The Alpine Garden Society, Avonbank, Pershore, Worcestershire.

Hardy, D. (text) & Fabian, A. (paintings), 1992. *Succulents of the Transvaal.* Southern Book Publishers (Pty) Ltd, Halfway House.

Hargreaves, B.J., 1990. *The succulents of Botswana. An Annotated Checklist.* National Museums, Monuments and Art Gallery Botswana, place of publication not stated [likely Gaborone].

Hargreaves, D. & Hargreaves, B., 1972. *African blossoms.* Hargreaves Company, Inc, Kailua, Hawaii.

Harley, R., 1991. The greasy pole syndrome. In: C.R. Huxley & D.F. Cutler (Eds), *Ant-plant interactions.* Oxford University Press, Oxford, pp. 430–433.

Harvey, W., 1838. *The genera of South African plants: arranged according to the natural system.* A.S. Robertson, Cape Town, pp. 1–429.

Harvey, W.H., 1862. Order LIII. Crassulaceae, D.C. VIII. *Kalanchoe,* Adans. In: W.H. Harvey & O.W. Sonder, *Flora capensis* [being a systematic description of the plants of the Cape Colony, Caffraria, & Port Natal] 2, L. Reeve & Co., Ltd, Kent, pp. 378–380.

Haworth, A.H., 1812. *Synopsis plantarum succulentarum, cum descriptionibus, synonymis, locis; observationibus anglicanis, culturaque.* Typis Richardi Taylor et Socii, Shoe-Lane, London.

Haworth, A.H., 1819. *Supplementum plantarum succulentarum, sistens plantas novas vel nuper introductas, sive omissas in* Synopse plantarum succulentarum: *cum observationibus variis anglicanis.* Impensis J. Harding, St James's Street, Londini.

Haworth, A.H., 1821. *Revisiones plantarum succulentarum.* In Ædibus R. et A. Taylor, Londoni. [*Kalanchoe* on pp. 23–24].

Haworth, A.H., 1824. XXXI. Decas novarum Plantarum Succulentarum. ("Mr Haworth's Decade of new Succulent Plants."). *The Philosophical Magazine and Journal* 64, 184–191.

Haworth, A.H., 1825 (06 July 1825). III. Decas quarta novarum Plantarum Succulentarum ("Mr Haworth's *Fourth Decade of new Succulent Plants.*"). *The Philosophical Magazine and Journal* 66, 27–33.

Haworth, A.H., 1829. XLVII. A new account of the genus *Kalanchoë. The Philosophical Magazine and Annals* 6, 301–305.

Heath, P.V., 1997. Notes on crassulaceous names (part 1). *Calyx* 5 (4), 129–132.

Henkel, J.S., 1934. *A field book of the woody plants of Natal and Zululand, being a key for the identification of woody plants based on leaf characters.*

Natal University Development Fund Committee, Durban & Pietermaritzburg.

Henning, M.W., 1926. Krimpsiekte. *Report on Veterinary Research, Union of South Africa* 12, 331–365.

Herrera, I. & Nassar, J.M., 2009. Reproductive and recruitment traits as indicators of the invasive potential of *Kalanchoe daigremontiana* (Crassulaceae) and *Stapelia gigantea* (Apocynaceae) in a Neotropical arid zone. *Journal of Arid Environments* 73, 978–986.

Heyes, J.A., 1989. Crassulacean acid metabolism in Zimbabwean succulents. *Excelsa* 14, 14–20.

Heyneman, A.J., 1983. Optimal sugar concentrations of floral nectars—dependence on sugar intake efficiency and foraging costs. *Oecologia* 60, 198–213.

Hiern, W.P., 1896. *Catalogue of the African Plants collected by Dr Friedrich Welwitsch in 1853–61* […] 1.1. Printed by order of the Trustees [of the British Museum], London, pp. 1–326.

Hooker, W.J., 1859. *Bryophyllum proliferum.* Proliferous Bryophyllum. Nat. Ord. Crassulaceae.—Octandria Monogynia. *Curtis's Botanical Magazine* 85, Tab. 5147. [see: https://www.biodiversitylibrary.org/item/14225#page/175/mode/1up].

Hooker, W.J., 1899. *Kalanchoe thyrsiflora.* Native of South Africa. Nat. Ord. Crassulaceae. *Curtis's Botanical Magazine ser. 3,* 55 (125 of the whole series), Tab. 7678. [see: https://www.biodiversitylibrary.org/item/14253#page/189/mode/1up].

Horvath, B., 2014. *The plant lover's guide to sedums.* Timber Press, Portland, Oregon, and London.

Hulme, M., 1954. *Wild flowers of Natal.* Shuter & Shooter, Pietermaritzburg.

Human, H. & Nicolson, S.W., 2008. Flower structure and nectar availability in *Aloe greatheadii* var. *davyana:* an evaluation of a winter nectar source for honeybees. *International Journal of Plant Sciences* 169, 263–269.

Humbert, H., 1933. *Kalanchoe* (Crassulacées) nouveaux ou peu connus de Madagascar. *Bulletin du Muséum National d'Histoire Naturelle, Ser. II.* 5, 163–170.

Hutchings, A., 1996. *Zulu medicinal plants: an inventory.* University of Natal Press, Pietermaritzburg.

Hutchings, A., Scott, A.H., Lewis, G. & Cunningham, A.B., 1996. *Zulu medicinal plants. An inventory.* University of Natal Press, Pietermaritzburg.

Hutchinson, J., 1946. *A botanist in southern Africa.* P.R. Gawthorn Ltd, London.

Hutchinson, J. & Dalziel, J.M., 1954. *Flora of West Tropical Africa.* 2nd edition, revised by R.W.J. Keay. Part 1. Volume 1. Crown Agents for Overseas Governments and Administrations, London.

Ihlenfeldt, H.-D., 1985. Lebensformen und Überlebensstrategien bei Sukkulenten. *Berichte der Deutschen Botanischen Gesellschaft* 98, 409–423.

Ingram, D.S., Vince-Prue, D., Gregory, P.J. (Eds), 2016. *Science and the garden: the scientific basis of horticultural practice.* 3rd edition. John Wiley & Sons, in association with the Royal Horticultural Society, Chichester (UK).

Iqbal, S.M., Jamil, Q., Jamil, N., Kashif, M., Mustafa, R. & Jabeen, Q., 2016. Antioxidant, antibacterial and gut modulating activities of *Kalanchoe laciniata. Acta Poloniae Pharmaceutica* 73, 1221–1227.

Irwin, R.E. & Brody, A.K., 1999. Nectar-robbing bumble bees reduce the fitness of *Ipomopsis aggregata* (Polemoniaceae). *Ecology* 80, 1703–1712.

Jacobsen, H., 1970. *Lexicon of succulent plants. Short descriptions, habitats and synonymy of succulent plants other than Cactaceae.* 2nd edition. Blandford Press, Poole, Dorset.

Jacobsen, H., 1986. *A handbook of succulent plants. Descriptions, synonyms and cultural details for succulents other than Cactaceae. Volume II. Ficus to Zygophyllum.* Blandford Press, Poole, Dorset.

Jacot Guillarmod, A.[F.M.G.], 1971. *Flora of Lesotho (Basutoland).* Verlag von J. Cramer, Lehre.

Jacquin, N.J., 1804. *Plantarum rariorum Horti Caesarei Schoenbrunnensis [...] Volume IV*. C. F. Wappler, Prostant Viennae; B. et J. White, Londini; S. et J. Luchtmans. Lugduni Batavorum.

Jadin, F. & Juillet, A., 1912. Recherches anatomiques sur trois espèces de *Kalanchoe* de Madagascar donnant des résines parfumées dans leurs écorces. *Annales du Musée Colonial de Marseille* 10, 137–156.

Jeppe, B., 1975. *Natal wild flowers*. Purnell (S.A.) (Pty) Ltd, Cape Town.

Johnson, A.T. & Smith, H.A., 1982. *Plant names simplified*. 2nd revised edition, sixth impression, fourth reprint. Landsmans Bookshop Ltd, Buckenhill, Bromyard, Herefordshire.

Johnson, M.A., 1934. The origin of the foliar pseudo-bulbils in *Kalanchoe daigremontiana*. *Bulletin of the Torrey Botanical Club* 61, 355–366.

Jürgens, N., 1995. Chapter 8. Contributions to the phytogeography of *Crassula*. In: H. 't Hart & U. Eggli (Eds), *Evolution and systematics of the Crassulaceae*. Backhuys Publishers, Leiden, pp. 136–172.

Kamboj, A., Rathour, A. & Kaur, M., 2013. Bufadienolides and their medicinal utility: a review. *International Journal of Pharmacy and Pharmaceutical Sciences* 5, 20–27.

Kamgang, R., Mboumi, R.Y., Fondjo, A.F., Tagne, M.A.F., N'dillé, G.P.R. M. & Yonkeu, J.N., 2008. Antihyperglycaemic potential of the water–ethanol extract of *Kalanchoe crenata* (Crassulaceae). *Journal of Natural Medicines* 62, 34–40.

Kapitany, A., 2007. *Australian succulent plants. An introduction*. Kapitany Concepts, Boronia, Victoria, Australia.

Kato, J. & Mii, M., 2012. Production of interspecific hybrids in ornamental plants. In: V. Loyola-Vargas & N. Ochoa-Alejo (Eds), *Plant cell culture protocols. Methods in molecular biology (methods and protocols)*. 877, Humana Press, Totowa (NJ), pp. 233–245.

Keay, R.W.J., 1954. Crassulaceae. In: *Flora of West Tropical Africa*. 2nd edition. Volume 1 (1), pp. 114–119.

Keissler, C., 1900. Crassulaceae. In: A. Zahlbruckner, Plantae Pentherianae. Pars I. *Annalen des K.K. Naturhistorischen Hofmuseums* 15, 36–41.

Kellerman, T.S., 2009. Poisonous plants. *Onderstepoort Journal of Veterinary Research* 76, 19–23.

Kellerman, T.S., Coetzer, J.A.W., Naude, T.W. & Botha, C.J., 2005. *Plant poisonings and mycotoxicoses of livestock in southern Africa*, 2nd edition. Oxford University Press, Cape Town.

Kellerman, T.S., Naudé, T.W. & Fourie, N., 1996. The distribution, diagnoses and estimated economic impact of plant poisonings and mycotoxicoses in South Africa. *The Onderstepoort Journal of Veterinary Research* 63, 65–90.

Kemp, E.S., 1981 [June 1981]. *Additions and name changes for the Flora of Swaziland*. The Swaziland National Trust Commission, place of publication not stated.

Kemp, E., 1983. *A flora checklist for Swaziland*. Occasional Paper No. 2. The Swaziland National Trust Commission, Lobamba.

Kerr, L.D., Gravatt, D.A. & Wiggers, R.J., 2018. The effects of ultraviolet light on anthocyanin accumulation in the adventitious roots of *Sedum wrightii* (Crassulaceae). *Annals of Biological Sciences* 6, 1–7.

Killick, D. [illustrated by Rosemary Holcroft], 1990. *A field guide to the flora of the Natal Drakensberg*. Jonathon Ball and Ad Donker Publishers, Johannesburg.

Killick, D.J.B. & Du Plessis, E., n.d., Index to volumes 1–49. / Indeks tot volumes 1–49. *The Flowering Plants of Africa / Die Blomplante van Afrika*. Botanical Research Institute, Department of Agriculture and Water Supply, place of publication not stated, likely Pretoria.

Kirby, G., 2013. *Wild flowers of Southeast Botswana*. Struik Nature, Cape Town.

Kluge, M. & Brulfert, J., 1996. Crassulacean acid metabolism in the genus *Kalanchoë*: ecological, physiological and biochemical aspects. In: K. Winter & J.A.C. Smith, (Eds), *Crassulacean acid metabolism: biochemistry, ecophysiology and evolution*. (Ecological Studies 114). Springer, Berlin, pp. 324–335.

Kluge, M., Brulfert, J., Ravelomanana, D., Lipp, J. & Ziegler, H., 1991. Crassulacean acid metabolism in *Kalanchoë* species selected at various climatic zones of Madagascar: a survey by $\delta^{13}C$ analysis. *Oecologia* 88, 407–414.

Kluge, M., Brulfert, J., Ravelomanana, D., Lipp, J. & Ziegler, H., 1993. A comparative study by $\delta^{13}C$-analysis of crassulacean acid metabolism (CAM) in *Kalanchoë* (Crassulaceae) species of Africa and Madagascar. *Botanica Acta* 106, 320–324.

Kluge, M. & Ting, I.P., 1978. *Crassulacean acid metabolism: analysis of an ecological adaptation*. (Ecological Studies 30). Springer, Berlin.

Knapp, U., 1994. Skulptur der Samenschale und Gliederung der Crassulaceae. *Botanische Jahrbücher fur Systematik Pflanzengeschichte und Pflanzengeographie* 116, 157–187.

Kobisi, K., 2005. *Preliminary checklist of the plants of Lesotho*. Southern African Botanical Diversity Network Report No. 34. Southern African Botanical Diversity Network [SABONET], Pretoria, and National University of Lesotho Herbarium, Roma.

Koch, K. & Ensikat, H.J., 2008. The hydrophobic coatings of plant surfaces: epicuticular wax crystals and their morphologies, crystallinity and molecular self-assembly. *Micron* 39, 759–772.

Köhlein, F., 1984. *Saxifrages and related genera*. B.T. Batsford Ltd, London.

Kolodziejczyk-Czepas, J. & Stochmal, A., 2017. Bufadienolides of *Kalanchoe* species: an overview of chemical structure, biological activity and prospects for pharmacological use. *Phytochemistry Reviews* 16, 1155–1171.

Koorders, S.H., 1918–1920. Beitrag sur Kenntniss der Flora von Java n° 18. *Bulletin du Jardin botanique. Buitenzorg* 3 (1), 169–180, figures 14 & 15.

Korevaar, L.C., van Donk, E., Peters, M., Deumer, P.V., Pullen, A.B., Smeenk, D.J., Theunissen, J., Waal, P.V.D., van Keppel, J.C., Uil, G.M., Noltee, F. & de Graaf, A., 1983. *Wat betekent die naam. Een alfabetische lijst van botanischen namen van succulente en aanverwante planten met hun verklaring*. Buining-fonds, Nederlands-Belgische vereniging van liefhebbers van cactussen en andere vetplanten, place of publication not stated.

Kotschy, T. & Peyritsch, J., 1867. *Plantae tinneanae [...]*. Carolis Gerold Filii., Vindobonae [Wien].

Krenn, L. & Kopp, B., 1998. Bufadienolides from animal and plant sources. *Phytochemistry* 48, 1–29.

Kroon, D.M., 1999. *Lepidoptera of southern Africa: host-plants & other associations: a catalogue*. Lepidopterists' Society of Africa & D.M. Kroon, Jukskei Park & Sasolburg.

Kuligowska, K., Lütken, H., Christensen, B., Skovgaard, I., Linde, M., Winkelmann, T. & Müller, R., 2015. Evaluation of reproductive barriers contributes to the development of novel interspecific hybrids in the *Kalanchoë* genus. *BMC plant biology* 15 (1), 15.

Kuntze, O., 1891. *Revisio generum plantarum vascularium omnium atque cellularium multarum secundum leges nomenclaturae internationales cum enumeratione plantarum exoticarum in itinere mundi collectarum. Pars I. Mit erläuterungen von Dr. Otto Kuntze, ordentlichem, ausländischem und ehren-mitgliede mehrerer gelehrter gesellschaften*. Commissionen. Arthur Felix, Leipzig; U. Hoepli, 37 Corso Vittore Emanuele, Milano; Dulau & Co., 37 Soho Square, London; Gust. E. Stechert, 828 Broadway, New York; and Charles Klincksieck, 11 Rue de Lille, Paris. [see: https://www.biodiversitylibrary.org/item/216222#page/392/mode/1up].

Kurz, W.S., 1871. On some new or imperfectly known Indian Plants, continuation from Journal, Volume XXXIX, Part II, pp. 61–91. *The Journal of the Asiatic Society of Bengal. [Part 2. (Natural History, &c.)]*. Calcutta 40 (2), 45–78.

Kwembeya, E.G. & Takawira, R., n.d. *A checklist of Zimbabwean vernacular plant names*. R&SS [Research & Scientific Services], National Herbarium and Botanic Garden, Harare, Zimbabwe.

Kwembeya, E.G. & Takawira, R., 2002. *A checklist of Zimbabwean vernacular plant names*. 2nd edition. National Herbarium and Botanic Garden, Harare, Zimbabwe.

Lamarck, J., 1786 [16 October 1786]. *Encyclopédie méthodique. Botanique* 2 (1). Panckoucke, Paris & Plomteux, Liège. [see: https://www.biodiversitylibrary.org/item/15260#page/143/mode/1up].

Landi, M., Tattini, M. & Gould, K.S., 2015. Multiple functional roles of anthocyanins in plant-environment interactions. *Environmental and Experimental Botany* 119, 4–17.

Landrum, L.R. & Bonilla, J., 1996. Anther glandularity in the American Myrtinae (Myrtaceae). *Madroño* 43, 58–68.

Lange, J.M.C., 1872. *Index semimum in horto academico hauniensi collectorum.* Hauniae, Hafniae, Copenhagen, Denmark.

Lange, Joh [M.C.], 1878. Udvalg af de i Københavns botaniske haves frøfortegnelser for 1854–75 beskrevne nye arter, på ny gennemgåede og forsynede med afbildninger, (Hertil tavle II–V). *Botanisk Tidsskrift* [Tredie række. Andet Bind] *Ser.* 3, Part 1, 131–142.

Lauzac-Marchal, M., 1974. TAXONOMIE VÉGÉTALE.—Réhabilitation du genre *Bryophyllum* Salisb. (Crassulacées Kalanchoïdées). [Note (*) de Mme Marguerite Lauzac-Marchal, présentée par M. André Aubréville.]. *Comptes Rendus Hebdomadaires des Séances de l'Académie des Sciences* 278 (20), 2505–2508.

Lawlor, D.W., 2001. *Photosynthesis.* 3rd edition. Springer, New York.

Lawrence, G.H.M., 1951. *Taxonomy of vascular plants.* The Macmillan Company, New York.

Lebrun, J.-P. & Stork, A.L., 2003. *Kalanchoe. Tropical African flowering plants. Ecology and distribution. Volume 1—Anonaceae–Balanitaceae.* Ville de Genève, Éditions des Conservatoire et Jardin botaniques de la Ville de Genève, Genève, pp. 212, 214–221.

Leistner, O.A. & Morris, J.W., 1976. Southern African place names. *Annals of the Cape Provincial Museums* 12, 1–565. Published jointly by the Cape Provincial Museums at the Albany Museum, Grahamstown.

Lemaire, A.[-]C. [CH.], 1852. Genre nouveau de la familie des Crassulacées. *Adromischus. Le Jardin Fleuriste, journal général des progrès et des Intérèts horticoles et botaniques.* Volume 2. Miscellanées. Chez les Éditeurs F. et E. Gyselynck, Gand [Ghent], pp. 58–60. [see: https://archive.org/stream/lejardinfleuris02lemagoog#page/n510/mode/2up].

Letty, C., 1962. *Veldblomme van Transvaal.* Die Kuratore, Boekefonds Veldblomme van Transvaal, place of publication not stated.

Leuenberger, B.E., 2004. Edith Raadts (1914–2004). *Willdenowia* 34, 323–325.

Lev-Yadun, S., 2001. Aposematic (warning) coloration associated with thorns in higher plants. *Journal of Theoretical Biology* 210, 385–388.

Lev-Yadun, S., 2017. How monocarpic is *Agave*? *Flora* 230, 12–13.

Linder, H.P., Dlamini, T., Henning, J. & Verboom, G.A., 2006. The evolutionary history of *Melianthus* (Melianthaceae). *American Journal of Botany* 93, 1052–1064.

Linnaeus, C., 1738. *Hortus cliffortianus, plantas exhibens, quas in hortis tam vivis quam siccis, Hartecampi in Hollandia, coluit […].* Amsterdam.

Linnaeus, C., 1753. *Species plantarum, exhibentes plantas rite cognitas, ad generarelatas, cum differentiis specificis, nominibus trivialibus, synonymis selectis, locis natalibus, secundum systema sexuale digestas.* Volume 1. Salvius, Stockholm. [see: https://www.biodiversitylibrary.org/item/13829#page/441/mode/1up].

Linnaeus, C. [filius], 1782 ("1781", published in April 1782). *Supplementum plantarum systematis vegetabilium editionis decimae tertiae, generum plantarum editiones sextae, et specierum plantarum editionis secundae.* Editum a Carolo a Linné. Impensis Orphanotrophei, Brunsvigae [Braunschweig]. [see: https://www.biodiversitylibrary.org/item/10321#page/206/mode/1up].

Lipscombe Vincett, B.A., 1984. *Golden days in the desert. Wild flowers of Saudi Arabia.* Immel Publishing, Jeddah and London.

Little, R.J. & Jones, C.E., 1980. *A dictionary of botany.* Van Nostrand Reinhold Company Inc., New York.

Long, C., 2005. Swaziland's Flora—siSwati names and uses. In: Swaziland's Flora Database. [see: http://www.sntc.org.sz/ Accessed November 2017].

Lopes Coelho, L., Kuligowska, K., Lütken, H. & Müller, R., 2015. Photoperiod and cold night temperature in control of flowering in *Kalanchoë*. *Acta Horticulturae* 1087, 129–134.

Lorenzo-Cáceres, J.M. Sánchez de, 2015. *Epónimos del género Kalanchoe* Adanson (Crassulaceae). Internet publication. pp. 1–12.

Lucas, A. & Pike, B., 1971. *Wild flowers of the Witwatersrand.* C. Struik Publishers, Cape Town.

Lüttge, U., 2004. Ecophysiology of crassulacean acid metabolism (CAM). *Annals of Botany* 93, 629–652.

Lüttge, U., 2015. Crassulacean acid metabolism. In: *eLS.* John Wiley & Sons, Chichester. http://www.els.net [see: https://doi.org/10.1002/9780470015902.a0001296.pub3].

Lyndon, R.F., 1962. Nitrogen metabolism of detached *Kalanchoe* leaves in the dark in relation to acidification, deacidification and O_2 uptake. *Journal of Experimental Botany* 13, 20–35.

Mabberley, D.J., 2017. *Mabberley's plant-book. A portable dictionary of plants, their classification and uses utilizing Kubitzki's* The families and genera of vascular plants *(1990–) and current botanical literature; arranged according to the principles of molecular systematics.* 4th edition, completely revised, with some 1400 additional entries. Cambridge University Press, Cambridge, United Kingdom.

Mabogo, D.E.N., 1990. *The ethnobotany of the Vhavenda.* MSc thesis. University of Pretoria, Pretoria.

Males, J., 2017. Secrets of succulence. *Journal of Experimental Botany* 68, 2121–2134.

Males, J. & Griffiths, H., 2017. Stomatal biology of CAM plants. *Plant Physiology* 174, 550–560.

Manan, M., Hussain, L., Ijaz, H. & Qadir, M.I., 2016. Antimicrobial activity of *Kalanchoe laciniata.* *Pakistan Journal of Pharmaceutical Sciences* 29, 1321–1324.

Mannheimer, C., Marais, A. & Schubert, S., 2012. *Toxic plants of veterinary importance in Namibia.* 2nd edition. Ministry of Agriculture, Water & Forestry, Windhoek.

Manning, J., Batten, A. & Bokelmann, H., 2001. *Eastern Cape. South African wild flower guide 11.* Botanical Society of South Africa in association with the National Botanical Institute, place of publication not stated.

Marion, L., 1973. Raymond-Hamet. 1890–1972. *Revue Cannadienne de Biologie* 32 (1), 1–2.

Mastenbroek, C., 1985. Cultivation and breeding of *Digitalis lanata* in the Netherlands. *British Heart Journal* 54, 262–268.

Masvingwe, C. & Mavenyengwa, M., 1997. *Kalanchoe lanceolata* poisoning in Brahman cattle in Zimbabwe: the first field outbreak: case report. *Journal of the South African Veterinary Association* 68, 18–20.

McCracken, D.P. & McCracken, P.A., 1990. *Natal. The Garden Colony. Victorian Natal and the Royal Botanic Gardens, Kew.* Frandsen Publishers (Pty) Ltd, Sandton, Johannesburg.

McGregor, M., 2008. *Saxifrages. A definitive guide to the 2000 species, hybrids & cultivars.* Timber Press, Portland, Oregon.

McKenzie, R.A., Franke, F.P. & Dunster, P.J., 1987. The toxicity to cattle and bufadienolide content of six *Bryophyllum* species. *Australian Veterinary Journal* 64, 298–301.

McNeill, J., Barrie, F.R., Buck, W.R., Demoulin, V., Greuter, W., Hawksworth, D.L., Herendeen, P.S., Knapp, S., Marhold, K., Prado, J., Prud'homme van Reine, W.F., Smith, G.F., Wiersema, J.H. & Turland, N.J. (Eds), 2012. *International Code of Nomenclature for algae, fungi, and plants (Melbourne Code) adopted by the Eighteenth International Botanical Congress Melbourne, Australia, July 2011.* Koeltz Scientific Books, Königstein [Regnum Vegetabile 154].

Mendelson, C., 2016. *Rhapsody in green. A novelist, an obsession, a laughably small excuse for a vegetable garden.* Kyle Books, London.

Mendonça, F.S., Nascimento, N.C., Almeida, V.M., Braga, T.C., Ribeiro, D.P., Chaves, H.A., Silva Filho, G.B. & Riet-Correa, F., 2018. An outbreak of poisoning by *Kalanchoe blossfeldiana* in cattle in northeastern Brazil. *Tropical Animal Health and Production* 50, 693–696.

Mes, T.H.M., 1995. *Origin and evolution of the Macaronesian Sempervivoideae (Crassulaceae).* PhD dissertation ("Proefschrift"), Faculteit Biologie, Universiteit Utrecht, Utrecht, The Netherlands, pp. 1–215.

Metcalfe, C.R. & Chalk, L., 1950. *Anatomy of the dicotyledons.* Volume 1. Clarendon Press, Oxford.

Meyer, N.L., Mössmer, M. & Smith, G.F. (Eds), 1997. *Taxonomic literature of southern African plants*. Strelitzia 5. National Botanical Institute, Pretoria, pp. 1–164.

Mii, M., 2009. Breeding of ornamental plants through interspecific hybridization using advanced techniques with a special focus on *Dianthus, Primula, Cosmos* and *Kalanchoe. Acta Horticulturae* 836, 63–72.

Milad, R., El-Ahmady, S. & Singab, A.N., 2014. Genus *Kalanchoe* (Crassulaceae): a review of its ethnomedicinal, botanical, chemical and pharmacological properties. *European Journal of Medicinal Plants* 4, 86–104.

Milburn, T.R., Pearson, D.J. & Ndegwe, N.A., 1968. Crassulacean acid metabolism under natural tropical conditions. *New Phytologist* 67, 883–897.

Moffett, R., 2014. *Biographical dictionary of contributors to the natural history of the Free State and Lesotho*. SUN MeDIA, Bloemfontein.

Mogg, A.O.D., 1958. An annotated checklist of the flowering plants and ferns of Inhaca Island, Mozambique. [Crassulaceae. *Kalanchoe*]. In: W. Macnae & M. Kalk (Eds), *A natural history of Inhaca Island, Moçambique*. Witwatersrand University Press, Johannesburg, pp. 139–156.

Molisch, H., 1928. Rote Wurzelspitzen. *Berichte der Deutschen Botanischen Gesellschaft* 46, 311–317.

Mooney, H.A., Troughton, J.H. & Berry, J.A., 1977. Carbon isotope measurements of succulent plants in southern Africa. *Oecologia* 30, 295–306.

Moreira, N.S., Nascimento, L.B.S., Leal-Costa, M.V. & Tavares, E.S., 2012. Comparative anatomy of leaves of *Kalanchoe pinnata* and *K. crenata* in sun and shade conditions, as a support for their identification. *Revista Brasileira de Farmacognosia* 22, 929–936.

Morsy, N., 2017. Cardiac glycosides in medicinal plants. In: H.A. El-Shemy (Ed.) *Aromatic and medicinal plants: back to nature*. IntechOpen, London, pp. 29–44.

Mort, M.E., Soltis, D.E., Soltis, P.S., Francisco-Ortega, J. & Santos-Guerra, A., 2001. Phylogenetic relationships and evolution of Crassulaceae inferred from *mat*K sequence data. *American Journal of Botany* 88, 76–91.

Mucina, L., Hoare, D.B., Lötter, M.C., Du Preez, J., Rutherford, M.C., Scott-Shaw, C.R., Bredenkamp, G.J., Powrie, L.W., Scott, L., Camp, K.G.T., Cilliers, S.S., Bezuidenhout, H., Mostert, T.H., Siebert, S.J., Winter, P.J.D., Burrows, J.E., Dobson, L., Ward, R.A., Stalmans, M., Oliver, E.G.H., Siebert, F., Schmidt, E., Kobisi, K. & Kose, L., 2006. Grassland Biome. In: L. Mucina & M.C. Rutherford (Eds), *The vegetation of South Africa, Lesotho and Swaziland*. Strelitzia 19, South African National Biodiversity Institute, Pretoria, pp. 349–436.

Mulroy, T.W., 1979. Spectral properties of heavily glaucous and non-glaucous leaves of a succulent rosette-plant. *Oecologia* 38, 349–357.

Naudé, T.W., 1977. The occurrence and significance of South African cardiac glycosides. *Journal of the South African Biological Society* 18, 7–20.

Newton, L.E. & Mbugua, P.K., 1993. A check-list and identification key for succulent plants in general cultivation in Nairobi. *Journal of the East Africa Natural History Society and National Museum* 82 (201), 43–53.

Neyland, M., Ng, Y.L. & Thimann, K.V., 1963. Formation of anthocyanin in leaves of *Kalanchoe blossfeldiana*—a photoperiodic response. *Plant physiology* 38, 447–451.

Nguelefack, T.B., Nana, P., Atsamo, A.D., Dimo, T., Watcho, P., Dongmo, A.B., Tapondjou, L.A., Njamen, D., Wansi, S.L. & Kamanyi, A., 2006. Analgesic and anticonvulsant effects of extracts from the leaves of *Kalanchoe crenata* (Andrews) Haworth (Crassulaceae). *Journal of Ethnopharmacology* 106, 70–75.

Nicolson, S.W. & Fleming, P.A., 2003. Nectar as food for birds: the physiological consequences of drinking dilute sugar solutions. *Plant Systematics and Evolution* 238, 139–153.

Nicolson, S.W. & Van Wyk, B.-E., 1998. Nectar sugars in Proteaceae: patterns and processes. *Australian Journal of Botany* 46, 489–504.

Notaguchi, M., Abe, M., Kimura, T., Daimon, Y., Kobayashi, T., Yamaguchi, A., Tomita, Y., Dohi, K., Mori, M. & Araki, T., 2008. Long-distance, graft-transmissible action of *Arabidopsis* FLOWERING LOCUS T protein to promote flowering. *Plant and Cell Physiology* 49, 1645–1658.

Oken, L., 1841. *Allgemeine Naturgeschichte fur alle Stande, von Professor Oken*. Volume 3 (3). Hoffmann'sche Verlags-Buchhandlung, Stuttgart, pp. 1459–2142; Index ("Register"), pp. 1–44. [see: https://www.biodiversitylibrary.org/item/180203#page/524/mode/1up].

Onderstall, J., 1984. *Transvaalse Laeveld en Platorand insluitende die Nasionale Krugerwildtuin. Veldblomgids van Suid-Afrika* 4. Botaniese Vereniging van Suid-Afrika, place of publication not stated.

Onderstall, J., 1996. *Sappi wild flower guide Mpumalanga & Northern Province*. DynamicAd, Nelspruit.

Osmond, C.B., 1978. Crassulacean acid metabolism: a curiosity in context. *Annual Review of Plant Physiology* 29, 379–414.

Pandurangan, A., Kaur, A. & Sharma, D., 2015. *Bryophyllum calycinum* (Crassulaceae)—an overview. *International Bulletin of Drug Research* 5, 51–63.

Patel, S., 2016. Plant-derived cardiac glycosides: role in heart ailments and cancer management. *Biomedicine & Pharmacotherapy* 84, 1036–1041.

Penzig, O., 1893. Piante raccolte in un viaggio botanico fra I Bogos edi Mensa, nell'Abissinia settentrionale. *Atti del Congresso Botanico Internazionale di Genova* 1892, 310–368.

Percival, M.S., 1961. Types of nectar in angiosperms. *New Phytologist* 60, 235–281.

Perrier de la Bâthie, H., 1928. Observations nouvelles sur le genre *Kalanchoe. Archives de Botanique II, Bulletin Mensuel* 2 (2), 17–31.

Persoon, C.H., 1805 [1 April–15 June 1805]. *Synopsis plantarum, seu enchiridium botanicum, complectens enumerationem systematicam specierum hucusque cognitorum*. Pars prima. Apud Carol. Frid. Cramerum, Parisiis lutetiorum [Paris] et apud J.G. Cottam, Tubingæ [London]. [see: https://www.biodiversitylibrary.org/item/11125#page/458/mode/1up].

Pertuit Jr., A.J., 1977. Influence of temperatures during long-night exposures on growth and flowering of 'Mace', 'Thor', and 'Telstar' *Kalanchoe. HortScience* 12, 48–49.

Pertuit Jr., A.J., 1992. *Kalanchoe*. In: R.A. Larson (Ed.), *Introduction to floriculture*. 2nd edition. Academic Press, San Diego, pp. 429–450.

Peters, W.C.H., 1862. *Naturwissenschaftliche Reise nach Mossambique auf befehl seiner Majestät des Königs Friedrich Wilhelm IV in den jahren 1842 bis 1848 ausgeführt. Botanik. I. Abtheilung. [Mit acht und vierzig Tafeln]*. Druck und Verlag von Georg Reimer, Berlin.

Pienaar, K. & Smith, G.F., 2011. *The southern African what flower is that?* 5th edition. Random House Struik, Cape Town.

Pilbeam, J., Rodgerson, C. & Tribble, D., 1998. *Adromischus*. The Cactus File Handbook 3. Cirio Publishing Services Ltd, Holbury, Southampton.

Pitzschke, A., 2013. From bench to barn: plant model research and its applications in agriculture. *Advances in Genetic Engineering* 2, 1–9.

Plukenet, L., 1692. *Phytographia [...]. Pars tertia*. Sumptibus auctoris, Londini.

Plukenet, L., 1696. *Almagestum botanicum*. Sumptibus auctoris, Londini.

Pole Evans, I.B. (Ed.), 1929. *Kalanchoe thyrsiflora*. Cape Province, Basutoland, Transvaal. Crassulaceae. *The Flowering Plants of South Africa* 9, plate 341.

Pooley, E., 1978. A checklist of plants of Ndumo Game Reserve, northeastern Zululand. *Journal of South African Botany* 44, 1–54.

Pooley, E., 1998. *A field guide to the wildflowers of KwaZulu-Natal and the eastern region*. Natal Flora Publications Trust, Durban.

Pooley, E., 2003. *Mountain flowers. A field guide to the flora of the Drakensberg and Lesotho*. The Flora Publications Trust, Durban.

Pott, R., 1920. Addendum to the first check-list of the flowering plants and ferns of the Transvaal and Swaziland. *Annals of the Transvaal Museum* 6 (4), 119–135.

Powrie, L., 2004. *Common names of Karoo plants*. G. Germishuizen & E. du Plessis, (Eds). Strelitzia 16. National Botanical Institute, Pretoria, pp. 1–199.

Proctor, M., Yeo, P. & Lack, A., 1996. *The natural history of pollination*. Timber Press, Portland, Oregon.

Purveur, S. [Comp.] & Harbour, E. [Illustrations], 1996. *Plant the name!* Michael Joseph Ltd, London.

Pyšek, P., Hulme, P.E., Meyerson, L.A., Smith, G.F., Boatwright, J.S., Crouch, N.R., Figueiredo, E., Foxcroft, L.C., Jarošík, V., Richardson, D.M., Suda, J. & Wilson, J.R.U., 2013. Hitting the right target: taxonomic challenges for, and of, plant invasions. *AoB Plants* 5, [see: https://doi.org/10.1093/aobpla/042. plt042].

Raadts, E., 1977. The genus *Kalanchoe* (Crassulaceae) in Tropical East Africa. *Willldenowia* 8 (1), 101–157.

Raadts, E., 1979. Rasterelektronenmikroskopische und anatomische Untersuchungen an Konnektivdrüsen von *Kalanchoë* (Crassulaceae). *Willldenowia* 9, 169–175.

Raadts, E., 1984. Cytotaxonomische Untersuchungen an *Kalanchoë* (Crassulaceae) 1. *Kalanchoë marmorata* Baker und 2 neue *Kalanchoë*-Arten aus Ostafrika. *Willldenowia* 13 (2), 373–385.

Raadts, E., 1985. Cytotaxonomische Untersuchungen an *Kalanchoe* (Crassulaceae). 2. Chromosomenzahlen in intermediärer Formen. *Willldenowia* 15, 157–166.

Raadts, E., 1989. Cytotaxonomische Untersuchungen an *Kalanchoë* (Crassulaceae) 3. Chromosomenzahlen Ostafrikanischer *Kalanchoë*-Sippen. *Willldenowia* 19 (1), 169–174.

Rabakonandrianina, E. & Carr, G.D., 1987. Chromosome numbers of Madagascar plants. *Annals of the Missouri Botanical Garden* 74, 123–125. [see: https://www.biodiversitylibrary.org/item/87376#page/127/mode/1up].

Raper, P.E., 1987. *Dictionary of southern African place names*. Lowry Publishers cc, Rivonia, Johannesburg.

Raper, P.E., 1989. *(A) Dictionary of southern African place names*, 2nd edition. Jonathan Ball Publishers, Parklands, Johannesburg.

Raper, P.E., 2004. *New dictionary of South African place names*. Jonathan Ball Publishers, Jeppestown, Johannesburg.

Raper, P.E., Möller, L.A. & Du Plessis, L.T., 2014. *Dictionary of Southern African place names*. Jonathan Ball Publishers, Jeppestown, Johannesburg.

Rauh, W., 1992. *Kalanchoe gastonis-bonnieri*. North-western Madagascar. Crassulaceae. *The Flowering Plants of Africa* 52 (1), Plate 2051.

Rauh, W., 1995. *Succulent and xerophytic plants of Madagascar*. Volume 1. Strawberry Press, Mill Valley, California.

Rauh, W., 1998. *Succulent and xerophytic plants of Madagascar*. Volume 2. Strawberry Press, Mill Valley, California.

Raymond-Hamet, 1910. Sur la présence dans la région de l'Usambara d'une plante considérée jusqu'ici comme une endémique malgache. *Bulletin de la Société Botanique de France* 88, 488–492.

Raymond-Hamet, 1916. Sur quelques Crassulacées nouvelles. *Journal of Botany (British and Foreign)* 54 (supplement 1), 1–33.

Raymond-Hamet, 1948. Plantes nouvelles, rares ou critiques des serres du Muséum. *Bulletin du Muséum national d'histoire naturelle, sér 2* 20 (5), 465–467.

Raymond-Hamet, 1955. Sur deux *Kalanchoe* de l'herbier de Dahlem-Berlin, l'un nouveau, l'autre peu connu. *Bulletin de la Societé Botanique de France* 10 (2), 239–240.

Raymond-Hamet, 1956. *Crassulacearum icones selectae*. Fascicle 2. Published by the author, Paris.

Raymond-Hamet, 1960. *Crassulacearum icones selectae*. Fascicle 4. Published by the author, Paris.

Raymond-Hamet, 1963. *Crassulacearum icones selectae*. Fascicle 5. Published by the author, Paris.

Raymond-Hamet & Marnier-Lapostolle, J., 1964. Le genre *Kalanchoe* au Jardin Botanique "Les Cèdres". *Archives du Muséum National d'Histoire Naturelle Septième Série*, Tome VIII, 1–110, Plates I–XXXVII.

Raymond-Hamet & Perrier de la Bâthie, J.M.H.A., 1914. Nouvelle contribution à l'étude des Crassulacées malgaches. *Annales du Musée Colonial de Marseille, sér. 3*, 2, 113–207. [see: https://www.biodiversitylibrary.org/item/23004#page/551/mode/1up].

Raymond-Hamet & Perrier de la Bâthie, H., 1915. Troisième contribution à l'étude des Crassulacées Malgaches. *Annales de l'Institut Botanico-Geologique Colonial de Marseille, série 3*, 3, 63–117.

Resende, F., 1956. Híbridos intergenericos e interespecíficos em Kalanchoideae 1. *Boletim da Sociedade Portuguesa de Ciências Naturais* 6, 241–244.

Resende, F. & Sobrinho, L.G., 1952. A new species of *Kalanchoe*. *Revista da Faculdade de Ciências, Universidade de Lisboa, 2ª Série, C- Ciências Naturais* 2 (2), 199–200, 1 plate.

Retief, E. & Herman, P.P.J., 1997. *Plants of the northern provinces of South Africa: keys and diagnostic features*. Strelitzia 6, National Botanical Institute, Pretoria, pp. 1–681.

Retief, E. & Meyer, N.L., 2017. *Plants of the Free State: inventory and identification guide*. Strelitzia 38. South African National Biodiversity Institute, Pretoria.

Richard, A., 1848. *Florae Abyssinicae seu enumeratio plantarum […]*. Volume I. Bertrand, Paris.

Riedel, M., Eichner, A. & Jetter, R., 2003. Slippery surfaces of carnivorous plants: composition of epicuticular wax crystals in *Nepenthes alata* Blanco pitchers. *Planta* 218, 87–97.

Riley, H.P., 1963. *Families of flowering plants of southern Africa*. University of Kentucky Press, place of publication not stated.

Ross, J.H., 1972. *The flora of Natal*. Memoirs of the Botanical Survey of South Africa 39, Department of Agricultural Technical Services, Botanical Research Institute, Pretoria, pp. 1–418.

Roth, A.W., 1821. *Alberti Guilielmi Roth, novae plantarum species praesertim Indiae orientalis. Ex collectione Doct. Benj. Heynii. Cum descriptionibus et observationibus*. Sumptibus H. Vogleri, Halberstadii. [see: https://www.biodiversitylibrary.org/item/41813#page/221/mode/1up].

Rourke, J.P., 1976. Prof. R.H. Compton's Flora of Swaziland. *Forum Botanicum* 14 (9), 57–58.

Rowley, G.D., 1978. *The illustrated encyclopedia of succulents. A guide to the natural history and cultivation of cacti and cactus-like plants*. A Salamander Book. Published by Leisure Books, place of publication not stated.

Rowley, G., 2003. Crassula. *A grower's guide*. Cactus & Co. libri, place of publication not stated.

Rowley, G., 2010. Worthwhile hybrid succulents no. 11: from *Kalanchoe blossfeldiana* Poelln. to the Double Flaming Katy Group. *CactusWorld* 28, 31–37.

Rowley, G.D., 2017. *Succulents in cultivation—breeding new cultivars*. The British Cactus & Succulent Society, Hornchurch, Essex.

Roxburgh, W., 1814. *Hortus bengalensis, or a catalogue of the plants growing in the Honourable East India Company's Botanical Garden at Calcutta*. Printed at the Mission Press, Serampore. [on p. 34 the name *Cotyledon rhizophylla* only]. [see: https://www.biodiversitylibrary.org/item/173186#page/56/mode/1up].

Roxburgh, W., 1832. *Flora indica; or, descriptions of Indian plants*. Volume II. Printed for W. Thacker and Co. Calcutta, Serampore; and Parbury, Allen and Co., London. [see: https://www.biodiversitylibrary.org/item/10529#page/459/mode/1up].

Rutherford, M.C., Powrie, L.W., Lötter, M.C., Von Maltitz, G.P., Euston-Brown, D.I.W., Matthews, W.S., Dobson, L. & McKenzie, B., 2006. Afrotemperate, Subtropical and Azonal Forests. In: L. Mucina & M.C. Rutherford (Eds), The vegetation of South Africa, Lesotho and Swaziland. *Strelitzia* 19, 584–614.

Rycroft, H.B., 1979. Professor R.H. Compton. *Veld & Flora* 65, 74–75.

Sajeva, M. & Costanzo, M., 1994. *Succulents. The new illustrated dictionary*. Cassell, London.

Sajeva, M. & Costanzo, M., 2000. *Succulents II. The new illustrated dictionary*. Timber Press, Portland, Oregon.

Salisbury, F.B. & Ross, C.W., 1992. *Plant physiology*. 4th edition. Wadsworth Publishing Company, Belmont (CA).

Salisbury, R.A., 1805 [1 June 1805]. *Bryophyllum calycinum*. Calyculated Bryophyllum. Ordo Naturalis. Sempervivae. Jusss. Gen. p. 307. *The Paradisus Londinensis: or coloured figures of plants cultivated in the vicinity of the metropolis* 1, t. 3. Published by William Hooker, No. 6, Frith Street, city not stated, presumably London. [see: https://www.biodiversitylibrary.org/item/113616#page/7/mode/1up].

Sapieha, T., n.d. *Wayside flowers of Kenya*. Wayside Flowers of Kenya, Nairobi, Kenya.

Schinz, H., 1888. Beiträge zur Kenntnis der Flora von Deutsch-Südwest-Afrika und der angrenzenden Gebiete II. *Verhandlungen des Botanischen Vereins der Provinz Brandenburg* 30, 138–186.

Schinz, H., 1912. Crassulaceae. In: H. Schinz (Ed.), Beiträge zur Kenntnis der afrikanischen Flora (XXV). *Vierteljahrsschriftder Naturforschenden Gesellschaft in Zürich* 57, 556–558.

Schinz, H. & Junod, H., 1900. Beiträge zur Kenntnis der Afrikanischen Flora (Neue Folge): I. Zur Kenntnis der Pflanzenwelt der Delagoa-Bay. *Mémoires de l'Herbier Boissier* 10, 25–75.

Schlechter, R., 1897. Decades plantarum novarum Austro-Africanarum. *The Journal of Botany* 35, 340–345.

Schmitt, H., 1974 (28 September 1974). Raymond Hamet (1890–1972). *Nouvelle Presse Medicale* 3 (32), 2042.

Schönland, S., 1891. Crassulaceae. 6. *Kalanchoe* Adans. In: A. Engler & K. Prantl (Eds), *Die natürlichen Pflanzenfamilien nebst ihren Gattungen und wichtigeren Arten insbesondere den Nutzpflanzen bearbeitet unter Mitwirkung zahlreicher hervorragender Fachgelehrten von A. Engler und K. Prantl*. III. Teil. 2. Abteilung a. [Podostemaceae von E. Warming; Crassulaceae von S. Schönland; Cephalotaceae, Saxifragaceae, Cunoniaceae von A. Engler; Myrothamnaceae von F. Niedenzu; Pittosporaceae von F. Pax; Hamamelidaceae, Bruniaceae, Platanaceae von F. Niedenzu]. Verlag von Wilhelm Engelmann, Leipzig, pp. 23–38. [see: https://www.biodiversitylibrary.org/item/100217#page/31/mode/1up].

Schönland, S., 1903a. On some new and some little-known species of South African plants. *Records of the Albany Museum* 1, 48–60.

Schönland, S., 1903b. A list of South African species of *Crassula* described or re-named during recent years. *Records of the Albany Museum* 1, 60–68.

Schonland, S., 1907. On some new and some little known species of South African plants belonging to the genera *Aloe, Gasteria, Crassula, Cotyledon* and *Kalanchoe*. *Records of the Albany Museum* 2, 137–155.

Schonland, S., 1919. *Phanerogamic flora of the Divisions of Uitenhage and Port Elizabeth*. Botanical Survey of South Africa Memoir No. 1. The Government Printing and Stationery Office, Pretoria, pp. 1–118.

Schonland, S., 1929. Materials for a critical revision of Crassulaceae. (The South African species of the genus *Crassula* L. (emend. Schonl.).). *Transactions of the Royal Society of South Africa* 17 (3), 151–293.

Schremmer, K., 1991. *A glossary of the names and terms used in connection with cacti & other succulent plants*. Published by the author, Kensington, Australia.

Schrire, B.D., 1983. Centenary of the Natal Herbarium, Durban, 1882–1982. *Bothalia* 14, 223–236.

Schulte, A.J., Koch, K., Spaeth, M. & Barthlott, W., 2009. Biomimetic replicas: transfer of complex architectures with different optical properties from plant surfaces onto technical materials. *Acta Biomaterialia* 5, 1848–1854.

Schwabe, W.W., 1971. Physiology of vegetative reproduction and flowering. In: F.C. Steward (Ed.), *Plant physiology: a treatise*. Volume 6A. Academic Press, New York, pp. 233–411.

Scott, P., 2008. *Physiology and behaviour of plants*. John Wiley & Sons, Chichester.

Scott-Elliot, G.F., 1890. Ornithophilous flowers in South Africa. *Annals of Botany* 4, 265–280.

Scott-Elliot, G.F., 1891. New and little-known Madagascar plants, collected and enumerated by G.F. Scott Elliot. *The Journal of the Linnean Society, Botany* 29, 1–67, 12 Plates.

Setshogo, M.P., 2005. *Preliminary checklist of the plants of Botswana*. Southern African Botanical Diversity Network Report No. 37. Southern African Botanical Diversity Network (SABONET), Pretoria and Gaborone.

Sharma, A.K. & Ghosh, S., 1967. Cytotaxonomy of Crassulaceae. *Biologische Zentralblatt* 86, 313–336. Supplement.

Shaw, J.M.H., 2008. An investigation of the cultivated *Kalanchoe daigremontiana* group, with a checklist of *Kalanchoe* cultivars. *Hanburyana* 3, 17–79.

Shaw, J.M.H., 2018. Raymond Hamet (25th March 1890–2nd Oct 1972)—a little known *Sedum* enthusiast. *Sedum Society Newsletter* 125, 72–80.

Siems, K., Jas, G., Arriaga-Giner, F.J., Wollenweber, E. & Dörr, M., 1995. On the chemical nature of epicuticular waxes in some succulent *Kalanchoe* and *Senecio* species. *Zeitschrift für Naturforschung C* 50, 451–454.

Silva, V., Figueiredo, E. & Smith, G.F., 2015. Alien succulents naturalised and cultivated on the central west coast of Portugal. *Bradleya* 33, 58–81.

Silva, M.C., Izidine, S. & Amude, A.B., 2004. *A preliminary checklist of the vascular plants of Mozambique./Catálogo provisório das plantas superiores de Moçambique*. Southern African Botanical Diversity Network Report No. 30. SABONET, Pretoria.

Sims, J., 1811. *Bryophyllum calycinum*. Pendulous-flowered Bryophyllum. Class and Order. Octandria Monogynia. *Curtis's Botanical Magazine* 34, Tab. 1409. [see: https://www.biodiversitylibrary.org/item/14320#page/93/mode/1up].

Sims, J., 1812. *Cotyledon crenata*. Scollop-leaved navel-wort. Class and Order. Decandria Pentagynia. *Curtis's Botanical Magazine* 35, Tab. 1436. [see: https://www.biodiversitylibrary.org/item/14321#page/53/mode/1up].

Skead C.J., 1967. *The sunbirds of southern Africa. Also the sugarbirds, the white-eyes and the spotted creeper*. Published for the Trustees of the South African Bird Book Fund by A.A. Balkema, Cape Town.

Smith, C.A., 1966. *Common names of South African plants*. E.P. Phillips & E. van Hoepen (Eds), Memoirs of the Botanical Survey of South Africa, 35, Department of Agricultural Technical Services, Botanical Research Institute, Pretoria, pp. 1–642.

Smith, G., 2004. *Kalanchoe* species poisoning in pets. *Veterinary Medicine* 99, 933–936.

Smith, G.F., 2005. *Gardening with succulents. Horticultural gifts from extreme environments*. Struik Publishers, Cape Town.

Smith, G.F., 2018. Notes on the fire ecology of *Petrosedum* Grulich and *Sedum* L. (Crassulaceae) in central continental Portugal. *Bradleya* 36, 61–69.

Smith, G.F. & Crouch, N.R., 2009. *Guide to succulents of southern Africa*. Struik Nature, Cape Town.

Smith, G.F., Crouch, N.R. & Figueiredo, E., 2016a. Reinstatement of *Kalanchoe montana* Compton (Crassulaceae), a distinctive species from the Barberton Center of Endemism, eastern southern Africa. *Haseltonia* 22, 64–72.

Smith, G.F., Crouch, N.R. & Figueiredo, E. 2017b. *Field guide to succulents in southern Africa*. Struik Nature, an imprint of Penguin Random House South Africa (Pty) Ltd, Century City, Cape Town.

Smith, G.F., Crouch, N.R. & Steyn, E.M.A., 2003. Notes on the distribution and ethnobotany of *Kalanchoe paniculata* (Crassulaceae) in southern Africa. *Bradleya* 21, 21–24.

Smith, G.F. & Figueiredo, E., 2011. *Umbilicus rupestris*: an interesting member of the Crassulaceae in Portugal. *Cactus & Succulent Journal (U.S.)* 83, 232–235.

Smith, G.F. & Figueiredo, E., 2013. The family Crassulaceae in continental Portugal. *Bradleya* 31, 76–88.

Smith, G.F. & Figueiredo, E., 2017a. The taxonomy of *Kalanchoe brachyloba* Welw. ex Britten (Crassulaceae). *Bradleya* 35, 2–14.

Smith, G.F. & Figueiredo, E., 2017b. The taxonomy of *Kalanchoe longiflora* Schltr. ex J.M.Wood (Crassulaceae), an endemic of South Africa's Maputaland-Pondoland Region of Endemism. *Bradleya* 35, 122–128.

Smith, G.F. & Figueiredo, E., 2017c. *Kalanchoe fedtschenkoi* Raym.-Hamet & H. Perrier (Crassulaceae) is spreading in South Africa's Klein Karoo. *Bradleya* 35, 80–86.

Smith, G.F. & Figueiredo, E., 2017d. The identity of *Kalanchoe paniculata* Harv. (Crassulaceae), a common species in southeastern and south-tropical Africa. *Haseltonia* 23, 72–78.

Smith, G.F. & Figueiredo, E., 2017e. Notes on *Kalanchoe rotundifolia* (Haw.) Haw. (Crassulaceae) in southern Africa: the taxonomy of a species complex. *Haseltonia* 23, 57–71.

Smith, G.F. & Figueiredo, E., 2018a. The infrageneric classification and nomenclature of *Kalanchoe* Adans. (Crassulaceae), with special reference to the southern African species. *Bradleya* 36, 162–172.

Smith, G.F. & Figueiredo, E., 2018b. Nomenclatural notes on *Kalanchoe pinnata* (Lam.) Pers. (Crassulaceae). *Bradleya* 36, 220–223.

Smith, G.F., Figueiredo, E. & Crouch, N.R., 2017a. *Kalanchoe leblanciae*. Crassulaceae. South Africa, Mozambique. *Flowering Plants of Africa* 65, 56–66, plate 2328.

Smith, G.F., Figueiredo, E. & Crouch, N.R., 2018a. The taxonomy of *Kalanchoe hirta* Harv. and *K. crenata* (Andrews) Haw. (Crassulaceae), and reinstatement of *K. hirta* as a distinctive, endemic species from southern and south-tropical Africa. *Haseltonia* 24, 40–50.

Smith, G.F., Figueiredo, E. & Mort, M.E., 2017c. Taxonomy of the three arborescent crassulas, *Crassula arborescens* (Mill.) Willd. subsp. *arborescens*, *C. arborescens* subsp. *undulatifolia* Toelken, and *Crassula ovata* (Mill.) Druce (Crassulaceae) from southern Africa. *Bradleya* 35, 87–105.

Smith, G.F., Figueiredo, E. & Mottram, R., 2018b. *Kalanchoe* 'Margrit's Magic' (Crassulaceae), a new cultivar from South Africa. *Bradleya* 36, 227–232.

Smith, G.F., Figueiredo, E. & Silva, V., 2015a. Geographical distribution range extension for *Sedum acre* L. (Crassulaceae) in central Portugal. *Bradleya* 33, 34–40.

Smith, G.F., Figueiredo, E. & Silva, V., 2015b. *Kalanchoe ×houghtonii* (Crassulaceae) recorded near Lisbon, Portugal. *Bouteloua* 20, 97–99.

Smith, G.F., Figueiredo, E. & Silva, V., 2016b. Notes on the geographical distribution range of *Sedum mucizonia* (Ortega) Raym.-Hamet (Crassulaceae), a miniature, annual succulent, in continental Portugal. *Bradleya* 34, 133–141.

Smith, G.F., Van Jaarsveld, E.J., Arnold, T.H., Steffens, F.E., Dixon, R.D. & Retief, J.A. (Eds), 1997. *List of southern African succulent plants.* Umdaus Press, Pretoria.

Smith, G.F. & Van Wyk, A.E., 1989. Biographical notes on James Bowie and the discovery of *Aloe bowiea* Schult. & J.H. Schult (Alooideae: Asphodelaceae). *Taxon* 38, 557–568.

Smith, G.F. & Van Wyk, B.[A.E.], 2008a. *Aloes in southern Africa.* Struik Nature, Cape Town.

Smith, G.F. & Van Wyk, B-E. [Ben-Erik], 2008b. *Guide to garden succulents.* Briza Publications, Arcadia, Pretoria.

Smith, G.F. & Willis, C.K. (Eds), 1997. *Index herbariorum: southern African supplement.* Southern African Botanical Diversity Network Report No. 2. Southern African Botanical Diversity Network (SABONET), Pretoria, pp. 1–55.

Smith, G.F. & Willis, C.K., 1999. *Index herbariorum: southern African supplement, 2nd edition.* Southern African Botanical Diversity Network Report No. 8. Southern African Botanical Diversity Network (SABONET), Pretoria, pp. 1–181.

Smith, P.P., 1997. A preliminary checklist of the vascular plants of the North Luangwa National Park, Zambia. *Kirkia* 16 (2), 205–245.

Soga, J.F., 1891. Disease 'Nenta' in goats. *Agricultural Journal of the Cape of Good Hope* 3, 140–142.

Spalding, G.H. (Comp.), 1953. Index. A botanist in southern Africa. *Lasca Miscellanea* 1, 1–53. Los Angeles State and County Arboretum, California.

Sprague, T.A., 1923. 1688. *Kalanchoe connata* Sprague [Crassulaceae]. *Bulletin of Miscellaneous Information, Kew* 1923 (5), 183–184.

Sprengel, C., 1825. *Systema vegetabilium*. Editio decima sexta. Volume II. Sumtibus Librariae Dieterichianae, Gottingae. [see: https://www.biodiversitylibrary.org/item/15253#page/260/mode/1up].

Springate, L.S., 1995. Kalanchöe. In: J. Cullen, J.C.M. Alexander, A. Brady, C.D. Brickell, P.S. Green, V.H. Heywood, P.-M. Jörgensen, S.L. Jury, S.G. Knees, A.C. Leslie, V.A. Matthews, N.K.B. Robson, S.M. Walters, D.O. Wijnands & P.F. Yeo (Eds), *European Garden Flora. Dicotyledons (Part II).* Volume 4. Cambridge University Press, Cambridge, United Kingdom, pp. 178–185.

Stafleu, F. & Cowan, R., 1983. *Taxonomic Literature: a selective guide to botanical publications and collections with dates, commentaries and types. Volume IV: P–Sak.* 2nd edition. Bohn, Scheltema & Holkema, Utrecht/Antwerpen; dr. W. Junk b.v., Publishers, The Hague/Boston. [see: http://www.sil.si.edu/DigitalCollections/tl-2/index.cfm].

Stearn, W.T., 1965. Biographical and bibliographical introduction. In: *Adrian Hardy Haworth. Complete works on succulent plants.* Volume 1. Gregg Press, place of publication not stated, pp. 9–80.

Stearn, W.T., 1971. The history of the discovery and botanical introduction of the Mesembryanthemaceae, with appropriate biographical notes. Adrian Hardy Haworth. In: H. Herre (Ed.), *The genera of Mesembryanthemaceae. Including a full set of botanical drawings by the artists of the Bolus Herbarium of the University of Cape Town, and others, and also distribution maps for each genus, identification keys, a scientific system, a contribution concerning the poisonous genus, and also a history of the introduction, with notes on and portraits of the various scientific workers from the beginning to the present day.* Tafelberg-Uitgewers Beperk, Cape Town, pp. 42–43.

Steenkamp, Y., Van Wyk, B., Victor, J., Hoare, D., Smith, G., Dold, T. & Cowling, R., 2004. Maputaland-Pondoland-Albany. In: R.A. Mittermeier, P. Robles Gil, M. Hoffman, J. Pilgrim, T. Brooks, C. Goetsch Mittermeier, J. Lamoreux & G.A.B. da Fonseca (Eds), *Hotspots revisited: earth's biologically richest and most endangered ecoregions.* Cemex, S.A. De C.V, Mexico City, pp. 218–229.

Stephenson, R., 1994. Sedum *cultivated stonecrops.* Timber Press, Portland, Oregon, pp. 1–335.

Stevens, J.F., 1995. *The systematic and evolutionary significance of phytochemical variation in the Eurasian Sedoideae and Sempervivoideae (Crassulaceae).* PhD dissertation ("Proefschrift"), in de Wiskunde en Natuurwetenschappen, Rijksuniversiteit Groningen, The Netherlands, pp. 1–210.

Stevens, P.F., 2001 onwards. Angiosperm Phylogeny Website. Version 14, July 2017 [and more or less continuously updated since]. [see: http://www.mobot.org/MOBOT/research/APweb/ Accessed 8 June 2018].

Steyn, D.G., 1949. *Vergiftiging van mens en dier.* Van Schaik, Pretoria.

Steyn, D.G. & Van der Walt, S.J., 1941. Recent investigations into the toxicity of known and unknown poisonous plants in the Union of South Africa XI. *Onderstepoort Journal of Veterinary Science and Animal Industry* 16, 121–147.

Steyn, P.S. & Van Heerden, F.R., 1998. Bufadienolides of plant and animal origin. *Natural Product Reports* 15, 397–413.

Storey, F.W. & Wright, K.M., 1916. *South African botany.* Longmans, Green and Co, London.

Stoudt, H.N., 1938. Gemmipary in *Kalanchoe rotundifolia* and other Crassulaceae. *American Journal of Botany* 25, 106–110.

Stuessy, T.F., Crawfod, D.J., Soltis, D.E. & Soltis, P.S., 2014. *Plant systematics: the origin, interpretation, and ordering of plant biodiversity.* Koeltz Scientific Books, Königstein.

Swain, T., 1977. Secondary compounds as protective agents. *Annual Review of Plant Physiology* 28, 479–501.

Symes, C.T., Nicolson, S.W. & McKechnie, A.E., 2008. Response of avian nectarivores to the flowering of *Aloe marlothii*: a nectar oasis during dry South African winters. *Journal of Ornithology* 149, 13–22.

't Hart, H., 1978. *Biosystematic studies in the Acre-group and the series Rupestria Berger of the genus* Sedum L. *(Crassulaceae).* Drukkerij Elinkwijk BV, Utrecht.

't Hart, H., 1995. Chapter 10. Infrafamilial and generic classification of the Crassulaceae. In: H. 't Hart & U. Eggli (Eds), *Evolution and systematics of the Crassulaceae*. Backhuys Publishers, Leiden, pp. 159–172.

't Hart, H. & Eggli, U. (Eds), 1995. *Evolution and systematics of the Crassulaceae*. Backhuys Publishers, Leiden.

Teeri, J.A., 1984. Seasonal variation in crassulacean acid metabolism in *Dudleya blochmanae* (Crassulaceae). *Oecologia* 64, 68–73.

Teixeira, L.B.C., Tostes, R.A., Andrade, S.F., Sakate, M. & Laurenti, R.F., 2010. Intoxicação experimental por *Kalanchoe blossfeldiana* (Crassulaceae) em cães. *Ciência Animal Brasileira* 11, 955–961.

Thiede, J. & Eggli, U., 2007. Crassulaceae. In: K. Kubitzki (Ed.), *The families and genera of vascular plants. IX. Flowering Plants. Eudicots*. Springer-Verlag, Berlin, pp. 83–118.

Thulin, M., 1993 (updated 2008). Crassulaceae. In: M. Thulin (Ed.), *Flora of Somalia*. Volume I. Royal Botanic Gardens, Kew, pp. 87–93. [see: http://plants.jstor.org/].

Tölken, H.R., 1977a. A revision of the genus *Crassula* in southern Africa. Part 1. In: E.A.C.L.E. Schelpe (Ed.), *Contributions from the Bolus Herbarium* 8, 1, 1–331. The Bolus Herbarium, University of Cape Town, Rondebosch.

Tölken, H.R., 1977b. A revision of the genus *Crassula* in southern Africa. Part 2. In: E.A.C.L.E. Schelpe (Ed.), *Contributions from the Bolus Herbarium* 8, 2, 332–595. The Bolus Herbarium, University of Cape Town, Rondebosch.

Tölken, H.R., 1978. Two new species and a new combination in the genus *Kalanchoe*. *Journal of South African Botany* 44, 89–91. [see: https://archive.org/stream/journalofsouthaf44unse#page/92/mode/2up].

Tölken, H.R., 1982. *Kalanchoe lanceolata*. Southern Africa, Tropical Africa, Arabia and India. Crassulaceae. *The Flowering Plants of Africa* 47 (1), plate 1848.

Tölken, H.R., 1983. *Kalanchoe sexangularis*. Natal, Swaziland, Transvaal and Zimbabwe. Crassulaceae. *The Flowering Plants of Africa* 47 (2), plate 1878.

Tölken, H.R., 1985. Crassulaceae. In: O.A. Leistner (Ed.), *Flora of southern Africa*. Volume 14, Department of Agriculture and Water Supply, Botanical Research Institute, Pretoria, pp. 1–244.

Tölken, H.R. & Leistner, O.A., 1986. *Bryophyllum delagoense*. Madagascar and naturalized in many countries. Crassulaceae. *The Flowering Plants of Africa* 49 (1), plate 1938.

Trager, J., 2001. The Huntington Botanical Gardens presents the 2001 offering of International Succulent Introductions [ISI 2001-37 *Kalanchoe luciae* subsp. *luciae*; ISI 2001-38 *Kalanchoe thyrsiflora*]. *Cactus & Succulent Journal (U.S)* 73, 87–101.

Trattinnick, L., 1821. *Auswahl vorzüglich schöner, seltener, berühmter, und ... merkwürdiger Gartenpflanzen in getreuen Abbildungen*. Volume 1. Wien.

Trimen, H., 1873. Friedrich Welwitsch. *Journal of Botany (British and Foreign)* 11, 1–11.

Turck, F., Fornara, F. & Coupland, G., 2008. Regulation and identity of florigen: FLOWERING LOCUS T moves center stage. *Annual Review of Plant Biology* 59, 573–594.

Turton, L., 1988. *Some flowering plants of south-eastern Botswana*. Botswana Society, Gaborone.

Uhl, C.H., 1948. Cytotaxonomic studies in the subfamilies Crassuloideae, Kalanchoideae, and Cotyledonoideae of the Crassulaceae. *American Journal of Botany* 35, 695–697.

Urton, N. [illustrations by D. Page], 1993. *Plants of the Swartkops valley bushveld*. Swartkops Trust, place of publication not stated.

Vahl, M., 1791. *Symbolae botanicae, sive plantarum, tam earum, quas in itinere, inprimis orientali, Collegit Petrus Forskål, quam aliarum, recentius detectarum, exactiores descriptiones, nec non observationes circa quasdam plantas dudum cognitas, auctore Martino Vahl. Pars Secunda. cum tabulis XXV aeri incisis*. Impensis Auctoris, Excudeant Nicolaus Möller et Filius, Aulae Regiae typographi, Hauniae. [see: https://www.biodiversitylibrary.org/item/118639#page/205/mode/1up].

Vahrmeijer, J., 1981. *Poisonous plants of southern Africa that cause stock losses*. Tafelberg Publishers, Cape Town.

Van der Schijff, H.P. & Schoonraad, E., 1971. The flora of the Mariepskop complex. *Bothalia* 10, 461–500.

Van der Walt, S.J. & Steyn, D.G., 1941. Recent investigations into the toxicity of known and unknown poisonous plants in the Union of South Africa XII. *Onderstepoort Journal of Veterinary Science and Animal Industry* 17, 211–223.

Van Ham, R.C.H.J., 1994. *Phylogenetic implications of chloroplast DNA variation in the Crassulaceae*. PhD dissertation ("Proefschrift"), Faculteit Biologie, Universiteit Utrecht, Utrecht, The Netherlands, pp. 1–153.

Van Ham, R.C.H.J. & 't Hart, H., 1998. Phylogenetic relationships in the Crassulaceae inferred from chloroplast DNA restriction-site variation. *American Journal of Botany* 85, 123–134.

Van Jaarsveld, E.[J.], 2017. *Kalanchoe waterbergensis*, a new *Kalanchoe* species from the Waterberg, Limpopo Province, South Africa. *Bradleya* 35, 166–170.

Van Jaarsveld, E.[J.] & Koutnik, D. [with illustrations by E. Bodley & L. Strachan], 2004. *Cotyledon and Tylecodon*. Umdaus Press, Hatfield, Pretoria.

Van Maarseveen, C. & Jetter, R., 2009. Composition of the epicuticular and intracuticular wax layers on *Kalanchoe daigremontiana* (Hamet et Perr. de la Bathie) leaves. *Phytochemistry* 70, 899–906.

Van Maarseveen, C., Han, H. & Jetter, R., 2009. Development of the cuticular wax during growth of *Kalanchoe daigremontiana* (Hamet et Perr. de la Bathie) leaves. *Plant, Cell & Environment* 32, 73–81.

Van Nieuwerburgh, M., Boterdaele, D., Coppens, A., De Meyer, H., De Meyer, J., Imschoot, A., Trivier, L. & Van Hulle, M. (Eds), n.d. *De Belgische sierteelt / L'horticulture Belge / Belgian horticulture/Der Belgische Zierpflanzenbau*. Printing office of Het Volk N.V., Ghent.

Van Voorst, A. & Arends, J.C., 1982. The origin and chromosome numbers of cultivars of *Kalanchoe blossfeldiana* Von Poelln.: their history and evolution. *Euphytica* 31 (3), 573–584.

Van Wyk, A.E. [Braam] & Smith, G.F., 2001. *Regions of floristic endemism in southern Africa. A review with emphasis on succulents*. Umdaus Press, Hatfield, Pretoria.

Van Wyk, B.[A.E.] & Malan, S., 1997. *Field guide to the wild flowers of the Highveld. Also useful in adjacent grassland and bushveld*. Struik Publishers (Pty) Ltd, Cape Town.

Van Wyk, B-E. [Ben-Erik] & Gericke, N., 2000. *People's plants: a guide to useful plants of southern Africa*. Briza Publications, Pretoria.

Van Wyk, B-E. [Ben-Erik] & Van Heerden, F. & Van Oudtshoorn, B., 2002. *Poisonous plants of South Africa*. Briza Publications, Pretoria.

Vanderplank, H.J., 1998. *Wildflowers of the Port Elizabeth area. Swartkops to Sundays Rivers*. Bluecliff Publishing, Hunters Retreat [Port Elizabeth].

Ventenat, E.P., 1804. *Jardin de la Malmaison*. Imprimerie de Crapelet, Paris & Chez l'auteur.

Venter, H.J.T., 1971. A preliminary check-list of the Pteridophyta and Spermatophyta of the grassland and swamp communities of the Ngoye Forest Reserve, Zululand. *Journal of South African Botany* 37, 103–108.

Verdoorn, I.C., 1942. '*n Inleiding tot plantkunde en tot enige Transvaalse veldblomme*. J.L. van Schaik, Bpk, Pretoria, pp. 1–146.

Verdoorn, I.C., 1946. *Kalanchoe crundallii*. Transvaal. Crassulaceae. *The Flowering Plants of Africa* 25, plate 967.

Verpoorte, R., 1998. Exploration of nature's chemodiversity: the role of secondary metabolites as leads in drug development. *Drug Discovery Today* 3, 232–238.

Victor, J.E., Smith, G.F. & Van Wyk, A.E., 2016. History and drivers of plant taxonomy in South Africa. *Phytotaxa* 269 (3), 193–208. [see: https://doi.org/10.11646/phytotaxa.269.3.3].

Villaseñor, J.L. & Espinosa-Garcia, F.J., 2004. The alien flowering plants of Mexico. *Diversity and Distributions* 10, 113–123.

Vogel, S.T., 1954. *Blütenbiologische Typen als Elemente der Sippengliederung*. Botanische Studien 1, 1–338. VEB Gustav Fischer, Jena.

Von Ahlefeldt, D., Crouch, N.R., Nichols, G., Symmonds, R., McKean, S., Sibiya, H. & Cele, M.P., 2003. *Medicinal plants traded on South Africa's eastern seaboard*. Porcupine Press, Durban.

Von Poellnitz, [J.]K.[L.A.], 1934. XV. *Kalanchoe Blossfeldiana*. *Repertorium specierum novarum regni vegetabilis*. Centralblatt für Sammlung und Veröffentlichung von Einzeldiagnosen neuer Pflanzen 35, 159–160. [see: http://bibdigital.rjb.csic.es/ing/Libro.php?Libro=6015].

Von Vest, L.C., 1820 [14 Jul 1820]. *Physocalycium* Vest. *Flora oder Botanische Zeitung* 3 (2) no. 26, 409–410.

Von Willert, D.J., Eller, B.M., Werger, M.J.A., Brinckman, E. & Ihlenfeldt, H-D., 1992. *Life strategies of succulents in deserts with special reference to the Namib desert*. Cambridge Studies in Ecology. Cambridge University Press, Cambridge.

Walker, C.C., 2017. Introducing the genus *Phedimus*. *New Zealand Cactus and Succulent Journal* 70 (2), 4–7.

Walters, M., 2011. Crassulaceae J. St.-Hil. (Stonecrop, Orpine or Houseleek family; *Plakkiefamilie*). In: M. Walters, E. Figueiredo, N.R. Crouch, P.J.D. Winter, G.F. Smith, H.G. Zimmermann & B.K. Mashope, *Naturalised and invasive succulents of southern Africa*. ABC Taxa 11. The Belgian Development Cooperation, Brussels, pp. 232–259.

Walters, M., Figueiredo, E., Crouch, N.R., Winter, P.J.D., Smith, G.F., Zimmermann, H.G. & Mashope, B.K., 2011. *Naturalised and invasive succulents of southern Africa*. ABC Taxa 11. The Belgian Development Cooperation, Brussels.

Wang, C.Y., Xiao, H.G., Liu, J. & Zhou, J.W., 2017. Differences in leaf functional traits between red and green leaves of two evergreen shrubs *Photinia × fraseri* and *Osmanthus fragrans*. *Journal for Forestry Research* 28, 473–479. [see: https://link.springer.com/article/10.1007%2Fs11676-016-0346-7].

Wang, Z.-Q., Guillot, D., Ren, M.-X. & López-Pujol, J., 2016. *Kalanchoe* (Crassulaceae) as invasive aliens in China—new records, and actual and potential distribution. *Nordic Journal of Botany* 34, 349–354.

Warburg, O., 1903. *Kunene-Sambesi-Expedition. H Baum*. Kolonial-Wirtschaftliches Komitee, Berlin.

Ward, D.B., 2008. Keys to the flora of Florida: 18. *Kalanchoe* (Crassulaceae). *Phytologia* 90 (1), 41–46.

Wareing, P.F. & Phillips, I.D.J., 1970. *The control of growth and differentiation in plants*. Pergamon Press, Oxford.

Watt, J.M. & Breyer-Brandwijk, M.G., 1962. *The medicinal and poisonous plants of southern and eastern Africa*, 2nd edition. E. & S. Livingstone Ltd, Edinburgh and London.

Weberling, F., 1989. *Morphology of flowers and inflorescences*. (Transl. by R.J. Pankhurst.) Cambridge University Press, Cambridge.

Welwitsch, F., 1856. Relação das plantas vivas, etc., da Flora Angolense que foram remetidas de Luanda, em 21 de Agosto de 1854, ao Conselho Ultramarino, pelo Dr. Welwitsch, com destino a serem enviadas ao Jardim Botânico de Coimbra, e a outros estabelecimentos hortículas de Lisboa. *Annaes do Conselho Ultramarino Parte Não Official, Série I*, 251–253.

Wickens, G.E., 1982. Miscellaneous notes on *Crassula*, *Bryophyllum* and *Kalanchoe*. Studies in the Crassulaceae for the 'Flora of Tropical East Africa': III. *Kew Bulletin* 36 (4), 665–674.

Wickens, G.E., 1987. Crassulaceae. In: R.M. Polhill (Ed.), *Flora of Tropical East Africa: Crassulaceae*. Published on behalf of the East African Governments by A.A. Balkema, Rotterdam, pp. 1–66.

Wild, H., Biegel, H.M. & Mavi, S., 1972. *A Rhodesian botanical dictionary of African and English plant names*. National Herbarium, Department of Research and Specialist Services, Ministry of Agriculture, Salisbury, Rhodesia.

Willdenow, C.L., 1799. *Caroli a Linné Species plantarum exhibentes plantas rite cognitas ad genera relatas cum differentiis specificis, nominibus trivialibus, synonymis selectis, locis natalibus, secundum systema sexuale digestas. Editio quarta, post Reicherdianum quinta adjectis vegetabilibus hucusque cognitis curante Carolo Ludovico Willdenow*. Tomus II. Pars I. Impensis G.C. Nauk, Berolini. [see: https://www.biodiversitylibrary.org/item/14556#page/472/mode/1up].

Willdenow, C.L., 1809. *Enumeratio plantarum horti regii botanici berolinensis, continens descriptiones omnium vegetabilium in horto dicto cultorum*. Pars I. Scholae Reale, Berlin. [see: http://bibdigital.rjb.csic.es/ing/Libro.php?Libro=1684].

Williams, V.L., Balkwill, K. & Witkowski, E.T.F., 2001. A lexicon of plants traded in the Witwatersrand *umuthi* shops, South Africa. *Bothalia* 31, 71–98.

Wilman, M., 1946. *Preliminary check-list of the flowering plants and ferns of Griqualand West*. Cambridge University Press, Cambridge.

Wingate, J.L. & Yeatts, L., 2003. *Alpine flower finder. The key to Rocky Mountain wild flowers found above treeline*. 2nd edition. Johnson Books, Boulder.

Wink, M. (Ed.), 1999. *Functions of plant secondary metabolites and their exploitation in biotechnology*. Annual Plant Reviews. Volume 3. Sheffield Academic Press, Sheffield.

Wink, M., 2010. Mode of action and toxicology of plant toxins and poisonous plants. *Julius-Kühn-Archiv* 421, 93–112.

Wink, M. & Van Wyk, B.-E., 2008. *Mind-altering and poisonous plants of the world: an illustrated scientific guide*. Briza Publications, Pretoria.

Winter, K. & Smith, J.A.C., 1996. An introduction to crassulacean acid metabolism. Biochemical principles and ecological diversity. In: K. Winter & J.A.C. Smith (Eds), *Crassulacean acid metabolism: biochemistry, ecophysiology and evolution*. (Ecological Studies 114). Springer, Berlin, pp. 1–13.

Winter, K., Holtum, J.A. & Smith, J.A.C., 2015. Crassulacean acid metabolism: a continuous or discrete trait? *New Phytologist* 208, 73–78.

Witt, A.B.R., 2004. Initial screening of the stem-boring weevil *Osphilia tenuipes*, a candidate agent for the biological control of *Bryophyllum delagoense* in Australia. *BioControl* 49, 197–209.

Witt, A.B.R., McConnachie, A.J. & Stals, R., 2004. *Alcidodes sedi* (Col.; Curculionidae), a natural enemy of *Bryophyllum delagoense* (Crassulaceae) in South Africa and a possible candidate for the biological control of this weed in Australia. *Biological Control* 31, 380–387.

Wood, J. Medley, 1899. *Kalanchoe rotundifolia*, Harv. Natural Order, Crassulaceae. *Natal Plants* Volume 1, 76, t. 94.

Wood, J. Medley, 1903. *Kalanchoe longiflora*, Schlechter, MSS. Natural Order, Crassulaceae. *Natal Plants* volume 4, part I, Plate 320 [lacking page numbers].

Wood, J. Medley, 1907. *A handbook to the flora of Natal*. Bennet & Davis Printers, Smith and Gardiner Streets, Durban, Natal.

Wood, J. Medley, 1909. Revised list of the flora of Natal. *Transactions of the South African Philosophical Society* 18, 121–280.

Wood, J. Medley & Evans, M., 1899. *Kalanchoe thyrsiflora* Harv. *Natal plants*. Volume 1, 43, t. 52.

World Health Organization, 2017. WHO Model List of essential medicines, 20th List – Rev. March 2017. [see: http://www.who.int/medicines/publications/essentialmedicines/20th_EML2017.pdf?ua=1 Accessed June 2018].

Yang, X., Hu, R., Yin, H., Jenkins, J., Shu, S., Tang, H., Liu, D., Weighill, D.A., Yim, W.C., Ha, J., Heyduk, K., Goodstein, D.M., Guo, H.-B., Moseley, R.C., Fitzek, E., Jawdy, S., Zhang, Z., Xie, M., Hartwell, J.A., Grimwood, J., Abraham, P.E., Mewalal, R., Beltrán, J.D., Boxall, S.F., Dever, L.D., Palla, K.J., Albion, R., Garcia, T., Mayer, J.A., Lim, S.D., Wai, C.M., Peluso, P., Van Buren, R., De Paoli, H.C., Borland, A.M., Guo, H., Chen, J.-G., Muchero, W., Yin, Y., Jacobson, D.A., Tschaplinski, T.J., Hettich, R.L., Ming, R., Winter, K., Leebens-Mack, J.C., Smith, J.A.C., Cushman, J.C., Schmutz, J. & Tuskan, G., 2017. The *Kalanchoë* genome provides insights into convergent evolution and building blocks of crassulacean acid metabolism. *Nature Communications* 8, 1899. https://doi.org/10.1038/s41467-017-01491.7.

Young, T.P. & Augspurger, C.K., 1991. Ecology and evolution of long-lived semelparous plants. *Trends in Ecology & Evolution* 6, 285–289.

Zotz, G. & Hietz, P., 2001. The physiological ecology of vascular epiphytes: current knowledge, open questions. *Journal of Experimental Botany* 52, 2067–2078.

Glossary

Close-up of the densely flowered thyrse of *Kalanchoe winteri*, a southern African endemic. Photograph: Gideon F. Smith.

To increase the flow and readability of the main text of this book and to make it user-friendly, terms are defined where they are introduced, sometimes in more than one place. In this 'Glossary', we therefore deliberately restrict the terms defined to those most often encountered.

Terms that are denoted with an asterisk when used in a definition are defined elsewhere in the 'Glossary'.

Useful sources for definitions of botanical terms, including those applicable to succulent plants, include Beentje (2010), Blackmore & Tootill (1984), Eggli (1993), Lawrence (1951), and Little & Jones (1980).

Apex   Distal end or tip.

Apiculate   Terminated by a short, flexible, usually sharply pointed tip.

Biennial   Applied to a plant that completes its life cycle, from seed germination to death, during two years (two growing seasons), of which the first is often a vegetative phase and the second a reproductive one. See also multiannual*.

Cordate   Heart-shaped, for example in the case of a leaf where the bases on both sides of the petiole* extend beyond the point of attachment of the petiole* to the lamina* and the lamina* terminates in a sharp or rounded tip.

Corymb   A type of inflorescence* with indeterminate* growth where the lowermost or outer flowers open first. A corymb is short, broad, and ± flat-topped as the flower stalks are of different lengths but all reach the same level.

Cuneate   Wedge-shaped, for example in the case of a leaf where the bases on both sides of the lamina* taper towards the tip. See also cordate*.

Cyme   A type of inflorescence* that is more or less broad and flat-topped and the main axis determinate* therefore where the central flowers (at the top) open first.

Determinate   A type of inflorescence* where the central or terminal flower opens first, so preventing the elongation of the central inflorescence* axis.

Dichasium   A type of inflorescence* with determinate* growth, like a cyme*, where the central, older flower, which is situated between two younger, lateral flowers, opens first. This yields an inflorescence* with a symmetrical appearance.

Follicle   A type of dry, dehiscent fruit that opens on one side only.

Hapaxanthic   Flowering in a single vegetative shoot rather than a whole, often multi-stemmed, plant.

Indeterminate   A type of inflorescence* where the bottommost flower or lateral flowers open first, so allowing the elongation of the central inflorescence* axis. The terminal flower is therefore the last to open.

Inflorescence   Any type of arrangement of flowers along a central axis, including all its bracts and branches.

Lamina   The usually expanded portion of a leaf. Also called the blade.

Monocarpic (syn. semelparous*)   Plants in which all resources are utilised for one episode of reproduction, followed by the death of the entire plant.

Mucronate   Where a structure, for example a corolla lobe, is terminated abruptly by an often harmless, spiny tip.

Multiannual   Applied to plants that grow vegetatively for some years, typically more than two, and die following a flowering event. See also biennial*.

Obcordate   The reverse of cordate*. Lobed at the apex*, rather than basally.

Opposite   A leaf arrangement where the leaves at a node are on opposite sides of an axis.

Ovate   With the broadest part below the middle, that is, shaped like the egg of a chicken.

Panicle   A type of branched inflorescence* with indeterminate* growth in which the branches of the main axis are racemes* and the flowers pedicellate (see pedicel*).

Pedicel   Stalk of an individual flower.

Peduncle   Stalk of an inflorescence* or of an individual flower, if that flower is the remainder of an inflorescence*.

Petiole   Stalk of a leaf.

Pleonanthy   Where a vegetative axis bears lateral flowers and remains indeterminate*.

Raceme   A type of unbranched inflorescence* with indeterminate* growth in which the flowers are pedicellate (see pedicel*) and the terminal flowers open last.

Semelparous (syn. monocarpic*)   Plants in which all resources are utilised for one episode of reproduction, followed by the death of the entire plant.

Thyrse   A type of inflorescence* where the main axis is indeterminate* and the lateral axes (side branches) determinate* (e.g. a dichasium* or cyme*), either of the latter usually being a compact flower cluster.

Index

Printed in the United States
By Bookmasters